Exclusive Reactions
at
High Momentum Transfer

Exclusive Reactions
at
High Momentum Transfer

Elba, Italy June 24– 26, 1993

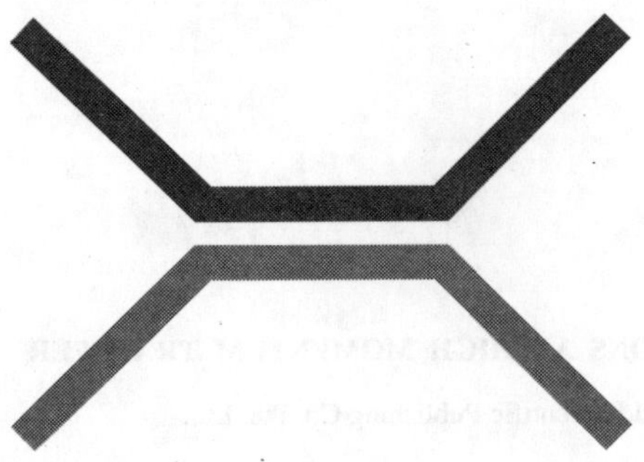

Editors

Carl Carlson
Physics Department, Coll. of William and Mary, Williamsburg, USA

Paul Stoler
Physics Department, Rensselaer Poly. Inst., Troy, USA

Mauro Taiuti
INFN, Genoa, Italy

World Scientific
Singapore • New Jersey • London • Hong Kong

Published by

World Scientific Publishing Co. Pte. Ltd.

P O Box 128, Farrer Road, Singapore 9128

USA office: Suite 1B, 1060 Main Street, River Edge, NJ 07661

UK office: 73 Lynton Mead, Totteridge, London N20 8DH

EXCLUSIVE REACTIONS AT HIGH MOMENTUM TRANSFER

ISBN 981-02-1768-4

Printed in Singapore.

CONTENTS

Preface — ix

The Challenges of Exclusive Processes in QCD — 1
    S. J. Brodsky

Can Sudakov Save Us? — 16
    C. E. Carlson

The Neutron Elastic Form Factor at High Q^2 — 24
    P. E. Bosted et al.

Nucleon Form Factors in PQCD and their Dependence on the
Nucleon Quark Distribution Amplitude — 32
    R. Eckerdt, J. Hansper and M. F. Gari

Baryon Resonance Form Factors at High Momentum Transfer — 36
    P. Stoler

The $\Delta(1232)$ Resonance Form Factor — 44
    L. M. Stuart et al.

Logarithmic Corrections to Baryon Form Factors — 48
    C. E. Carlson and N. C. Mukhopadhyay

Meson Form Factors at Large Q^2 — 55
    J. Milana

Measuring the Charged Pion Form Factor at CEBAF — 62
    D. J. Mack

The Pion Form Factor: Role of Intrinsic Transverse Momentum — 66
    R. Jakob and P. Kroll

Form Factors in the Timelike Region — 70
    K. Seth

On the Proton Form Factor in the Time-Like Region   84
S. M. Bilinsky, C. Giunti and V. Wataghin

On Exclusive Reactions in the Time-like Region   88
M. Schuermann

Strong Form Factors   92
A. Szczurek, H. Holtmann and J. Speth

Photon Absorption on Complex Nuclei   107
V. Muccifora et al.

Inclusive Electron Scattering from an Oxygen and Argon Jet Target   115
M. Anghinolfi et al.

Hadron Structure and Hard Exclusive Processes   120
V. Chernyak

Fixed Angle High Energy Elastic Scattering and Problem
of Helicity Conservation   134
M. P. Chavleishvili

On Proton and Delta Wave Functions   137
N. G. Stefanis and M. Bergmann

Evolution Effects on the Nucleon Distribution Amplitude   146
M. Bergmann and N. G. Stefanis

Probing Minimal Fock Space Components in Meson in Diffractive
Hadron-Nucleon Processes   150
L. Frankfurt and M. Strikman

Strangeness and Charm Components in Heavy Meson Production   162
J. M. Laget

Two Pion Decay of Electroproduced Baryon Resonances   171
M. Ripani and V. Burkert

Signature for Hybrids   179
T. Barnes

Semi-Inclusive Deep-Inelastic Scattering on A = 1, 2   191
   A. E. L. Dieperink and G. D. Bosveld

Quark Correlation Functions in Semi-Inclusive Hard Processes   196
   P. J. Mulders and J. Levelt

Exclusive Charmonium Decays in Perturbative QCD   201
   M. Anselmino

Spin-1 Form Factors   206
   C. R. Ji

Deuteron Photodistintegration: Data and Plans   214
   J. Napolitano et al.

Deuteron Photodisintegration at Small t or u   222
   L. A. Kondratyuk et al.

Nucleon and Baryon Components of Six-Quark Deuteron
Wave Functions   226
   L. Ya. Glozman

Relativistic Light-Cone Approach to the Study of Deuteron
Properties   230
   L. Kondratyuk, M. M. Giannini and P. Saracco

Color Transparency   233
   B. K. Jennings and G. A. Miller

How to Test Experimentally for Color Transparency and
Nuclear Filtering   241
   P. Jain and J. Ralston

Exclusive Hadronic Interaction Mechanism and Color
Transparency   249
   S. Heppelmann

Color Transparency and the $(e, e'p)$ Reaction   256
   A. Lung

Future Tests of Color Transparency Using Leptons as a Probe	264
  J. Jourdan

Fermi Momentum Bias of Color Transparency	272
  B. Z. Kopeliovich

Hadron Dynamics of Color Transparency	280
  A. Bianconi, S. Boffi and D. E. Kharzeev

Physics with a 15 to 30 GeV Electron Accelerator (ELFE)	284
  B. Frois and B. Pire

PREFACE

During the past decade a considerable literature has appeared on the topic of exclusive reactions at high momentum transfer. Two important recent developments offer the possibility of experimentally testing the theoretical ideas in a systematic way. These are the potential CEBAF energy upgrades and the proposed 15–30 GeV European Electron Accelerator, ELFE. The time was right to have a meeting to collect the ideas in this field with a view to supporting the physics programs of these new facilities.

We met on the Island of Elba for three days, 24–26 June 1993, under the heading "International Workshop on Exclusive Reactions at High Momentum Transfer." The field has now many workers united in the goal of using exclusive reactions to better understand the structure of hadrons. The spectrum of interested physicists is broad and includes persons who are nuclear physicists by background as well as particle physicists, and the number of us who met was about sixty-five.

These proceedings of the "Workshop" record the formal talks given there and thus, we hope, also record a reasonably full idea of the state of the art in studying exclusive reactions at high momentum transfers and establish a common body of knowledge to be shared by all. There is some emphasis on the use of perturbative calculations to connect hadron structure to experimental measureables, and other topics, including color transparency, the use of QCD sum rules, and experimental data. To all our attendees, speakers, and authors, go many thanks.

Thanks are also due to the people who worked hard—and sucessfully—to ensure that our stay on Elba was pleasant and productive. Allow us to mention especially the conference secretaries in Genoa and on the Island of Elba, Laura Opisso, Antonella Sapere, and Daniella Vatinno, and the Local Organizing Committee centered in Genoa and Frascati, consisting of Marco Anghinolfi, Valeria Muccifora, and Mauro Taiuti.

Finally, we thank the Istituto Nazionale di Fizica Nucleare (INFN), the National Science Foundation, and CEBAF for supporting the workshop.

Carl Carlson
Paul Stoler
Mauro Taiuti

THE CHALLENGES OF EXCLUSIVE PROCESSES IN QCD

STANLEY J. BRODSKY

Stanford Linear Accelerator Center, Stanford University

Stanford, CA 94309 USA

ABSTRACT

I review some of the outstanding theoretical and experimental issues confronting the application of perturbative quantum chromodynamics to large momentum transfer exclusive reactions.

1. Introduction

The analysis of exclusive hadronic amplitudes such as form factors, electroweak transition matrix elements, and two-body scattering amplitudes has remained among the most challenging computational problems in quantum chromodynamics. The physics of exclusive amplitudes clearly depends on the fundamental relativistic structure of the hadrons as well as the dynamics governing quark and gluon propagation, QCD vacuum structure, Regge behavior, and color confinement. Numerical predictions for exclusive processes involving low momentum transfer are beginning to be obtained from lattice gauge theory and QCD sum rules. However, the most interesting insights into hadron structure at the amplitude level and the most transparent connections to the underlying QCD physics emerges at high momentum transfer where perturbative analyses for the leading twist contributions to exclusive processes can be combined with non-perturbative hadron wavefunction information.

The least-complicated exclusive amplitudes to analyze from first principles in QCD are the space-like electromagnetic form factors of hadrons. An elastic form factor is the probability amplitude for a hadron to remain intact after absorbing momentum q by its local quark current. If one uses light-cone quantization in the $q^+ = q^0 + q^z = 0$ frame with $\vec{q}_\perp^{\,2} = -q^2 = Q^2$, then vacuum fluctuation contributions to the j^+ current can be avoided. Nevertheless, the computation of an elastic form factor requires knowledge of all of the hadron's light-cone Fock state wavefunctions. For example, the helicity-conserving form factor has the form[1]

$$F(Q^2) = \langle p + q | j^+ | p \rangle \,/\, 2p^+$$

$$= \sum_{n,\lambda_i} \sum_a e_a \int \overline{\prod_i} \frac{dx_i\, d^2\vec{k}_{\perp i}}{16\pi^3} \psi_n^{(\Lambda)*}(x_i, \vec{\ell}_{\perp i}, \lambda_i)\, \psi_n^{(\Lambda)}(x_i, \vec{k}_{\perp i}, \lambda_i).$$

2

The constituents in the initial state have longitudinal light-cone momentum fractions $x_i = (k^0 + k^z)_i/(p^0 + p^z)$, relative transverse momentum, $\vec{k}_{\perp i}$, and helicities λ_i. Here e_a is the charge of the struck quark, $\Lambda^2 \gg \vec{q}_\perp^2$, and the transverse momenta in the final state are

$$\vec{\ell}_{\perp i} \equiv \begin{cases} \vec{k}_{\perp i} - x_i \vec{q}_\perp + \vec{q}_\perp & \text{for the struck quark} \\ \vec{k}_{\perp i} - x_i \vec{q}_\perp & \text{for all other partons.} \end{cases}$$

In principle, one can obtain all of the required Fock State wavefunctions by diagonalizing the light-cone QCD Hamiltonian.[2] This has in fact been done for meson and baryon wavefunctions in the case of QCD in one-space and one-time dimensions, but the corresponding task appears to be formidable for QCD(3+1).

Fortunately, because of asymptotic freedom and the point-like behavior of quark and gluon interactions at short distances, the computation of exclusive amplitudes in QCD becomes much simpler at large momentum transfer. The primary ingredient in the analysis is factorization: the non-perturbative dynamics of the bound states can be isolated in terms of process-independent distribution amplitudes, and the dynamics of the momentum transfer to the hadrons can be isolated in terms of perturbatively-calculable hard-scattering quark and gluon subprocesses. Thus general properties of exclusive reactions at large momentum transfer can be derived without explicit knowledge of the non-perturbative structure of the theory.[3]

The most characteristic feature of an exclusive amplitude in QCD is that it falls off slowly with momentum transfer, not as an exponential or a Gaussian, but as an inverse power of $Q = p_T$ which is directly related to the degree of complexity of the scattering hadrons. The nominal power-law fall-off[4] $\mathcal{M} \sim Q^{4-n}$ of an exclusive amplitude at large momentum transfer reflects the elementary scaling of the lowest-order connected quark and gluon tree graphs obtained by replacing each of the external hadrons by its respective collinear quarks. Here n is the total number of initial state and final state lepton, photon, or quark fields entering or leaving the hard scattering subprocess. The empirical success of the dimensional counting rules for the power-law fall-off of form factors and general fixed center-of-mass angle scattering amplitudes gave early and important evidence for the scale-invariance of quark and gluon interactions at short distances.

Thus only the valence-quark Fock components of the hadron wavefunctions contribute to the leading power-law fall-off of an exclusive amplitude. In particular, since the internal momentum transfer at the quark level is required to be large, one can obtain the basic scaling and helicity structure of the hadron amplitude by simply iterating the gluon-exchange term in the effective potential for the light-cone wavefunctions. The result is that exclusive amplitudes at high momentum transfer Q^2 can be written in a factorized form as a convolution of process-independent "distribution amplitudes" $\phi(x_i, Q)$, one for each hadron involved in the amplitude, with a hard-scattering amplitude T_H describing the scattering of the valence quarks from the initial to final state.[5,6]

The distribution amplitude is the fundamental gauge invariant wavefunction which describes the fractional longitudinal momentum distributions of the valence quarks in a hadron integrated over transverse momentum up to the scale Q.[5] For example, the pion's electromagnetic form factor can be written as[5,6,7]

$$F_\pi(Q^2) = \int_0^1 dx \int_0^1 dy \, \phi_\pi^*(y, Q) \, T_H(x, y, Q) \, \phi_\pi(x, Q) \left(1 + \mathcal{O}\left(\frac{1}{Q}\right)\right).$$

Here T_H is the scattering amplitude obtained when pions replaced by collinear $q\bar{q}$ pairs. This factorized form is the prototype for the factorization of general exclusive amplitudes in QCD at high momentum transfer. All of the non-perturbative dynamics is factorized into the distribution amplitudes,[5] $\phi_B(x_i, \lambda_i, Q)$, for the baryons with $x_1 + x_2 + x_3 = 1$, and $\phi_M(x_i, \lambda_i, Q)$, for the mesons with $x_1 + x_2 = 1$ which sum all internal momentum transfers up to the scale Q^2. On the other hand, all momentum transfers higher than Q^2 appear in T_H, which can be computed perturbatively in powers of the QCD running coupling constant $\alpha_s(Q^2)$. The distribution amplitudes are thus the process-independent hadron wavefunctions which interpolate between the QCD bound state and their valence quarks at transverse separation $b_\perp \simeq 1/Q$. The pion's distribution amplitude, for example, is directly related to its valence light-cone wavefunction:

$$\phi_\pi(x, Q) = \int \frac{d^2 \vec{k}_\perp}{16\pi^3} \, \psi_{q\bar{q}/\pi}^{(Q)}(x, \vec{k}_\perp)$$

$$= P_\pi^+ \int \frac{dz^-}{4\pi} \, e^{ix P_\pi^+ z^-/2} \left\langle 0 \left| \bar{\psi}(0) \frac{\gamma^+ \gamma_5}{2\sqrt{2n_c}} \psi(z) \right| \pi \right\rangle^{(Q)} \Bigg|_{z^+ = \vec{z}_\perp = 0}.$$

The $\vec{k}_\perp$ integration is cut off by the ultraviolet cutoff $\Lambda = Q$ implicit in the wavefunction; thus only valence Fock states with invariant mass squared $\mathcal{M}^2 \leq Q^2$ contribute.

Given the factorized structure of exclusive amplitudes at large momentum transfer, one can read off a number of general features of the PQCD predictions: the dimensional counting rules, hadron helicity conservation, and color transparency.[3] QCD also predicts calculable corrections to the nominal dimensional counting power-law behavior due to the running of the strong coupling constant, higher order corrections to the hard scattering amplitude, Sudakov effects, pinch singularities, as well as the evolution of the hadron distribution amplitudes, $\phi_H(x_i, Q)$.

Evolution equations for the meson and baryon distribution amplitudes can be derived and employed in analogy to the evolution of structure functions.[3,8] If one can calculate the distribution amplitude at an initial scale Q_0 using QCD sum rules or lattice gauge theory,[8] then one can determine $\phi(x_i, Q)$ at higher momentum scales via

4

evolution equations in $\log Q^2$ or equivalently, the operator product expansion.[9] Empirical constraints on the hadron distribution amplitudes can be obtained from the normalization and scaling of form factors at large momentum transfer and the angular dependence of two body scattering amplitudes.

Perhaps the most surprising feature of the QCD predictions for exclusive processes in QCD is "color transparency",[10] which reflects the fact that only the small transverse separation $b_\perp \sim 1/Q$ valence wavefunction can contribute to exclusive amplitude at large momentum transfer. Since these color-singlet states have small color-dipole moments, they will have small initial and final state interactions. In particular if the large momentum transfer occurs as a quasi-elastic process within a nucleus, there will be minimal initial state or final state absorption—in striking contrast to the standard picture of strong absorption predicted in Glauber theory. A careful treatment of color transparency requires consideration of the expansion time and coherence length of the small size configurations.[11]

2. A Detailed Example: Compton Scattering in Perturbative QCD

Exclusive reactions involving two real or virtual photons provide a particularly interesting testing ground for QCD because of the relative simplicity of the couplings of the photons to the underlying quark currents, and the absence of significant initial state interactions—any remnant of vector-meson dominance contributions is suppressed at large momentum transfer, and the photon enters the amplitude as a direct point-like coupling.

The simplest example of a two-photon exclusive process is the $\gamma^*(q)\gamma \to M^0$ process which is measurable in tagged $e^+e^- \to e^+e^-M^0$ reactions. The photon to neutral meson transition form factor $F_{\gamma \to M^0}(Q^2)$ is predicted to fall as $1/Q^2$—modulo calculable logarithmic corrections from the evolution of the meson distribution amplitude. This QCD prediction reflects the elementary scaling of the quark propagator at high momentum transfer, the same scale-free behavior which leads to Bjorken scaling of the deep inelastic lepton-nucleon cross sections. The existing data from TPC/$\gamma\gamma$ are consistent with the predicted scaling and normalization of the transition form factors for the π^0, η_0, and η'.

The angular distributions for the hadron pair production processes $\gamma\gamma \to H\overline{H}$ are sensitive to the x_i dependence of the hadron distribution amplitudes.[12] Lowest order predictions for meson pair production in two photon collisions using this formalism are given in Refs. 12 and 8; the analysis of the $\gamma\gamma$ to meson pair process has been carried out to next-to-leading order in $\alpha_s(Q^2)$ by Nizic.[13] The Mark II and TPC/$\gamma\gamma$ measurements of $\gamma\gamma \to \pi^+\pi^-$ and $\gamma\gamma \to K^+K^-$ reactions are also consistent with PQCD expectations. A review of this work is given in Ref. 14.

Compton scattering $\gamma p \to \gamma p$ at large momentum transfer and its s-channel crossed reactions $\gamma\gamma \to \overline{p}p$ and $\overline{p}p \to \gamma\gamma$ are classic tests of the perturbative QCD formalism for exclusive reactions. At leading twist, each helicity amplitude has the factorized form,[3]

$$\mathcal{M}_{hh'}^{\lambda\lambda'}(s,t) = \sum_{d,i} \int [dx][dy]\phi_i(x_1,x_2,x_3,\widetilde{Q}) T_i^{(d)}(x,h,\lambda;y,h',\lambda';s,t)\phi_i(y_1,y_2,y_3;\widetilde{Q}) \,.$$

The index i labels the three contributing valence Fock amplitudes at the renormalization scale $\widetilde{Q}$. The index d labels the 378 connected Feynman diagrams which contribute to the eight-point hard scattering amplitude $qqq\gamma \to qqq\gamma$ at the tree level; i.e. at order $\alpha\alpha_s^2(\widehat{Q})$. The arguments $\widehat{Q}$ of the QCD running coupling constant can be evaluated amplitude by amplitude using the method of Ref. 15. The evaluation of the hard scattering amplitudes $T_i^{(d)}(x,h,\lambda;y,h',\lambda';s,t)$ has now been done by several groups.[16,17,18,19]

An important simplification of Compton scattering in PQCD is the fact that pinch singularities are readily integrable and do not change the nominal power-law behavior of the basic amplitudes.[18] Physically, the pinch singularities correspond to the existence of potentially on-shell intermediate states in the hard scattering amplitudes. This leads to a non-trivial phase structure of the Compton amplitude. Such phases can in principle be measured by interfering the virtual Compton process in $e^{\pm}p \to e^{\pm}p\gamma$ with the purely real Bethe-Heitler bremsstrahlung amplitude.[20] A careful analytic treatment of the integration over the on-shell intermediate states has been given by Kronfeld and Nizic.[18]

The most characteristic feature of the PQCD predictions is the scaling of the differential Compton cross section at fixed t/s or θ_{CM}

$$s^6 \frac{d\sigma}{dt}(\gamma p \to \gamma p) = F(t/s).$$

The power s^6 reflects the fact that 8 elementary fields enter or leave the hard scattering subprocess.[4] The scaling of the existing data[21] is remarkably consistent with the PQCD power-law prediction, but measurements at higher energies and momentum transfer are needed to test the predicted logarithmic corrections to this scaling behavior and determine the angular distribution of the scaled cross section over as large a range as possible.

The predictions for the normalization of the Compton cross section and the shape of its angular distribution are sensitive to the shape of the proton distribution amplitude $\phi_p(x_i,Q)$. The forms predicted for the proton distribution amplitude from QCD sum-rule constraints[8] by Chernyak, Oglobin, and Zhitnitskii, and King and Sachrajda, appear to give a reasonable representation of the existing data. More recent QCD sum rule analyses of the proton distribution amplitude are given in Ref. 22. These distributions, which predict that approximately 65% of the proton's momentum is carried by the u quark with helicity parallel to the proton's helicity also provide empirically consistent predictions for the normalization of the proton's form factor and the $J/\psi \to p\overline{p}$ decay rate. The crossing behavior from spacelike Compton scattering to the timelike annihilation channels will also provide important tests and constraints on the PQCD formalism and the shape of the proton distribution amplitudes. Predictions for the time-like processes have been made by Farrar *et al.*,[16] Millers and Gunion[17], and Hyer.[19]

The theoretical uncertainties from finite nucleon mass corrections, the magnitude of the QCD running coupling constant, and the normalization of the proton distribution amplitude largely cancel out in the ratio of differential cross sections

$$R_{\gamma\gamma/e^+e^-}(s,\theta_{cm}) = \frac{d\sigma(\overline{p}p \to \gamma\gamma)/dt}{d\sigma(\overline{p}p \to e^+e^-)/dt},$$

which is predicted by QCD to be essentially independent of s at large momentum transfer. If this scaling is confirmed, then the center-of-mass angular dependence of $R_{\gamma\gamma/e^+e^-}(s,\theta_{cm})$ will be one of the best ways to determine the shape of $\phi_p(x_i, Q)$.

2. Lepto-Production of Vector Mesons as a Test of PQCD and Color Transparency

The study of real and virtual photoproduction of vector mesons on protons and nuclei provides an elegant illustration of the emergence of perturbative QCD features in the large momentum transfer domain.[23,24]

1. At small momentum transfer and high energy where the coherence length $2\nu/(\mathcal{M}^2 + Q^2)$ is large compared to the target size, the incident photon is expected to act as a coherent sum of vector mesons with mass squared $\mathcal{M}^2 \leq \mathcal{O}(Q^2)$. This is the generalized vector meson dominance picture of photon interactions. In addition, $s-$channel helicity conservation predicts that the vector meson will be dominantly produced with transverse polarization equal to that of the incident photon.

2. At small momentum transfer where photon interactions are dominantly hadron-like, the cross section for vector meson photoproduction on a nucleus should have the same nuclear properties as meson-nucleon scattering. Due to the optical theorem, the forward high energy coherent nuclear amplitude $\gamma^* A \to V^0 A$ must then scale with the nuclear size the same as the total hadron-nucleus cross section; *i.e.* $A^{2/3}$. The $t-$dependence of the coherent nuclear cross section is of the form $d\sigma/dt \sim \exp^{b_A t}$ where $b_A \propto R_A^2$ and R_A is the nuclear size. Thus the total coherent cross section $\sigma(\gamma^* A \to V^0 A)$ is predicted to scale with nuclear number as $A^{4/3}/R_A^2 \sim A^{2/3}$.

3. The predictions for $\gamma^* A \to V^0 A'$ are in striking contrast to the above results when Q^2 becomes large compared to Λ_{QCD}^2. The virtual quark loop connecting the photon to the vector meson is now highly virtual, and only the point-like

piece of the photon and the small transverse size of the valence $q\bar{q}$ light-cone wavefunction of the vector meson enter the exclusive amplitude. Thus at high Q^2 the nuclear absorption in the initial and final state should vanish, and the nuclear amplitude becomes additive: $M(\gamma^* A \to V^0 A') = A^1 M(\gamma^* N \to V^0 N')$. The integrated coherent cross section $\sigma(\gamma^* A \to V^0 A)$ is thus predicted to scale with nuclear number as $A^2/R_A^2 \sim A^{4/3}$. This contrasting nuclear dependence of the virtual photoproduction cross section provides a dramatic test of color transparency. Preliminary results from E665[25] for ρ lepto-production at Fermilab appear to confirm these QCD predictions.

4. Another important prediction of PQCD in the large Q^2 domain is that the vector meson should be produced with zero helicity since it is formed from a quark and antiquark with equal and opposite helicities.[26] The change-over from transverse to longitudinal vector meson polarization with increasing Q^2 also appears to be confirmed by the E665 data.

5. At large photon virtuality Q^2 the photon and vector meson will act as point-like systems, and thus the $t-$ dependence of the differential cross section $d\sigma/dt(\gamma^* p \to V^0 p')$ should only reflect the finite size of the scattered nucleon. At large t the form factors should reflect the underlying two-gluon exchange structure of the PQCD Pomeron.

6. At large momentum transfer $-t \gg \Lambda_{QCD}^2$, $-u \gg \Lambda_{QCD}^2$, PQCD predicts that the photoproduction cross section has the nominal fixed CM angle scaling: $d\sigma/dt(\gamma p \to V^0 p') \sim f(\theta_{CM})/s^7$. The dominant amplitudes will conserve hadron helicity: $\lambda_{p'} + \lambda_V = \lambda_p$.

7. At larger momentum transfers $-t > R_A^2$, one can study quasi-elastic lepto-production in the nucleus; $d\sigma/dt(\gamma^* A \to V^0 N' X)$ where X represents a sum over excited nuclear states, but without extra particle production. When $p_T^2 \gg \Lambda_{QCD}^2$, color transparency predicts the absence of initial or final state absorption of the incident photon and the outgoing meson and nucleon. Thus the quasi-elastic cross section should approach additivity in nuclear number at large momentum transfer.

4. When Do Leading-Twist Predictions for Exclusive Processes Become Applicable?

The factorized predictions for exclusive amplitudes are evidently rigorous predictions of QCD at large momentum transfer. However, it is important to understand the kinematic domain where the leading twist predictions become valid. The basic scales of QCD are set by the quark masses and the scale Λ_{QCD} which parameterizes the QCD running coupling constant. Thus one normally would expect that the leading power-law predictions should become dominant at momentum transfers exceeding these parameters. In the case of inclusive reactions, Bjorken scaling is already apparent at momentum transfers $Q \sim 1$ GeV or less.

In fact, the data for hadron form factors is consistent with the onset of PQCD scaling at momentum transfers of a few GeV. Stoler[27] has shown that the measurements of the transition form factors of the proton to the $N(1535)$ and $N(1680)$ resonances are consistent with the predicted PQCD Q^{-4} scaling to beyond $Q^2 = 20\ GeV^2$. The normalization is also in reasonable agreement with that predicted from QCD sum rule constraints on the nucleon distribution amplitudes, allowing for uncertainties from higher order QCD corrections. In the case of the proton to $\Delta(1232)$ transition, the form factor falls faster that Q^{-4}. This anomalous behavior is, in fact, predicted by QCD sum rule constraints, since unlike the proton, the Δ has a highly symmetric distribution amplitude which results in a small net coupling to the QCD hard scattering amplitude. The observed scaling pattern of the transition form factors gives strong support to the QCD sum rule predictions and PQCD factorization.

Isgur and Llewellyn Smith[28] and Radyushkin[29] have raised the concern that important contributions to exclusive processes could arise from the endpoint regions $x_i \rightarrow 1$; such behavior would imply the breakdown of PQCD factorization. For example, the denominator of the hard scattering amplitudes, $e.g.$, $T_H \propto \alpha_s/[(1 - x)(1 - y)Q^2]$ for the meson form factor becomes singular in the endpoint integration region at $x \sim 1$ and $y \sim 1$. Such endpoint regions are even further emphasized when one assumes the strongly asymmetric forms for the hadron distribution amplitudes derived from QCD sum rules. However, it is important to note that these endpoint regimes correspond to scattering processes where one quark carries nearly all of the proton's momentum and is at a fixed transverse separation $b_\perp$ from the spectator quarks.

When a quark which is isolated in space receives a large momentum transfer $x_i Q$, it will normally strongly radiate gluons into the final state due to the displacement of both its initial and final self-field, which is contrary to the requirements of exclusive scattering. For example, in QED the radiation from the initial and final state charged lines is controlled by the coherent sum $\sum_i \frac{\epsilon \cdot p_i}{k \cdot p_i} \eta_i q_i$ where q_i and p_i are the charges four-momenta of the charged lines, ϵ and k are polarization and four-momentum of the radiation, and $\eta_i = \pm 1$ for initial and final state particles, respectively. Radiation will occur for any finite momentum transfer scattering as long as the photon's wavelength is less than the size of the initial and final neutral bound states. The probability amplitude that radiation does not occur is given by rapidly falling Sudakov form factor, as first discussed by in Refs. 5 and 30. An elegant and much more complete discussion has now been given by Botts and Li and Sterman.[31] The radiation from the colored lines in QCD have similar coherence properties as in QED:[32] because of the destructive color interference of the radiators, the momentum of the radiated gluon in a QCD hard scattering process only ranges from k of order $1/b_\perp$, where color screening occurs, up to the momentum transfer $x_i Q$ of the scattered quarks. This analysis and unitarity allows one to compute the probability that no radiation occurs during the hard scattering.[31,19] It is given by a rapidly falling exponentiated Sudakov form factor $S = S(x_i Q, b_\perp, \Lambda_{QCD})$; thus at large Q and fixed impact separation, the Sudakov factor strongly suppresses the endpoint contribution.

On the other hand, when $b_\perp = \mathcal{O}(x_i Q)^{-1}$, the Sudakov form factor is of order 1, and the radiation leads to logarithmic evolution and contributions of higher order in $\alpha_s(Q^2)$, the corrections already contained in the PQCD predictions.[5,30,33] This is the starting point of the detailed analysis of the suppression of endpoint contributions to meson and baryon form factors and its quantitative effect on the PQCD predictions recently presented by Li and Sterman.[31] This analysis has now also been applied to two-photon reactions and the timelike proton form factor by Hyer.[19]

Thus the leading PQCD contributions to large momentum transfer exclusive reactions derive from wavefunction configurations where the valence quarks are at small transverse separation $b_\perp = \mathcal{O}(1/k_\perp) = \mathcal{O}(1/Q)$, the regime where there is no Sudakov suppression. Furthermore, as noted by Li and Sterman, the hard scattering amplitude loses its singular endpoint structure if one explicitly retains the valence quark transverse momenta in the denominators. For example, in the case of the pion form factor, the hard scattering amplitude is effectively modified to the form

$$T_H \propto \frac{\alpha_s}{(1-x)(1-y)Q^2 + (\mathbf{k}_1^\perp + \mathbf{k}_2^\perp)^2}.$$

The Sudakov effect thus ensures that the denominators are always protected at large momentum transfers. In their numerical studies, Li and Sterman find that the pion form factor becomes relatively insensitive to soft gluon exchange at momentum transfers beyond 20 Λ_{QCD}. In the case of the proton Dirac form factor, the corresponding analysis by Li[31] is in good agreement with experiment at momentum transfers greater than 3 GeV. Thus the leading twist QCD predictions based on the factorization of long and short distance physics appear to be self-consistent and valid for momentum transfers as low as a few GeV, thus accounting for the empirical success of quark counting rules in exclusive process phenomenology. The Sudakov effect suppression also enhances the QCD "color transparency" phenomena, since only small color singlet wavefunction configurations can scatter at large momentum transfer without radiation.[10]

The extension of the leading order PQCD analysis to higher orders including Sudakov effects is technically very challenging. Thus far, the next-to-leading $\alpha_s(Q^2)$ corrections to the hard scattering amplitudes T_H have been computed for only a few exclusive processes: the meson form factor, the photon-to-meson transition form factors, and $\gamma\gamma$ to meson pairs. There are many outstanding theoretical issues which are being resolves, such as how to extend these calculations to baryon processes, how to set the renormalization scale in α_s,[15] how to implement conformal symmetry and its breaking,[9,34] and how to formulate and solve the evolution equations for the hadron distribution amplitudes to next-to-leading order.

6. Other Applications of Large Momentum Transfer Exclusive QCD.

The factorization techniques used to derive the leading-twist behavior of exclusive amplitudes have general applicability to processes where hadron wavefunctions have to be evaluated at far off-shell configurations. In each of these applications, one can separate the perturbative quark and gluon dynamics from momentum transfer higher than a scale Q from the non-perturbative long-distance physics contained in the distribution amplitudes $\phi(x_i, Q)$. For example at $x \sim 1$ the struck quark in deep inelastic lepton-hadron scattering is kinematically far off shell and space-like. Thus the leading power law fall off in $(1 - x)$ is determined by iterating the gluon exchange kernel in the valence Fock state wavefunction. In this way one derives "spectator" counting rules for the nominal power law behavior [e.g. $G_{q/p}(x) \sim (1 - x)^3$] and helicity-retention rules at $x \to 1$. The resulting structure functions connect smoothly to the behavior of large momentum transfer elastic and inelastic transition form factors at fixed $\mathcal{M}^2$. In fact, when $(1 - x)Q^2$ is fixed, the usual evolution of the structure functions breaks down and there is no increase in the effective power beyond that given by the spectator counting rules. Further discussion may be found in Ref. 35.

Higher-twist corrections to inclusive reactions are of two types: coherent corrections which depend on the multiparticle structure of hadrons, and single particle corrections, such as mass and condensate insertions, which affect single quark or single gluon propagators. Exclusive processes represent the completely coherent limit of dynamical higher twist terms in inclusive reactions. At fixed $(1 - x)Q^2$, the multiquark higher twist contributions can be computed using the exclusive factorization analysis, and they contribute at the same order as the leading twist terms.[36,37] Strong higher-twist corrections are in fact observed in the angular and Q^2–dependence of Drell-Yan processes and in deep inelastic lepton scattering at $x \sim 1$.[38]

The factorization techniques used to derive the leading twist contributions to form factors can also be applied to the exclusive decays of heavy hadrons when large momentum transfers are involved. An interesting example of this analysis is "atomic alchemy",[39] *i.e.* the exclusive decays of muonic atoms to electronic atoms plus neutrinos. In this case the calculation requires the high momentum tail of the atomic wavefunctions, which in turn can be obtained via the iteration of the relativistic atomic bound-state equations. Again one obtains a factorization theorem for exclusive atomic transitions where the atomic wavefunction at the origin plays the role of the distribution amplitude.

7. Outstanding Phenomenological Issues in Exclusive Processes.

Although most large momentum transfer exclusive reactions appears to be empirically consistent with perturbative QCD expectations, there are a number of glaring exceptions where theory and experiment diverge. If one accepts that the underlying formalism for the leading twist behavior of exclusive reactions is reliable, then these exceptions provide important insights into new physical mechanisms within QCD.

What accounts for the structure in the spin correlations in pp **elastic scattering at large momentum transfer?** Measurements[40] of large angle pp elastic scattering at Argonne and Brookhaven show a dramatic spin-spin correlation

A_{NN} which reaches ~ 0.6 at $\sqrt{s} \sim 5$ GeV: *i.e.* the spin-analyzed cross section is four times larger if the protons scatter with their spins parallel and normal to the scattering plane compared to antiparallel. The explanation for this phenomena is far from settled. The most popular explanations[41] are based on the interference of Landshoff pinch singularities[42] with the quark interchange amplitude, but there is no understanding why the Landshoff contribution would itself have a large A_{NN}[43] or sufficient normalization[44] to explain this phenomena. Guy de Teramond and I have proposed[45] that the large spin correlations reflects inelastic channels corresponding to the production of charm at threshold. This effect leads to enhancement in the $J = L = S = 1$ $pp \to pp$ partial wave which implies a large value of A_{NN} at the energies sufficient to produce open charm. This explanation would be confirmed by the observation of a sizeable charm production rate of order $1\mu bn$. A similar enhancement of A_{NN} is seen at the open strangeness threshold regime. The heavy quark explanation has received some support from the work of Luke, Savage, and Manohar,[46] who have shown that the interactions of $c\bar{c}$ systems at low relative velocity with hadrons is enhanced due to the QCD scale anomaly; in fact, the scalar exchange interaction is predicted to be strong enough to bind charmonium to heavy nuclei.[47]

Why does QCD color transparency appear to break down in quasielastic pp scattering? The Brookhaven measurements[48] of the transparency ratio for large angle quasi-elastic pp scattering increases with momentum transfer, as predicted by PQCD, but the ratio then appears to revert to normal absorption at $\sqrt{s} \sim 5$ GeV. This suggests that whatever is causing the structure in A_{NN} at the same energies and angles involves large transverse sizes and is far from perturbative in origin. The charm threshold effect is a candidate for this type of explanation.

The preliminary results for the SLAC color transparency experiment NE18 reported at this meeting[49] indicate that color transparency in quasi-elastic ep scattering is not a strong effect up to the accessible momentum transfers. Higher momentum transfers exceeding 5 GeV are needed for a decisive test. A sensitive test of color transparency is provided by measuring the sign of the derivative of the transparency ratio $d/dQ^2 \sigma(eA \to e'p(A-1))/Z\sigma(ep \to e'p)$. Perturbative QCD predicts a positive slope, whereas conventional Glauber theory predicts a negative derivative in the low Q^2 domain.

Why does the J/ψ decay copiously to $\rho\pi$? According to the principle of hadron helicity conservation[26] in exclusive decays, the J/ψ produced with $J_z = \pm 1$ in e^+e^- annihilation should not decay to vector plus pseudoscalar meson pairs. In fact, this is true for the ψ' and other S-state charmonium states, but in the case of the J/ψ, the $\rho\pi$ and KK^* psuedoscalar-vector meson channels are actually the dominant two-body hadronic decays. A possible explanation is that the J/ψ mixes with a nearby gluonic or hybrid $J = 1$ state $\mathcal{O}$ that favors vector plus pseudoscalar meson pair decay.[50] One can search for the $\mathcal{O}$ by looking for a $\rho\pi$ mass peak near the J/ψ in the decay $\psi' \to \pi\pi\mathcal{O} \to \pi\pi\rho\pi$.

Why do effective Reggeon trajectories flatten to values below $\alpha_R(t) = 0$ at large momentum transfer? A fundamental prediction of perturbative QCD is that the Reggeon trajectories $\alpha_\rho(t)$ and $\alpha_{A_2}(t)$ governing charge exchange reactions at high energies $s \gg -t$ monotonically approach zero at large spacelike momentum transfer.[51] More generally, the leading Reggeon in an exclusive process will reflect the minimal particle number exchange quantum numbers: two gluons in the case of the Pomeron, three gluons in the case of the Odderon, and quark plus anti-quark in the case of meson exchange trajectories. Because of asymptotic freedom the leading trajectory at large momentum transfer is thus simply $j_1 + j_2 - 1$ with corrections of order $\sqrt{\alpha_s}(-t)$. The asymptotic prediction $\lim_{-t \to \infty} \alpha_R(t) = 0$ reflects the fact that a weakly interacting quark-antiquark pair is exchanged in the $t-$channel.[51] Thus one expects that the effective ρ Reggeon should asymptote at $\alpha_\rho(t) \to 0$ at large $-t$. However, measurements of the inclusive processes $\pi^- p \to \pi^0 X$ at $s \simeq 300$ GeV^2 and $8 > -t > 2$GeV^2 indicate that the effective non-singlet ρ trajectory becomes negative at large $-t$.[52] Thorn, Tang and I have recently shown that the hard QCD part of the trajectory is weakly coupled and that its contribution may well be hidden until much higher energy.[53] Quark interchange[54] may thus be the dominant subprocess at presently accessible kinematic ranges. We also show that Reggeon contributions to exclusive and semi-inclusive mesonic exchange hadron reactions can be systematically studied in perturbative QCD.

Why is quark interchange the dominant mechanism for large-angle hadron-hadron scattering? The comprehensive measurements at BNL[55] of the relative normalization and angular dependence of a large set of exclusive hadron scattering channels strongly suggests that the dominant mechanism for scattering hadrons at large momentum transfer is quark interchange.[54] For example, if gluon exchange were the dominant mechanism, then the differential cross sections for $K^+ p \to K^+ p$ and $K^- p \to K^- p$ at large p_T would be roughly equal in magnitude and angular shape. In fact they have grossly different magnitudes and shapes. The $K^+ p \to K^+ p$ cross section has the approximate form predicted by the exchange of their common u quark. A possible explanation of this fact is that quark interchange involves the least number of large momentum exchanges within the hadron scattering amplitude.

Acknowledgements

I wish to thank Carl Carlson and Paul Stoler and the other members of the organizing committee for organizing an outstanding meeting in Elba. This work was supported by the U.S. Department of Energy under Contract No. DE-AC03-76SF00515.

REFERENCES

[1] S. D. Drell and T. M. Yan, *Phys. Rev. Lett.* **24** (1970) 181.

[2] S. J. Brodsky and H. C. Pauli in *Recent Aspects of Quantum Fields*, H. Mitter and H. Gausterer, Eds.; Lecture Notes in Physics, Vol. 396, Springer-Verlag, Berlin, Heidelberg, (1991), and reference therein.

[3] For a review of the theory of exclusive processes in QCD and additional references see S. J. Brodsky and G. P. Lepage in *Perturbative Quantum Chromodynamics*, edited by A. Mueller (World Scientific, Singapore, 1989).

[4] S. J. Brodsky and G. R. Farrar, *Phys. Rev.* **D11** (1975) 1309.

[5] G. P. Lepage and S. J. Brodsky, *Phys. Rev.* **D22**, 2157 (1980); *Phys. Lett.* **87B** (1979) 359; *Phys. Rev. Lett.* **43** (1979) 545, 1625E.

[6] General QCD analyses of exclusive processes are given in Ref. 5, S. J. Brodsky and G. P. Lepage, SLAC-PUB-2294, presented at the *Workshop on Current Topics in High Energy Physics*, Caltech (Feb. 1979), S. J. Brodsky, in the *Proc. of the La Jolla Inst. Summer Workshop on QCD*, La Jolla (1978), A. V. Efremov and A. V. Radyushkin, *Phys. Lett.* **B94** (1980) 245, V. L. Chernyak, V. G. Serbo, and A. R. Zhitnitskii, *Yad. Fiz.* **31**, (1980) 1069, S. J. Brodsky, Y. Frishman, G. P. Lepage, and C. Sachrajda, *Phys. Lett.* **91B** (1980) 239, and A. Duncan and A. H. Mueller, *Phys. Rev.* **D21** (1980) 1636.

[7] QCD predictions for the pion form factor at asymptotic Q^2 have ben given by V. L. Chernyak, A. R. Zhitnitskii, and V. G. Serbo, *JETP Lett.* **26** (1977) 594, D. R. Jackson, Ph.D. Thesis, Cal Tech (1977), and G. Farrar and D. Jackson, *Phys. Rev. Lett.* **43** (1979) 246; and ref.6. See also A. M. Polyakov, *Proc. of the Int. Symp. on Lepton and Photon Interactions at High Energies*, Stanford (1975), and G. Parisi, *Phys. Lett.* **84B** (1979) 225. See also S. J. Brodsky and G. P. Lepage, in *High Energy Physics–1980, Proceedings of the XXth International Conference*, Madison, Wisconsin, edited by L. Durand and L. G. Pondrom (AIP, New York, 1981); p. 568. A. V. Efremov and A. V. Radyushkin, Rev. Nuovo Cimento **3**, 1 (1980); *Phys. Lett.* **94B** (1980) 245. V. L. Chernyak and A. R. Zhitnitskii, *JETP Lett.* **25** (1977) 11; M. K. Chase, *Nucl. Phys.* **B167** (1980) 125.

[8] V. L. Chernyak and A. R. Zhitnitskii, *Phys. Rept.* **112** (1984) 173; V. L. Chernyak, A. A. Oglobin, and I. R. Zhitnitskii, *Sov. J. Nucl. Phys.* **48** (1988) 536; I. D. King and C. T. Sachrajda, *Nucl. Phys.* **B297** (1987) 785; M. Gari and N. G. Stefanis, *Phys. Rev.* **D35** (1987) 1074; and references therein.

[9] S. J. Brodsky, Y. Frishman, G. P. Lepage and C. Sachrajda, *Phys. Lett.* **91B** (1980) 239. M. E. Peskin, *Phys. Lett.* **88B** (1979) 128.

[10] S. J. Brodsky and A. H. Mueller, *Phys. Lett.* **206B** (1988) 685, and references therein; G. Bertsch, S. J. Brodsky, A. S. Goldhaber, J.F. Gunion, *Phys. Rev. Lett.* **47** (1981) 297.

[11] See, for example, B. K. Jennings and G.A. Miller DOE-ER-40427-00-N93-11, (1993) and *Phys. Lett.*$\mathbf{B236}$ (1990) 209; G. R. Farrar, H. Liu, L. L. Frankfurt, and M. I. Strikman, *Phys. Rev. Lett.* $\mathbf{61}$ (1988) 686; N. N. Nikolaev and B. G. Zakharov, *Z. Phys.* $\mathbf{C49}$ (1991) 607; L. L. Frankfurt, M. I. Strikman, and M. B. Zhalov, preprint (1993); J. P. Ralston and B. Pire, *Phys. Rev. Lett.* $\mathbf{61}$ (1988) 1823, and in the *Proceedings of the 1989 24th Rencontre de Moriond* (1989).

[12] S. J. Brodsky and G. P. Lepage, *Phys. Rev.* $\mathbf{D24}$ (1981) 1808.

[13] B. Nizic, *Fizika* $\mathbf{18}$ (1986) 113.

[14] For a review of exclusive two-photon processes, see S. J. Brodsky, *Proceedings of the Tau-Charm Workshop*, Stanford, CA (1989).

[15] S. J. Brodsky, G. P. Lepage, and P. B. Mackenzie, *Phys. Rev.* $\mathbf{D28}$ (1983) 228.

[16] G. R. Farrar, *et al. Nucl. Phys.* $\mathbf{B311}$ (1989) 585.

[17] D. Millers and J. F. Gunion, *Phys. Rev.* $\mathbf{D34}$ (1986) 2657.

[18] A. N. Kronfeld and B. Nizic, *Phys. Rev.* $\mathbf{D44}$ (1991) 3445; B. Nizic, *Phys. Rev.* $\mathbf{D35}$(1987) 80.

[19] T. Hyer, *Phys. Rev.* $\mathbf{D47}$ (1993) 3875.

[20] S. J. Brodsky, F. E. Close, J. F. Gunion, *Phys. Rev.* $\mathbf{D6}$ (1972) 177.

[21] M. A. Shupe, *et al.*, *Phys. Rev.* $\mathbf{D19}$ (1979) 1921.

[22] M. Bergmann and N. G. Stefanis, Bochum preprints RUB-TPH-36/93, RUB-TPH-46/93, and RUB-TPH-47/93.

[23] S. J. Brodsky, in the *Proceedings of the Topical Conf. on Electronuclear Physics with Internal Targets*, Stanford, (1989); S. J. Brodsky, B. T. Chertok, *Phys. Rev.* $\mathbf{D14}$ (1976) 3003.

[24] B. Z. Kopeliovich, J. Nemchick, N. N. Nikolaev, B. G. Zakharov, *Phys. Lett.* $\mathbf{B309}$ (1993) 179.

[25] G. Fang, preliminary E665 results presented at the INT - Fermilab Workshop on *Perspectives of High Energy Strong Interaction Physics at Hadron Facilities* (1993).

[26] G. P. Lepage and S. J. Brodsky, *Phys. Rev.* $\mathbf{D24}$ (1981) 2848.

[27] P. Stoler, *Phys. Rev.* $\mathbf{D44}$ (1991) 73, *Phys. Rev. Lett.* $\mathbf{66}$ (1991) 1003.

[28] N. Isgur and C. H. Llewellyn Smith, *Phys. Rev. Lett.* $\mathbf{52}$ (1984) 1080; *Phys. Lett* $\mathbf{B217}$ (1989) 535.

[29] A. V. Radyushkin, *Nucl. Phys.* $\mathbf{A532}$ (1991) 141.

[30] A. Duncan, and A. H. Mueller, *Phys. Lett.* $\mathbf{90B}$ (1980) 159.

[31] J. Botts and G. Sterman, *Nucl. Phys.* $\mathbf{B325}$ (1989) 62; *Phys. Lett.* $\mathbf{B224}$ (1989) 201; J. Botts, J.-W. Qiu, and G. Sterman, *Nucl. Phys.* $\mathbf{A527}$ (1991) 577. H. N. Li and G. Sterman, *Nucl. Phys.* $\mathbf{B381}$ (1992) 129. H. N. Li, Stony Brook preprint ITP-SB-92-25 (1991).

[32] S. J. Brodsky and J. F. Gunion, *Phys. Rev. Lett.* $\mathbf{37}$ (1976) 402.

[33] A. Szczepaniak and L. Mankiewicz, *Phys. Lett.* **B266** (1991) 153.

[34] D. Mueller, SLAC-PUB (1993).

[35] S. J. Brodsky, I. A. Schmidt, *Phys. Lett.* **B234** (1990) 144, and references therein; S. J. Brodsky, in the *Proceedings of the International Symposium on High-Energy Spin Physics,* Nagoya, Japan, (1992).

[36] S. J. Brodsky, E. L. Berger, G. Peter Lepage, *Proc. of the Drell-Yan Workshop,* Fermilab (1982); E. L. Berger and S. J. Brodsky, *Phys. Rev. Lett.* **42** (1979) 940. For a recent analysis and additional references see S. S. Agaev, *Z. Phys.* **C57** (1993) 403.

[37] S. J. Brodsky, P. Hoyer, A. H. Mueller, W-K. Tang, *Nucl. Phys.* **B369** (1992) 519.

[38] See, *e.g.,* J. S. Conway *et al., Phys. Rev.* **D39** (1989) 92.

[39] S. J. Brodsky, C. Greub, C. Munger, and D. Wyler, to be published.

[40] For a summary of the spin correlation data see A. D. Krisch, *Nucl. Phys. B (Proc. Suppl.)* **25B** (1992) 285.

[41] See, for example, J. P. Ralston and B. Pire, *Phys. Rev. Lett.* **49** (1982) 1605; C. E. Carlson, M. Chachkhunashvili, F. Myhrer, *Phys. Rev.* **D46** (1992) 2891; G. P. Ramsey, D. Sivers, *Phys. Rev.* **D47** (1992) 93; and references therein.

[42] P. V. Landshoff, *Phys. Rev* **D10** (1974) 1024.

[43] S. J. Brodsky, C. E. Carlson, H. J. Lipkin, *Phys. Rev.* **D20** (1979) 2278.

[44] Presented at the INT - Fermilab Workshop on *Perspectives of High Energy Strong Interaction Physics at Hadron Facilities* (1993).

[45] S. J. Brodsky and G. F. de Teramond, *Phys. Rev. Lett.* **60** (1988) 1924.

[46] M. Luke, A. V. Manohar, M. J. Savage, *Phys. Lett.* **B288** (1992) 355.

[47] S. J. Brodsky, and G. F. de Teramond, and I. A. Schmidt, *Phys. Rev. Lett.* **64** (1990) 1011.

[48] S. Heppelmann, *Nucl. Phys. B, Proc. Suppl.* **12** (1990) 159, and references therein.

[49] A. Lung, these proceedings.

[50] S. J. Brodsky, G. Peter Lepage, S. F.Tuan, *Phys. Rev. Lett.* **59** (1987) 621, and references therein.

[51] R. Kirshner and L. N. Lipatov, *Sov. Phys. JETP* **56** (1982) 266; *Nucl. Phys.* **B213** (1983) 122.

[52] R. Blankenbecler, S. J. Brodsky, J. F. Gunion, and R. Savit, *Phys. Rev.* **D8** (1973) 4117.

[53] S. J. Brodsky, W-K. Tang, and C. B. Thorn, SLAC-PUB-6227 (1993).

[54] J. F. Gunion, S. J. Brodsky, and R. Blankenbecler, *Phys. Rev.* **D8** (1973) 287.

[55] A. Carroll, these proceedings; C. White *et al.,* BNL-49059, (1993); B. R. Baller *et al., Phys. Rev. Lett.* **60** (1988) 1118.

CAN SUDAKOV SAVE US?

CARL E. CARLSON

Physics Department, College of William and Mary
Williamsburg, VA 23187, USA

ABSTRACT

The question of the validity of perturbative QCD is examined from the viewpoint that any pertinent cutoffs should not be ad hoc but should be calculable as due to such things as Sudakov form factors or transverse momenta of quarks within a hadron.

1. Viewpoint

This talk will have two goals. One goal is to explain what a Sudakov form factor is. The phrase "Sudakov form factor" is heard often nowadays, and no–one should be in the dark about its meaning and origin. The other goal is to make a direct response to criticism of the use of perturbative QCD at experimentally feasible energies. The exclusive reactions program is motivated by a desire to probe the structure of hadrons and perturbative QCD is one of our calculational tools for relating observed quantities to hadron structure. We should know if its use is valid, and I shall give reason for believing that perturbative QCD is correctly giving a substantial fraction of the observed quantities at high but experimentally accessible momentum transfers.

Perturbative QCD (pQCD) has been applied to exclusive reactions at experimentally accessible energies with some apparent success. In particular, the scaling laws seem to have general success, and in some cases absolutely normalized calculations can be done with good results. In this contest, even one apparent failure of the pQCD scaling laws, namely the anomalously rapid falloff of the electromagnetic $N \to \Delta$ transition form factor, can be counted a success in that a normalized calculation can be done and it shows an unusually small coefficient of the term with the slowest falloff with increasing momentum transfer. In the absence of calculations of the next leading term in the falloff with momentum transfer of the form factor or whatever we are considering, the argument in favor of applying pQCD is roughly that when we try it, it seems to work.

Countering this are attempts to estimate the size of effects which may vitiate the expansion in inverse momentum transfer. Such estimates often make plausible ansatzes about how (say) some internal propagator will be modified from the approximation that leads to the leading order pQCD result, and then they see how much the leading order result is in fact affected. The results are dramatic in a very negative way. The critics show that the modifications of the propagators lead to great decreases in the results for the quantity under consideration. However, the propagator modification depends on some parameter that defines a mass or momentum scale for where the cutoff takes effect, and the value of this parameter is crucial.

From a modern viewpoint, one should calculate directly whatever is responsible for the correction to the leading order (in an expansion in 1/momentum transfer) and not make any ad hoc modifications to anything. Corrections to the leading order term come from Sudakov form factors, quark mass effects, quark transverse momenta, and perhaps other things.

2. Problem

We are discussing the use of pQCD to calculate exclusive reactions. For definiteness we will talk about the pion form factor calculation. The problem is in a plausible claim the calculation is unreliable, e.g., [1,2]. A synopsis of the argument is that the pQCD calculation of the form factor involves a range of internal gluon momenta, much of it at low momentum and thus long distance. Since the propagators used in a pQCD calculation are not correct at long distances, the calculation itself is not perturbative (i.e., perturbation theory does not give a good approximation to the correct answer).

To expand upon the negative argument, consider the usual pQCD calculation of F_π. In Fig. 1, we have a photon coming in "sideways" (meaning in this context $q^+ = q^0 + q^3 = 0$) and a single gluon, in lowest non–trivial order, carrying momentum from one quark to the other.

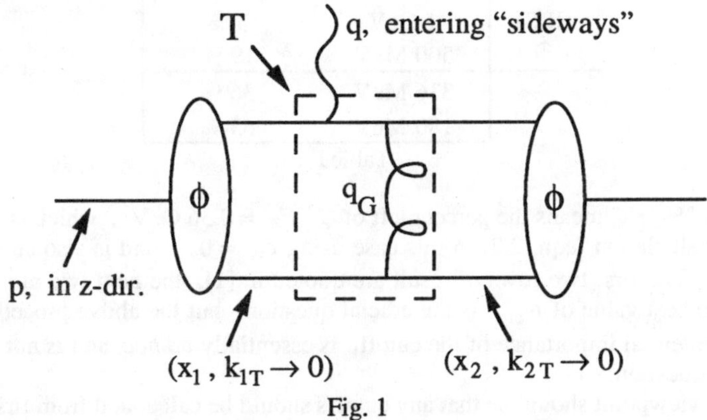

Fig. 1

In terms of the full wave function ψ and the full scattering amplitude T, the form factor is given by

$$F_\pi(Q^2) = \int dx_1\, dx_2\, d^2k_{1T}\, d^2k_{2T}\, \psi(x_2,k_{2T})\, T(x_2,k_{2T},x_1,k_{1T},Q)\, \psi(x_1,k_{1T}) \ , \qquad (2.1)$$

where $Q = q_T$ and $Q^2 = -q^2$. If Q is large enough we may neglect k_T in T and project the wave function into the distribution amplitude and get

$$F_\pi(Q^2) = \int dx_1\, dx_2\, \phi(x_2)\, T_H(x_2,x_1,Q)\, \phi(x_1) \ . \qquad (2.2)$$

But one may work out that with the neglect of k_T,

$$Q_G^2 = x_1 x_2 Q^2 \ , \qquad (2.3)$$

so that if x_1 (say) $\to 0$, then $Q_G^2 \to 0$, and pQCD is not applicable for the gluon propagator.

To make the non–applicability of pQCD quantitative, the critics take the viewpoint that the simple expression for Q_G^2 is modified by transverse momenta, by masses, by Sudakov form factors, and perhaps still other things. They approximate this by letting

$$Q_G^2 = x_1 x_2 Q^2 + m_{cut}^2 \tag{2.4}$$

and seeing how much the result is reduced. (For simplicity, they actually leave unchanged $Q_G^2 = x_1 x_2 Q^2$ and cut off the integrals if $Q_G^2 < m_{cut}^2$.) The result depends on the value of m_{cut} as shown in the table below, given for the Chernyak–Zhitnitsky wave function and for $Q^2 = 5\,\text{GeV}^2$.

	m_{cut}	% left
(ILS)	1 GeV	3 %
↑	500 MeV	19 %
	316 MeV	36 %
	180 MeV	63 %

Table I

The column "% left" means the percent left of $Q^2 F_\pi = 0.36\,\text{GeV}^2$, which is the result of the pQCD calculation (eqn. 2.2) in this case using $\alpha_S = 0.3$, and is also about what the data shows. The first two rows of result are quoted in [1], the next two are added here. Clearly, the best value of m_{cut} is the crucial question, but the above procedure, which shows the potential importance of the cutoff, is essentially *ad hoc* and is not designed to answer this question.

Today's viewpoint should be that any cutoffs should be calculated from first principles or as close thereto as we can come. The cutoffs are due to such things as Sudakov form factors or transverse momenta of the quarks in a hadron. We now turn to the first of these.

3. What did Sudakov do?

Sudakov calculated the form factor for free electrons [3]. (Sudakov's paper, incidentally, was submitted in 1954 though published in 1956, showing that people had problems even in the old days.) He included only the terms that were leading at very high Q^2, but included all orders of perturbation theory. For the photon–electron–electron vertex, with electron momenta p and p', his result was

$$\Gamma_\mu(p,p') = \gamma_\mu \, F(Q^2) \tag{3.2}$$

where

$$F(Q^2) = \exp-\left\{ \frac{e^2}{2\pi} \ln\frac{Q^2}{|p^2|} \ln\frac{Q^2}{|p'^2|} \right\} \equiv \exp-\left(\frac{e^2}{2\pi} \ln^2 Q^2 \right). \tag{3.1}$$

The form factor is thus small as $Q^2 \to \infty$, meaning that elastic scattering becomes unlikely. Elastic scattering must compete with inelastic scattering for event rate, and quoting directly from Sudakov, "Processes in which a large number of real photons are simultaneously emitted will occur with much greater probability."

Sudakov's result also works for QCD for hypothetical isolated quarks. The main difference is the running coupling, which leads to a simple change

$$\ln Q^2 \times \ln Q^2 \;\to\; \ln Q^2 \times \ln\left(\ln Q^2\right) . \tag{3.3}$$

The Sudakov form factor still squelches elastic scattering at high momentum transfer.

4. Effect and non-effect of Sudakov form factors on composites.

The effects of Sudakov form factors upon composite states was studied by Li and Sterman [4]. In a bound state there are two or more constituents, overall neutral in color. This will have the effect of canceling the Sudakov form factor when the constituents are close together. Think for a moment of gluon bremsstrahlung. If the constituents are far apart, judged compared to the wavelength of the emitted gluon, each constituent radiates independently and the Sudakov form factor for each of then is the same as for a free constituent. If they are close together, the gluon will see just a single color neutral object, and the gluon emission and hence the Sudakov form factor will not exist.

The effects of gluon bremsstrahlung are mirrored in the elastic process. If the quarks in the bound state are far apart, the Sudakov form factor suppresses high momentum transfer scattering as usual. However, if the quarks are near to each other, there are no Sudakov form factor effects. Thus a physical mechanism for the suppression of long distance contributions has been found, and we may find out, albeit with some calculational effort but without guesswork, how large the effect of the suppression is. Also, we are on our way to a self consistent perturbative calculation, since the Sudakov form factor removes what we cannot calculate reliably and leaves what we can.

Allow me to quote some details. Feynman diagrams such as those in Fig. 2 need to be calculated (and their leading terms extended to all orders). The discussion is easiest if we Fourier transform the transverse degrees of freedom, and discuss results in terms of a relative transverse separation b,

$$\psi(x,k_T) \to \psi(x,b) = \int d^2k_T \; \psi(x,k_T) \exp(-i\vec{k}_T \cdot \vec{b}) . \tag{4.1}$$

Ref. [4] casts the result for the Sudakov suppression as a factor for each quark (momentum fractions x_1 and $1-x_1$) multiplying the distribution amplitude of the bound state,

$$\psi(x_1,b_1) \to \phi(x_1)\, e^{-S(x_1,b_1,Q)-S(1-x_1,b_1,Q)} , \tag{4.2}$$

and give an expression for S. The quantity $\exp(-S)$ is 0 at long distance and essentially unity at short distance, which is taken to be below $\sqrt{2}/(xQ)$. Since x can be either x_1 or $(1-x_1)$, the condition for no suppression is

$$1 - \frac{\sqrt{2}}{bQ} < x_1 < \frac{\sqrt{2}}{bQ} \ . \tag{4.3}$$

Thus for small b, as expected, there is no x_1 suppressed, while at medium b, middle x_1 is unaffected, but there is a suppression of contributions from the end points. Incidentally, the physics of the $x_1 \to 0$ suppression is that the other quark is moving fast and hence is susceptible to bremsstrahlung.

So far only the diagrams in the first line of Fig. 2 are included [4]. That means not all of the phenomenon is yet included and in particular the *non*-suppression when the quarks are close to each other is not completely accounted for. Hence the true answer should be somewhat larger than the estimate given. However, the published results are the best currently available, and we proceed to show them.

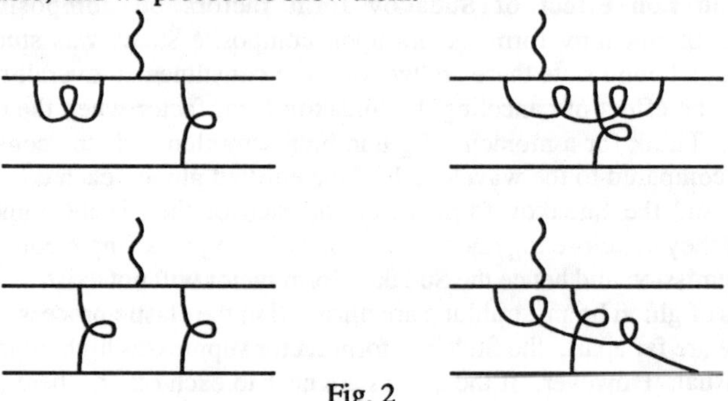

Fig. 2

The results are shown in Fig. 3, borrowed from Ref. [5]. The dashed line shows the result of Li and Sterman using the Chernyak–Zhitnitsky distribution amplitude [6] and a running α_S . The result is not in poor agreement with the data! We shall discuss the solid curve in the next section.

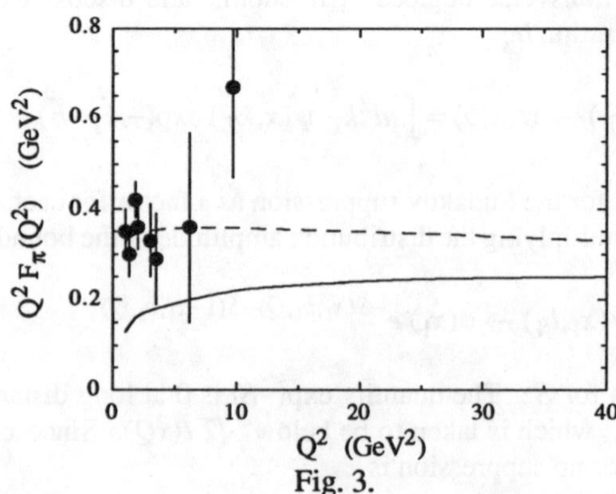

Fig. 3.

5. Transverse momentum effects

The quarks within the pion have some transverse momenta. We may neglect this fact at infinite momentum transfer and we have neglected it in the previous section. Now we shall estimate the effect of including transverse momentum at finite momentum transfers.

Transverse momentum effects give "higher twist" corrections, that is, corrections that add or subtract from leading order but are asymptotically smaller by powers of $1/Q^2$. There are other higher twist corrections, such as contributions from higher Fock components of the pion (e.g., $q\bar{q}G$). By and by, all the next higher twist contributions should be calculated, and we should view the transverse momentum effects in this context as a start in a longer project.

We will follow Jakob and Kroll [5]. We return to the earliest formula for F_π, Eq. (2.1) which was written in terms of the full $q\bar{q}$ wave function and the full T. For the wave function we use something like

$$\psi(x,k_T) = N\,\phi(x)\,e^{-\beta^2 k_T^2}$$

$$or \quad N\,\phi(x)\,(1+\beta^2 k_T^2)^{-n} \tag{5.1}$$

and we recall that the leading contribution at high momentum transfer will come from a term in T proportional to $1/Q_G^2$, and using an approximation of Li and Sterman,

$$Q_G^2 = x_1 x_2\, Q^2 + (k_{1T} - k_{2T})^2 \tag{5.2}$$

(see Fig. 1). Clearly for $x_1 x_2 Q^2 \gg (k_{1T} - k_{2T})^2$ we recover the old result. Keeping the transverse momenta and doing the integrals numerically leads to the solid curve shown in Fig. 3 when β is adjusted to give $\langle k_T \rangle = 300$ MeV. There is a noticeable but not, in my opinion, catastrophic decrease in the calculated size of $Q^2 F_\pi$.

Allow me a technical comment before ending this section. The approximation for the propagator above is inspired by Breit frame considerations, where the initial and final pions are moving in opposite directions. Writing four momentum components in the order $(+,-,T)$ where $k^\pm \equiv k^0 \pm k^3$, we have

$$k_1 = \left(x_1 Q, k_{1T}^2/x_1 Q, k_{1T}\right)$$

$$k_2 = \left(k_{2T}^2/x_2 Q, x_2 Q, k_{2T}\right). \tag{5.3}$$

Neglect of the k_T^2 terms directly above leads to the result for the propagator Eq. (5.2). Note that Q_G^2 from Eq. (5.2) is never zero and that including the transverse momenta always increases Q_G^2, thus always decreasing F_π.

Calculations are often best [7] done in the light cone frame, $q^+ = 0$, or

$$q = (0, 2m_N \, v, q_T) \ .$$

(5.4)

Now,

$$k_1 = \left(x_1 p^+, k_{1T}^2 / x_1 p^+, k_{1T}\right)$$

$$k_2 = \left(x_2 p^+, (x_2 q_T + k_{2T})^2 / x_2 p^+, x_2 q_T + k_{2T}\right) ,$$

(5.5)

and

$$Q_G^2 = \frac{1}{x_1 x_2}\left(x_1 x_2 q_T + x_1 k_{2T} - x_2 k_{1T}\right)^2 - i\eta$$

(5.6)

where η is a positive infinitesimal number. The propagator is (unsurprisingly) singular for the collinear case, but pQCD still applies, although that is a longer argument dependent on pQCD working at short range whatever the momentum. Do remember that whatever corrections come must be higher twist. Still, this result for Q_G^2 is different from the previous one, and one would like to know what effect this has. One might wish to work with the projection operator formalism shown in [8] where the singularity will not be so apparent.

To end this section on transverse momentum effects, we see that the result is a noticeable but not, in my opinion, huge suppression of the calculated form factor at experimental momentum transfers. The authors of [5] are more pessimistic, but to me it seems that pQCD is in the vicinity of the data. We do need to settle some questions of how best to do the calculations that include transverse momentum, and further need to pursue other corrections that are also next to leading twist.

6.   Higher order perturbative corrections

It is not only higher twist that we should worry about, but also simply higher orders in perturbation theory. It happens that the higher order perturbative corrections are very dependent on the wave function chosen. We may write the leading twist form factor as

$$F_\pi(Q) = F_\pi^{\mathrm{LO}}(Q,\mu)\left\{1 + \frac{\alpha_s(\mu^2)}{\pi} B(Q,\mu) + O(\alpha_s^2)\right\} ,$$

(6.1)

where "LO" stands for "lowest order." It is known that there are analogous situations where B can be as large as 20, making the perturbative expansion questionable. In the present case, we may write

$$B = -A \ln \frac{Q^2}{\mu^2} + C$$

(6.2)

where A and C involve integrals over the wave function and are otherwise Q^2 independent. There are four references to cite [9–12]. The first three give three different answers, the fourth provides the signal service of finding the errors. (In the notation of the references, I have written the result with $\mu_F = \mu_R = \mu$.)

If we use the asymptotic wave function and $\mu^2 = Q^2/9$, the same as the average Q_G^2 for the asymptotic wave function, then (Cf., [13])

$$B \approx 2 \, , \qquad (6.3)$$

an acceptable value. For broader wave functions, B is also small, but is apparently not small for the Chernyak–Zhitnitsky wave function.

7. Finis

We have concentrated on the pion form factor for simplicity, and have seen that the pQCD calculated value is a significant part of the pion form data at a few GeV^2 of momentum transfer. It is certainly much more than the 3% occasionally quoted! If one were to compare to results given in terms of an effective cutoff mass, then m_{cut} of 200–300 MeV would give the correct normalization.

There are further things that ought to be calculated or thought about. I may mention doing more complete calculations of the Sudakov form factor for composite states when the constituents are near each other, more thought on how to include the transverse momenta of the constituents in a high momentum transfer form factor calculation, and consideration of higher Fock components, which also contribute higher twist corrections to perturbativly calculated quantities. There is a real possibility that a full set of more accurately calculated higher twist corrections would show the corrections to be smaller rather than larger than the ones discussed in this report.

8. Acknowledgments

I thank S. Brodsky, V. Chernyak, J. Napolitano, A. Radyushkin, J. Ralston and P. Stoler for useful comments, some of them incorporated into the text, and also thank the National Science Foundation for support under grant PHY–9112173.

9. References

[1] N. Isgur and C.H. Llewellyn–Smith, Nucl. Phys. **B317**, 526 (1989).
[2] S.V. Mikhailov and A.V. Radyushkin, Sov. J. Nucl. Phys. **49**, 494 (1989).
[3] V.V. Sudakov, JETP **3**, 65 (1956).
[4] H. Li and G. Sterman, Nucl. Phys. **B381**, 129 (1992).
[5] R. Jakob and P. Kroll, CERN Report, CERN–TH–6900–93 (1993).
[6] V.L. Chernyak and A.R. Zhitnitsky, Nucl. Phys. B **201**, 492 (1982).
[7] M. Sawicki, Phys. Rev. D **46**, 474 (1992).
[8] G.P. Lepage and S.J. Brodsky, Phys. Rev. D **22**, 2157 (1980).
[9] F.–M. Dittes and A.V. Radyushkin, SJNP **34**, 293 (1981) {Yad. Fiz. **34**, 529 (1981)}.
[10] R.D. Field, R. Gupta, S. Otto, and L. Chang, Nucl. Phys. **B186**, 429 (1981).
[11] M.H. Sarmadi, University of Pittsburgh Report, Ph. D. thesis (1982).
[12] E. Braaten and S.–M. Tse, Phys. Rev. D **35**, 2255 (1987).
[13] A.V. Radyushkin, in *Proceedings of Baryons 92*, New Haven, CT, 1992, ed. by M. Gai (World Scientific, Singapore, 1993) p. 366.

THE NEUTRON ELASTIC FORM FACTOR AT HIGH Q^2

P. E. Bosted,[1] L. M. Stuart,[2,4,a] A. Lung,[1,b] L. Andivahis,[1]
J. Alster,[12] R. G. Arnold,[1] C. C. Chang,[5] F. S. Dietrich,[4] W. R. Dodge,[7,c]
R. Gearhart,[10] J. Gomez,[3] K. A. Griffioen,[8] R. S. Hicks,[6] C. E. Hyde-Wright,[13]
C. Keppel,[1] S. E. Kuhn,[11,d] J. Lichtenstadt,[12] R. A. Miskimen,[6] G. A. Peterson,[6]
G. G. Petratos,[9,a] S. E. Rock,[1] S. H. Rokni,[6,a] W. K. Sakumoto,[9]
M. Spengos,[1] K. Swartz,[13] Z. Szalata,[1] L. H. Tao[1]

[1] *The American University, Washington D.C. 20016*
[2] *University of California, Davis, California 95616*
[3] *CEBAF, Newport News, Virginia 23606*
[4] *Lawrence Livermore National Laboratory, Livermore, California 94550*
[5] *University of Maryland, College Park, Maryland 20742*
[6] *University of Massachusetts, Amherst, Massachusetts 01003*
[7] *National Institute of Standards and Technology, Gaithersburg, Maryland 20899*
[8] *University of Pennsylvania, Philadelphia, Pennsylvania 19104*
[9] *University of Rochester, Rochester, New York 14627*
[10] *Stanford Linear Accelerator Center, Stanford, California 94309*
[11] *Stanford University, Stanford, California 94305*
[12] *University of Tel-Aviv, Ramat Aviv, Tel-Aviv 69978, Israel*
[13] *University of Washington, Seattle, Washington 98195*

ABSTRACT

Quasielastic e–d cross sections have been measured over a large ϵ range for $Q^2 =$ 1.75, 2.50, 3.25 and 4.00 (GeV/c)^2. Rosenbluth separations have been made on the cross sections to obtain R_L and R_T and the neutron form factors, G_{En} and G_{Mn}, have been extracted via model dependent methods. The sensitivity of the form factor results to various model assumptions has been studied. The results for G_{Mn} are consistent with form factor scaling, while G_{En} is consistent with zero.

*Work supported in part by National Science Foundation grants PHY–87–15050 (AU), PHY–89–18491 (Maryland), PHY–88–19259 (U Penn), and PHY–86–58127 (UW); by Department of Energy contracts DE–AC03–76SF00515 (SLAC), W–7405–ENG–48 (LLNL), DE–FG02–88ER40415 (U Mass), DE–AC02–ER13065 (UR) and DE–FG06–90ER40537 (UW) ; and by the US–Israel Binational Science Foundation. Present addresses:

a) Stanford Linear Accelerator Center, Stanford, CA 94309

b) California Institute of Technology, Pasadena, CA 91125

c) George Washington University, Wash., D.C. 20052

d) Old Dominion University, Norfolk, VA 23529

1. Introduction

The neutron electromagnetic form factors, G_{En} and G_{Mn}, which reflect the charge and magnetization distributions within the neutron, are of fundamental importance for understanding nucleon structure, and are necessary for calculations of processes involving the electromagnetic interaction with complex nuclei. These quantities are functions of four-momentum transfer squared, Q^2. SLAC experiment NE11 has measured these form factors out to a Q^2 of 4.0 $(\text{GeV}/c)^2$ with high precision, and the results have been recently published.[1] This paper provides some additional details[2] on the extraction of G_{Mn} and G_{En}. The measurements of proton elastic form factors also made in the NE11 experiment have also been published,[3] and will not be discussed further in this paper.

Several approaches have been developed to understand the nucleon form factors. Vector Meson Dominance (VMD) models[4,5] are based on superpositions of photon couplings to various vector mesons. These models generally involve free parameters which are fit to form factor data at low Q^2, and are not expected to be valid at high Q^2. For asymptotically large Q^2, dimensional scaling methods and perturbative Quantum Chromodynamics (pQCD)[6] predict form factor behavior at large Q^2, but so far detailed calculations have been made only for $Q^2 > 10$ $(\text{GeV}/c)^2$. To describe the form factor behavior at intermediate values of Q^2, a hybrid model[7] by Gari and Krümpelmann (GK) uses VMD constraints at low Q^2 and pQCD constraints at high Q^2. Free parameters in the model are adjusted to fit existing form factor data. Other approaches include the use of QCD sum rules[8] to make absolute predictions, diquark models,[9] and relativistic constituent quark models.[10]

2. Experiment

Previous measurements of the elastic electron–neutron cross sections which depend on both G_{En} and G_{Mn} extend to $Q^2 = 10$ $(\text{GeV}/c)^2$, but separations of the two form factors have only been made up to $Q^2 = 2.7$ $(\text{GeV}/c)^2$ with large errors.[11] These results are consistent with dipole scaling:

$$\frac{G_{Mn}(Q^2)}{\mu_n} = G_D(Q^2) = \frac{1}{(1 + Q^2/0.71)^2} \tag{1}$$

where $\mu_n = -1.913$ nm is the neutron anomalous magnetic moment. The present experiment, NE11, has made significant improvement to the experimental precision of the measured proton[3] and neutron[1] form factors as well as increasing the measured Q^2 range. The Nuclear Physics Injector at SLAC provided beams with energies E ranging from 1.5 to 5.5 GeV at average currents from 0.5 to 10 μA. The beam angle and position were determined to within 0.05 mr and 1 mm, respectively using position sensitive resonant cavities and wire arrays. The total incident charge was measured by two independent toroidal charge monitors which agreed to within 0.2% and measured the absolute charge to 1%. The target consisted of a 15 cm long liquid deuterium cell which was 6.44 cm in diameter, with 0.1 mm thick aluminum walls and endcaps. The liquid was circulated through the targets at 2 m/sec so that local density changes were negligible. The average density was determined from platinum resistors and vapor pressure bulbs with a run-to-run precision of 0.2% and an overall normalization of 0.9%. A 1.8 mm thick aluminum target was used to measure endcap contributions.

Scattered electrons were measured simultaneously in two magnetic spectrometers. The SLAC 8 GeV/c spectrometer detected electrons at central scattering angles θ between 15° and 90°, and momentum between 0.5 and 7.5 GeV/c. The uncertainties in the 8 GeV spectrometer central momentum and angle were 0.05% and 0.005° respectively. The SLAC 1.6 GeV/c spectrometer was upgraded for this experiment with two 10Q18 quadrupole magnets in order to increase its solid angle by nearly a factor of four. It was fixed at 90° which allowed for the use of tungsten slits to shield from the target endcaps. It measured cross sections with central momenta E' between 0.5 and 0.8 GeV/c. The uncertainty in the 1.6 GeV/c spectrometer angle was 0.05°. The optics of the 8 GeV/c spectrometer were better understood than those of the 1.6 GeV/c spectrometer due to a precision wire float calibration.[12] Therefore, the cross sections in the 1.6 GeV/c spectrometer were normalized to the 8 GeV/c data using a single normalization factor of 1.013 ± 0.010.

In both spectrometers, gas threshold Čerenkov counters and lead glass shower counter arrays were used to identify electrons in the presence of a large background of pions. Wire chambers and scintillator hodoscopes were used to measure the particle trajectories and hence to determine the electron momenta and scattering angles.

3. Analysis

Monte Carlo simulations of the spectrometer properties were used to generate models of the acceptance as a function of relative momentum, δ, relative horizontal scattering angle, $\Delta\theta$, and vertical scattering angle, ϕ. The 8 GeV/c Monte Carlo was based on surveyed aperture information and on a magnetic field model designed to agree with floating-wire[12] measurements of the optical coefficients. The Monte Carlo was also used to model the momentum dependence of multiple scattering effects and for the change in effective target length as a function of spectrometer central angle. The 1.6 GeV/c spectrometer Monte Carlo ray-traced particles using measured field maps of the quadrupoles and calculations of the field of the dipole magnet. The δ-dependence of the acceptance functions was checked by comparing deuterium inelastic cross sections measured at the same beam energy and scattering angle, but with the central spectrometer momenta differing by a few percent.

The measured counts were corrected for electronics and computer dead time and for the detector inefficiencies. Quasielastic e–d spectra at each kinematic point were found as a function of E' at fixed θ by correcting for the small $\Delta\theta$ dependence of the cross section within the angular acceptance of the spectrometer using a model cross section. A correction of 0.85% was made to the cross sections due to hydrogen contamination in the deuterium target, and an average correction of 2% was made to the 8 GeV/c spectrometer cross sections for aluminum endcap contributions. Subtractions were also made for a background contamination of pions (typically 0.2%), and for electrons originating from pair-production in the target. The latter was measured in separate runs by reversing the polarity of the spectrometers, and was 3.5% in the worst case at $Q^2 = 4.0$ (GeV/c)^2 and $\theta = 90°$. Finally, radiative corrections were applied which were found using the peaking approximation formulas of Mo and Tsai.[13] The final radiative corrections were found using an iterative procedure where the input cross section model was adjusted after each iteration until convergence was obtained.

The measured e–d cross sections per nucleon, $\sigma(E, E', \theta)$, were converted to reduced cross sections, defined as:

$$\sigma_R = \epsilon(1 + \tau')\frac{\sigma(E, E', \theta)}{\sigma_{Mott}} = R_T + \epsilon R_L \tag{2}$$

where $\sigma_{Mott} = \alpha^2 \cos^2(\theta/2)/4E^2 \sin^4(\theta/2)$, $\epsilon = [1 + 2(1 + \tau')\tan^2(\theta/2)]^{-1}$ is the longitudinal polarization of the virtual photon, with $\tau' = \nu^2/Q^2$, and $\nu = E - E'$. Quasielastic spectra were measured over a large ϵ range (typically 0.2 to 0.9) for $Q^2 = 1.75, 2.50, 3.25$, and 4.00 (GeV/c)^2. There were four ϵ values for each of the two lowest Q^2 points, and three and two ϵ values for $Q^2 = 3.25$ and 4.00 (GeV/c)^2, respectively. The inelastic contribution at the quasielastic peak increased with Q^2 to a maximum of $\sim 15\%$ at $Q^2 = 4.00$ (GeV/c)^2. Rosenbluth separations were done using linear fits to the reduced cross sections for each W^2 value at each Q^2. A normalized longitudinal response function, R_L/G_D^2, was obtained from the slope, and a transverse response function, R_T/G_D^2, from the intercept.

From this point on in the analysis, the neutron form factor extraction is model dependent. A comprehensive study has been made of this model dependence, and a summary is given here while further details are available.[14,15] Three different form factor extraction methods were implemented including two "area" methods. The first "area" method was a least-squares simultaneous fit to all the reduced cross section spectra at a given Q^2. The second was a similar fit applied separately to the extracted R_L and R_T spectra. A "peak" method of extraction was also done which only used data in the quasielastic peak region. This method is less sensitive to the modeling of the quasielastic peak shape, but the statistics are significantly reduced.

The shape of the quasielastic peak was modeled with a non-relativistic Plane Wave Impulse Approximation (PWIA) calculation[16] where the Paris,[17] Bonn,[18] and Reid soft-core[19] deuteron wave functions were all studied. In the PWIA, the quasielastic portion of R_L is proportional to $(G_{Ep}^2 + G_{En}^2)$, and R_T is proportional to $(G_{Mp}^2 + G_{Mn}^2)$. In addition to this nonrelativistic model, two sets of relativistic corrections were studied by Keister[20] and Gross.[21] The inelastic tail which extends under the quasielastic peak was modeled using a fit to the measured proton resonance region data which was convoluted with the deuteron wavefunction using a variety of Fermi-smearing models.[22,23] The smeared cross sections were fit to the deuterium data in the resonance region assuming two parameters: the ratio of neutron and proton cross

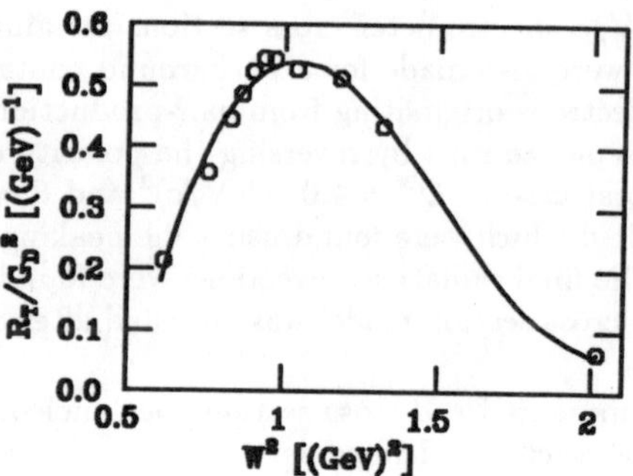

Figure 1. R_T^{MEC} due to calculations by Laget[26] at $Q^2 = 1.75$ $(\text{GeV}/c)^2$. The solid line is a fit to the MEC shape, used in extracting $G_{Mn}(Q^2)$ from the deuterium spectra.

sections, σ_n/σ_p, for resonance production, and for nonresonant background production. Several off-shell corrections which are applied to the input structure functions in the smearing models were also investigated.[24,25]

An effort was also made to estimate the effects due to meson-exchange currents (MEC). For the kinematic range of the NE11 data, no theoretical calculations were available for this effect. In lieu of these calculations, the MEC contribution was estimated using calculations made by Laget[26] for SLAC experiment NE4[27] at the largest Q^2 of 1.75 $(\text{GeV}/c)^2$. This Q^2 corresponds exactly to the lowest Q^2 NE11 point. The calculations included theoretical cross sections with and without contributions from MEC as well as final-state interactions (FSI), which were small compared to MEC. The difference in calculated reduced cross sections due to the MEC and FSI contributions was fit as a function of W^2 using a third degree polynomial fit. Results of this fit are shown in Fig. 1. The cross sections were assumed to be purely transverse so that $R_L = 0.0$ and $\sigma_R = R_T^{MEC}$. The fit shown in Fig. 1 was used for the shape of the (MEC + FSI) cross sections while the magnitude was a fit parameter.

4. Results

All fits to the data yielded a measurement of the sum of the squares of the proton and neutron form factors. The neutron form factors were determined by subtracting the proton form factors measured in this experiment.[3] The results using "area" fit method 1, the Paris wave function, Keister relativistic corrections, the first smearing method as given by Sargsyan, Frankfurt and Strikman,[23] the off-shell correction given by Kusno and Moravczik,[25] and no MEC are shown in Figure 2. This choice of models will hereafter be referred to as "standard". In Fig. 2 the inner error bars are statistical only, while the outer error bars include systematic errors. The point-to-point errors include the combined uncertainties in beam energy and scattering angle. The absolute systematic errors result from uncertainties in absolute values of the incident charge, radiative corrections, and solid angles of the spectrometers, as well as the absolute normalization of the proton form factors.

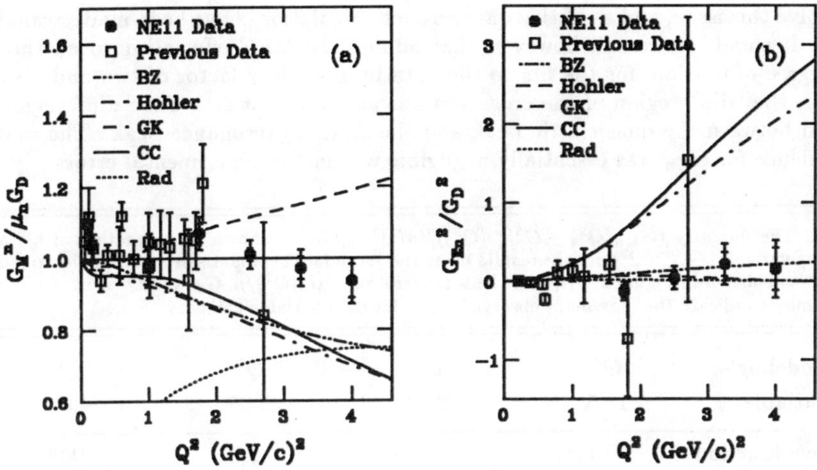

Figure 2. Results for a) $G_{Mn}/\mu_n G_D$ and b) G_{En}^2/G_D^2 versus Q^2 extracted using the "standard" model assumptions as described in text. The inner error bars are statistical, and the outer include systematic errors. Also shown are previous data and several theoretical curves: BZ,[5] Hohler,[4] GK,[7] CC,[10] and Rad.[8]

Figures 2a and 2b show $G_{Mn}/\mu_n G_D$ and G_{En}^2/G_D^2 respectively, along with previous data and various theoretical predictions. The new data are in good agreement with previous data where there is overlap. None of the theoretical curves agree with the results for G_{Mn}, but dipole scaling is in agreement. The GK hybrid model[7] (dash-dot) and the relativistic constituent quark model[10] (CC, solid) both predict $F_{1n} = 0$, or $G_{En} = \tau G_{Mn}$ where $\tau = Q^2/4M^2$. This prediction is in very poor agreement with the new data for G_{En}. All of the remaining curves are in reasonable agreement with the G_{En} data which is also consistent with zero for all the measurements.

A careful study of the model dependence of the form factor extraction indicates that most of the changes made to model variables produce negligible results. Table I summarizes the results of this study for G_{Mn} variations from the "standard" model assumptions, expressed as ratios to the total error on each point. In the approximate order of increasing influence we have deuteron wavefunctions, extraction fit methods, off-shell corrections, smearing methods, relativistic corrections and MEC effects. The first four of these are essentially negligible variations. The relativistic corrections of Gross[21] produce downward shifts of greater than one σ for G_{Mn}/μ_n, where σ is the total error as shown in Fig. 2. The MEC effects give large downward shifts on the order of two σ for G_{Mn}/μ_n. This indicates that further study on the MEC effect is warranted. It can be seen in Fig. 2 that shifts this large will move the data points for G_{Mn} down to agree with the theoretical curves BZ, CC, and GK. However, the simple method used here to estimate MEC effects could be giving anomalously large variations. Theoretical calculations for the kinematics of this experiment are needed to resolve this issue, and as of this date no such calculations have been made available to us. It should be noted, however, that adding the MEC effects improved the χ^2 per degree of freedom for the fits to the data by roughly a factor of two and visibly filled in the "dip" region of the cross sections at the lowest Q^2. The "dip" region is located between the quasielastic peak and the $\Delta(1232)$ resonance peak. The model dependence for G_{En} was essentially negligible within the experimental errors.

Table I. The quantity $(G_{Mn}(Q^2) - G_{Mn}^{Stan}(Q^2))/(\sigma(Q^2)\mu_n G_D)$ is shown as a function of Q^2 and modeling variable. $G_{Mn}^{stan}(Q^2)$ are the results using the "standard" comparison model as described in the text and shown in Figure 2, $\sigma(Q^2)$ is the total error on $G_{Mn}(Q^2)/\mu_n G_D$ and Q^2 is in $(GeV/c)^2$. The numbers indicate the maximum observed effect for each variable category

Modeling variable	Q^2 1.75	Q^2 2.50	Q^2 3.25	Q^2 4.00
D_2 wave function	-0.17	-0.15	-0.04	0.09
extraction method	-0.40	-0.03	0.06	0.27
off-shell corr.	-0.13	-0.20	-0.22	-0.25
smearing method	0.13	0.25	0.37	0.47
relativistic corr.	-0.66	-1.10	-1.24	-0.95
MEC effects	-1.98	-1.63	-2.45	-2.91

5. Conclusions

Quasielastic e–d cross sections have been measured and Rosenbluth separations used to obtain R_L and R_T, at $Q^2 = 1.75$, 2.50, 3.25 and 4.00 $(GeV/c)^2$. Using a PWIA model, values for G_{En} and G_{Mn} have been extracted which greatly increase the Q^2 range of previous data with significantly smaller error bars. Model studies indicate that there is some sensitivity to relativistic corrections and possibly a large sensitivity to MEC effects, but more work is needed for conclusive results. Assuming no MEC effects, the results for $G_{Mn}/\mu_n G_D$ are consistent with form factor scaling, and the results for G_{En}^2/G_D^2 are consistent with zero. None of the theoretical models agree well with both sets of form factor data. If the MEC effects are as big as these preliminary studies indicate, then the results for $G_{Mn}/\mu_n G_D$ are significantly shifted down for all four data points, and the data agrees better with theoretical calculations. However, there is still no single model which adequately describes both the neutron and proton form factors. It is possible that use of the new data to adjust free parameters may improve agreement for many of the models.

References

1. A. Lung *et al.*, Phys. Rev. Lett. **70**, 718 (1993).

2. L. Stuart *et al.*, SLAC-PUB-6235, 1993.

3. P. Bosted *et al.*, Phys. Rev. Lett. **68**, 3841 (1992).

4. G. Höhler *et al.*, Nucl. Phys. **B114**, 505 (1976). Fit 5.3.

5. S. Blatnik and N. Zovko, Acta. Phys. Austr. **30**, 62 (1974).

6. S. J. Brodsky and G. F. Farrar, Phys. Rev. D **11**, 1309 (1975); G. P. Lepage and S. J. Brodsky, Phys. Rev. Lett. **43**, 545 (1979); **43**, 1625 (1979).

7. M. Gari and W. Krümpelmann, Z. Phys. **A322**, 689 (1985).

8. A. V. Radyushkin, Acta. Phys. Pol. **B15**, 403 (1984).

9. P. Kroll, M. Schürmann, and W. Schweiger, Z. Phys. **A338**, 339 (1991).

10. P. L. Chung and F. Coester, Phys. Rev. D **44**, 229 (1991). Model with $M_q = 0.24$.

11. W. Bartel *et al.*, Nucl. Phys. **B58**, 429 (1973).

12. L. Andivahis *et al.*, SLAC preprint SLAC–PUB–5753 (1992).

13. L. W. Mo and Y. S. Tsai, Rev. Mod. Phys. **41**, 205 (1969); Y. S. Tsai, SLAC-PUB-848 Rev. (1971), Rev; Mod. Phys. **46**, 815 (1974).

14. L. M. Stuart, Ph.D. Thesis, University of California, Davis (1992), UCRL-LR-110897.

15. A. Lung, Ph.D. Thesis, The American University (1992).

16. L. Durand III, Phys. Rev. **123**, 1393 (1961); I. J. McGee, Phys. Rev. **161**, 1640 (1967); **158**, 1500 (1967) as given by W. Bartel, *et al.*, Nucl. Phys. **B58**, 429 (1973).

17. M. Lacombe *et al.*, Phys. Lett. **101B**, 139 (1981).

18. R. Machleidt *et al.*, Phys. Rep. **149**, 1 (1987).

19. R. Reid, Ann. Phys. **50**, 411 (1968).

20. B. D. Keister, private communication. See also Phys. Rev. C **37**, 1765 (1988).

21. F. Gross and J. W. Van Orden, private communication. See also Phys. Rev. C **43**, 1022 (1991).

22. W. Atwood and G. West, Phys. Rev. D **7**, 773 (1973).

23. M. M. Sargsyan, L. L. Frankfurt, and M. I. Strikman, Z. Phys. **A335**, 431 (1990).

24. A. Bodek, *et al.*, Phys. Rev. D, **20**, 1471 (1979).

25. D. Kusno and M. Moravczik, Phys. Rev. C **27**, 2173 (1983).

26. J. M. Laget, Can. J. Phys. **62**,1046 (1984); J. M. Laget, Phys. Lett. **B199**, 493 (1987).

27. R. G. Arnold, *et al.*, Phys. Rev. Lett. **61**, 806 (1988).

NUCLEON FORM FACTORS IN PQCD AND THEIR DEPENDENCE ON THE NUCLEON QUARK DISTRIBUTION AMPLITUDE

Robert Eckardt, Jörg Hansper and Manfred F. Gari

Institut für Theoretische Physik II, Ruhr-Universität Bochum
Postfach 10 21 48, D-44780 Bochum, Germany
E-Mail: QCD-MEP@gari.MEP.Ruhr-Uni-Bochum.DE

ABSTRACT

We start from the problems connected with the end-point dominance in the calculation of form factors in pQCD via the hard-scattering formalism. We discuss the resulting importance of the end-point behaviour of the quark distribution amplitude (QDA). In addition we look at the influence of the QDA on the Q^2-behaviour of form factors. It is shown that nucleon form factors can be reasonably reproduced in size and shape if an effective gluon mass is introduced into the hard-scattering amplitude and if the Q^2-dependence of α_S is taken into account as well as the Q^2-evolution of the QDA.

There has been some criticism of the perturbative approach[1] where the form factor is factorized into hard and soft contributions.[2,3] Since the gluon momenta – which look typically like $x_3 y_3 Q^2$ – are too small in the end-point region ($x_i \to 0$) perturbative QCD is not applicable.

One can introduce an effective mass m_g in the gluon propagator to suppress the end-point region (e.g. via $x_3 y_3 Q^2 + m_g^2$) to take only perturbative contributions into account. But one sees that form factors become too small compared to experiment as we checked in table 1 for some "old" amplitudes.[4,5,6,7]

TABLE 1: $Q^4 \cdot F_1^p(Q^2 = 30\,\mathrm{GeV}^2)/\mathrm{GeV}^4$ for some "old" amplitudes,[4,5,6,7] evaluated with $\bar{\alpha}_S \equiv 0.3$.

$\dfrac{m_g}{\mathrm{MeV}} =$	0	300	450	630
CZ[4]	0.890	0.151	0.094	0.059
COZ[5]	1.031	0.167	0.103	0.063
KS[6]	1.340	0.203	0.122	0.073
GS[7]	0.891	0.095	0.032	0.022

The question arises, whether this really indicates that perturbative QCD is not applicable.

We use a refined approach taking the full logarithmic dependence of α_S on the gluon momentum into account not only in the hard-scattering amplitude T_H but also in the QDA ϕ_N. For the coupling constant we use the well-known one-loop approximation $\alpha_S(t) = 4\pi/\beta \ln\left(\frac{t}{\Lambda_{\mathrm{QCD}}^2}\right)$ with $\beta = 11 - \frac{2}{3} n_f$ and $\Lambda_{\mathrm{QCD}} = 100 \ldots 400\,\mathrm{MeV}$. We put the "gluon mass" not only into the gluon propagator in T_H but also into α_S, i.e. we replace $t \mapsto t + m_g^2$.

Therefore, the QDA in the form factor becomes Q^2-dependent and α_S in T_H now depends on the gluon momentum.

$$F(Q^2) = \int [dx]\,[dy]\; \phi^*(y; \tilde{Q}_y^2)\, T_H(x, y; Q^2)\, \phi(x; \tilde{Q}_x^2) + \cdots \tag{1}$$

with $q := p' - p$, $Q^2 := -q^2$, $x_i := \frac{p_i^+}{P^+}$, $\int [dx] := \int\limits_0^1 dx_1 \int\limits_0^1 dx_2 \int\limits_0^1 dx_3\, \delta(1 - x_1 - x_2 - x_3)$

and $\tilde{Q}_x := \min\{x_1, x_2, x_3\} Q$. The general solution of the evolution equation[2] for ϕ_N is given by

$$\phi_N(x; \tilde{Q}^2) = 120\, x_1 x_2 x_3 \sum_{i=0}^{\infty} \left(\frac{\alpha_S(\tilde{Q}^2)}{\alpha_S(\mu_0^2)} \right)^{\gamma_i} b_i\, A_i(x) \qquad (2)$$

which gives the full logarithmic dependence of ϕ_N through the evolution factor $(\cdot\cdot)^{\gamma_i}$.

According to this ansatz, we determined a nucleon QDA $\phi_N^{(1)}$ up to 3rd order exclusively from QCD sum rule moments.[5] The form factor is given in table 2. (To verify the agreement with sum rule moments refer to ref. 8.) As we get in spite of the "gluon mass" still a large value for F_1^p we can conclude that PQCD is not ruled out!

But we note that the third order expansion coefficients[9] are large, which indicates a slowly converging Appell polynomial expansion. This leads us to problems in the construction of the QDA.

TABLE 2: $Q^4 \cdot F_1(Q^2 = 30\,\text{GeV}^2)/\text{GeV}^4$ for our new amplitudes,[8] evaluated with α_S depending on the gluon momenta.

$\frac{m_G}{\text{MeV}} =$	300	450	630
$Q^4\, F_1^p[\phi_N^{(1)}]$	6.631	1.574	0.482
$-Q^4\, F_1^n[\phi_N^{(1)}]$	3.175	0.746	0.225
$Q^4\, F_1^p[\phi_N^{(2)}]$	4.015	1.190	0.413
$-Q^4\, F_1^n[\phi_N^{(2)}]$	1.944	0.572	0.195

The true, full solution Φ_N is decomposed into a model distribution amplitude ϕ_N – the one that is calculated – and an error of missing higher order polynomials $\Delta\Phi_N$.

$$\Phi_N(x, \tilde{Q}_x^2) = \underbrace{\phi_N^{\text{model}}(x, \tilde{Q}_x^2)}_{\text{lower order}} + \underbrace{\Delta\Phi_N(x, \tilde{Q}_x^2)}_{\text{higher Appell polynomials}} \qquad (3)$$

Of course, ϕ_N behaves according to the evolution equation. But a truncated, i.e. finite, polynomial expansion is unreliable at the end-points. This is due to the slow convergence of the expansion (as seen in the case of $\phi_N^{(1)}$). But it is also caused by the very limited number of sum-rule moments that, in addition, have errors.

In the end-point region the contributions due to the error $\Delta\Phi_N$ are large even for high Q^2.

As can be seen from a typical gluon propagator $\sim \dfrac{1}{Q^2(xy + \varepsilon)}$, where $\varepsilon = \dfrac{m_g^2}{Q^2}$, at low Q^2 the contributions of the error $\Delta\Phi_N$ to the form factor are suppressed by the "gluon mass" m_g in the HSA (large ε). On the other hand, for large Q^2 ($\varepsilon \to 0$) the contributions of $\Delta\Phi_N$ are enhanced by the HSA and therefore cannot be neglected compared to those of the model distribution amplitude ϕ_N. Moreover, in the intermediate range, where the form factor stops increasing and begins to decrease (logarithmically), the form factor is very sensitive to the error $\Delta\Phi_N$.

The inclusion of the $\tilde{Q}^2$-evolution of ϕ_N^{model} in form factor calculations spoils the Q^2-behaviour of the form factors rather than improving it, especially due to the errors in the end-point region. The reason is that higher orders do not get suppressed by the evolution factor.

It seems that

a truncated (i.e. finite) pure Appell polynomial expansion of the QDA is unreliable.

To cure these problems we multiplied the truncated ansatz (Eq. (2) summed up to some $n_{\max}$) by an auxiliary function χ to account for higher order contributions in the expansion. This function is intended to smooth the QDA in the end-point region as one expects that the true QDA is some smooth function.

$$\phi_N(x;\tilde{Q}^2) = 120\, x_1 x_2 x_3 \sum_{i=0}^{9} \left(\frac{\alpha_S(\tilde{Q}^2 + m_g^2)}{\alpha_S(\mu_0^2 + m_g^2)}\right)^{\gamma_i} b_i\, A_i(x) \cdot \chi(x) \qquad (4)$$

$$\chi(x) := \begin{cases} 1 & \longrightarrow \quad \phi_N^{(1)} \\ \exp\left(-a(\sum_{i=1}^{3}\frac{1}{x_i} - 9)\right) & \longrightarrow \quad \phi_N^{(2)},\ \phi_N^{(3)} \\ \prod_{i=1}^{3} \theta(x_i - \delta) & \longrightarrow \quad \phi_N^{(4)} \end{cases}$$

To see how good this ansatz works we started with auxiliary functions of θ- and exp-type, but one can think of many others. The constants a and δ are taken to be constant in our first approximation, but to fulfill the evolution equation[2] for the QDA one has to solve the evolution equation for χ which then becomes $\chi(x, \tilde{Q}^2, n_{\max})$.

Since only a finite number of sum-rule moments, which also have errors, is known, the QDA is not uniquely fixed. Therefore one has several possibilities to construct model distribution amplitudes that all give excellent sum-rule agreement.[8,9]

First of all, we determined a model function of the exponential type ($\phi_N^{(2)}$) that gives a large form factor (table 2) in spite of the smoothing auxiliary function χ.

We also found an amplitude of exp-type $(\phi_N^{(3)})$[8] that gives a good Q^2-behaviour while providing at the same time a small form factor ratio $|F_1^n/F_1^p|$ as implied by older analyses of experimental data.[10]

Finally we obtained a QDA of the θ-type $(\phi_N^{(4)})$[9] which gives an excellent Q^2-behaviour of the form factor. It is shown in fig. 1 for different "gluon masses" in comparison to the experimental data of the proton magnetic form factor. It was, of course, calculated with evolution in

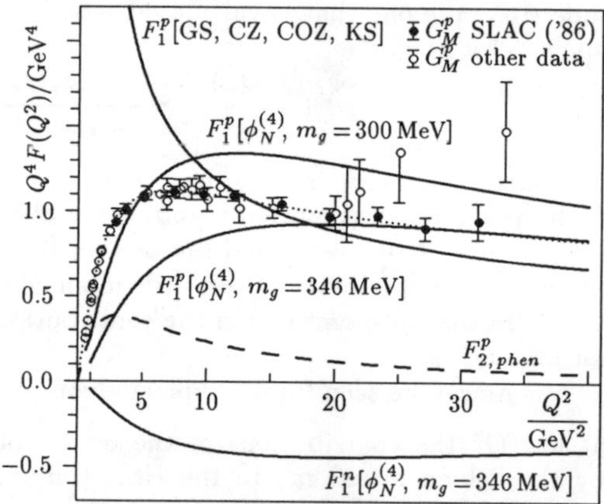

FIG. 1: Dirac form factor of our QDA $\phi_N^{(4)9}$ with different "gluon masses" compared to experimental data of G_M^p[10,11] and a typical representative of the "old" amplitudes.

the distribution amplitude and the gluon momenta as the arguments in α_S which implies that α_S was taken under the integral.

We also show a typical representative of the old model amplitudes evaluated without "gluon mass", without evolution in the QDA and with the coupling constant outside of the integral.

Whereas both procedures work fairly well for describing the nucleon form factor data at higher Q^2, which are known up to $30\,\mathrm{GeV}^2$, one sees that the old method fails to describe the region of the N-Δ form factor, where data is known only up to $10\,\mathrm{GeV}^2$.

But even when one evaluates the form factor for the old model amplitudes using the same procedure as for our new ones, one sees that they still increase at very high Q^2.[9] On the other hand our model amplitudes start decreasing at much lower Q^2 (e.g. at about $20\,\mathrm{GeV}^2$ for $\phi_N^{(4)}$), where it is indicated by the data. This cannot be solely caused by the introduction of the "gluon mass" as the representatives of the old model amplitudes were computed under the same conditions.

The reason for the different behaviour is that the error $\Delta\Phi_N$ becomes more important for intermediate Q^2. But the damping in the end-point region (by our function χ) keeps the error small. This improves the Q^2-behaviour.

- We consider the momentum dependence of α_S both in the hard-scattering amplitude as well as in the QDA.
- We saw that form factors can be large even with "gluon masses" included, while fulfilling at the same time the QCD sum rules!
- As we expect the true quark distribution amplitude to be some smooth function we need a means to correct the important but wrong end-point behaviour of a finite Appell expansion. One can take for example our auxiliary function χ of the exp- or θ-type.
- Due to the limited number and the errors of the QCD sum-rule moments one has still a broad freedom in finding model amplitudes. Other means (e.g. form factors) are needed to restrict the freedom.
- QDAs with our auxiliary functions were found, yielding a good Q^2-behaviour of F_1^p over a wide Q^2-range.
- In the first approximation, where our function χ is taken independent of Q^2, the error in the $\tilde{Q}^2$-evolution of ϕ_N is negligible in the $\tilde{Q}^2$-region considered. Nevertheless, one can restore the compliance with the evolution equation by solving an evolution equation for χ.

References

1. N. Isgur and C.H. Llewellyn Smith, Nucl. Phys. **B317** (1989) 526.
2. G.P. Lepage and S.J. Brodsky, Phys. Rev. **D22** (1980) 2157.
3. V.L. Chernyak and A.R. Zhitnitskiĭ, JETP Lett. **25** (1977) 510.
4. V.L. Chernyak and I.R. Zhitnitsky, Nucl. Phys. **B246** (1984) 52.
5. V.L. Chernyak, A.A. Ogloblin and I.R. Zhitnitsky, Z. Phys. **C42** (1989) 569.
6. I.D. King and C.T. Sachrajda, Nucl. Phys. **B279** (1987) 785.
7. M.F. Gari and N.G. Stefanis, Phys. Lett. **175B** (1986) 462.
8. J. Hansper, R. Eckardt and M.F. Gari, Z. Phys. **A341** (1992) 339.
9. R. Eckardt, J. Hansper and M.F. Gari, Z. Phys. **A343** (1992) 443.
10. M.F. Gari and W. Krümpelmann, Z. Phys. **A322** (1985) 689.
11. R.G. Arnold *et alii*, Phys. Rev. Lett. **57** (1986) 174.

BARYON RESONANCE FORM FACTORS
AT HIGH MOMENTUM TRANSFER

PAUL STOLER

Physics Department, Rensselaer Polytechnic Institute
Troy, New York 12180, USA

ABSTRACT

Baryon resonance transition form-factors at high Q^2 are examples of exclusive reactions which can yield information about the *non-perturbative* structure of hadrons. The subject is discussed in terms of what is currently known, and the potential for a program of measurements at new electron accelerators such as CEBAF and the ELFE.

1. Introduction

The description of hadrons and their excitations in terms of elementary quark and gluon constituents is one of the fundamental problems in physics. For small distances, inclusive reactions such as deep inelastic electron scattering have established that *perturbative QCD (PQCD)* is the correct description for Q^2 as low as a few GeV^2/c^2. These phenomena involve the virtual photon interacting with a single asymptotically free quark, followed by complicated hadronization processes. However, at large distances, on the hadron scale, the situation is much less satisfactory. How QCD evolves from the perturbative asymptotic regime to yield the properties of hadrons remains unsolved. It is a very difficult, but fascinating problem, and surely one that must be solved, since hadronic matter makes up most of our ambient universe.

The $non-perurbative$ structure of hadrons is too complex to study microscopically in its entirety. Exclusive reactions allow us to focus on simpler substructures and subprocesses which may be more tractable for studying current theoretical ideas about the $non-perurbative$ structure of hadrons. At low Q^2 the *constituent quark model (CQM)* has proved extremely successful in parameterizing the complexity of *non-perturbative QCD (NPQCD)* into a manageable number of variables involving three constituent quarks and their mutual interactions. However, this model is not fundamental, and one woud like to access the *non-perturbative* structure in terms of the current quarks and gluons. Exclusive reactions at high Q^2 provide selective windows on such substructures.

2. Accessing Non-Perturbative Physics via PQCD

With increasing Q^2 exclusive reactions select successively simpler components of the hadronic Fock state wave function, so that in the limit of very high Q^2 presumably only the simplest valence quark Fock states will be involved.

2.1 High Q^2 Form-Factors:

It was shown in Ch-77, Le-80, Br-81 that at high Q^2 baryon form-factors may written in the following factorized form:

$$G_T(Q^2) = F_1(Q^2) = \int_0^1 \int_0^1 dxdy \; \Phi_f(x)^* \; T_H \; \Phi_i(y). \qquad (1)$$

In eq. 1 the variables x and y stand for the momentum fractions of the three quarks in the initial and final states respectively.

The two main ingredients are the hard transition operator T_H involve *short-distance perturbative physics*, and the distribution amplitudes Φ_i and Φ_f, which are determined by the *long-range non-perturbative* physics.

2.1.1 Transition Operator

The transition operator is constructed *perturbatively* by considering the leading order Feynman diagrams, which involve the exchange of two gluons among the three valence quarks. All the leading order diagrams result in the following form of T_H:

$$T_H = \alpha_s(k_1)\alpha_s(k_2)\frac{f(x,y)}{k_1^2 k_2^2} \; \sim \; \frac{\alpha_s^2(Q^2,x,y)}{Q^4}f(x,y). \qquad (2)$$

This results in the well-known dimensional scaling rule in the limit of high Q^2, $F_1 \rightarrow G_T \propto Q^{-4}$.

2.1.2 Distribution Amplitudes:

The distribution amplitudes $\Phi(x)$ contain the physics of the baryon structure. They are determined primarily by the dominant $long - range \; non - perturbative$ interactions of the quarks among themselves and with the physical vacuum. A significant advance in determining these distributions has been made possible by the successful application of QCD *sum-rules* (Sh-79, Ch-84). Using sum-rule constraints Ch-84 *et seq.* had obtained model three-quark nucleon Fock state.

2.2 Range of validity of PQCD

Using eq. 1 with Φ's determined from sum-rules it has been possible to obtain reasonable fits to the Dirac proton elastic form-factor $F_1(Q^2)$ for Q^2 as low as about $5 \; GeV^2/c^2$. However the procedures have been the subject of strong criticism, since the transition amplitudes become large near the kinematic limits of x and y, so that most of the form-factor is due to contributions near these kinematic limits, where the gluon momentum is not necessarily large, and $PQCD$ would not be expected to be operative.

Recently it has been shown (Li-92a,b) that color radiative (Sudakov) effects strongly suppress the contributions to the form-factors from the *non-perturbative* domain of eq. 1., thus extending the validity of the $PQCD$ based calculations down to a lower range of Q^2 than would otherwise be the case. This is discussed in more detai in the contribution to these proceedings by C. Carlson.

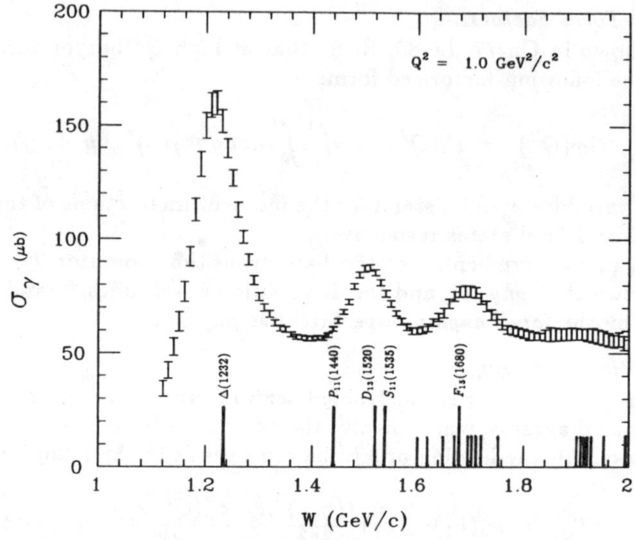

Figure 1. The virtual photon cross section for proton excitation at $Q^2 =$ 1 GeV^2/c^2. The data are reconstructed from an evaluation by Br-76. The known contributing states are indicated at the bottom by vertical lines. The largest contributing states are indicated by long lines and are spectroscopically denoted by $L_{2I,2J}(W)$.

3. Baryon Resonances

The study of transition form-factors to excited baryons at high Q^2 can have an important impact on our knowledge of hadronic structure. In eq. 1 Φ_i and Φ_f are the nucleon and excited baryon distribution amplitudes respectively. Such a program will be a significant component of experimental programs with new electron accelerators in the 1990's and beyond, such as CEBAF and the proposed ELFE.

Figure 1 shows the virtual photon cross section at $Q^2 = 1\ GeV^2/c^2$ as a function of baryon invariant mass W. For $W < 2\ GeV$, the most significant feature is the existence of three maxima, known as the first, second and third resonance regions. In this interval there are about 20 known resonances. The first maximum is due to the $\Delta(1232)$ resonance. The second resonance region is dominated by two strong negative parity states, the $D_{13}(1520)$ and the $S_{11}(1535)$. At low Q^2 ($< 1\ GeV^2/c^2$) the $D_{13}(1520)$ is dominant, whereas at higher Q^2 ($> 3\ GeV^2/c^2$) the $S_{11}(1535)$ dominates. In the third resonance region, the largest excitation at low Q^2 is the $F_{15}(1680)$ state. The relative strength of the other states is not well determined, especially at increasing Q^2. The Roper resonance, the $P_{11}(1440)$, has not yet been definitely observed at $Q^2 > 0$, but is of considerable interest since it may correspond to a radial nucleon excitation, or alternatively a $3Q - G$ *hybrid* (Li-92) in the CQM.

4. Resonance Form-Factors

At low Q^2 the excitations indicated in Figure 1 have been rather successfully described in terms of the CQM. The region of the breakdown of the CQM and the onset of $PQCD$ is of great interest. A large body of exclusive data at high Q^2 including inelastic resonant transition form-factors in addition to elastic form-factors is needed to obtain baryon distribution amplitudes. However, except for the $\Delta(1232)$ the resonances are largely overlapping even at relatively low excitation. The separation of the contributing electromagnetic multipoles will require measurement of exclusive reactions such as $(e, e'\pi)$ and $(e, e'\eta)$ to as high Q^2 as possible, with polarized beams and targets, a very difficult task made more difficult since the form-factors become small at high Q^2, and there is a significant contribution from non-resonant processes.

The current experimental situation is that exclusive $(e, e'\pi)$ and $(e, e'\eta)$ data exist only for Q^2 less than 3 GeV^2/c^2. Although there is a total absence of exclusive data above $Q^2{=}3$ GeV^2/c^2, there is a considerable amount of inclusive data in the resonance region obtained mostly at SLAC (see references in St-93). The available inclusive data have been fit with resonance functions (St-93) to extract information about the Q^2 dependence of resonance form-factors at higher Q^2 than is available from exclusive measurements. The purpose was to see whether any of the features predicted by $PQCD$ are observable. Although the statistical accuracy of the data diminishes with increasing Q^2, the second and third resonances remain prominent.

Transverse form-factors, $|G_T(Q^2)|^2$ were extracted for the three dominant peaks near $W = 1232$, 1535 and 1680 MeV were extracted as a function of Q^2 by fitting them with resonance cross sections and phenomanological non-resonant backgrounds, and assuming the resonance cross sections are purely transverse (see St-93 for details). The form-factors, are shown in Figure 2, relative to a dipole shape of the form $G(Q^2)_{dipole} = 3/(1 + Q^2/0.71)^2$ versus Q^2. The quantity $G_T(Q^2)/G(Q^2)_{dipole}$ is plotted rather than $Q^4 G_T(Q^2)$ in order to better display the low Q^2 behavior. For high Q^2 the quantity $Q^4 G_T(Q^2)_{dipole} \to$ constant. Also shown at lower Q^2 are form-factors extracted from data obtained from exclusive $(e, e', p)\pi^\circ$ and $(e, e', p)\eta$ experiments. The proton magnetic form-factor relative to $\mu_P/(1 + Q^2/0.71)^2$ is also shown in the top panel for comparison.

Figure 2 shows that the obtained form-factors for the second and third resonance regions are consistent with a Q^{-4} dependence, although with large statistical uncertainty. On the other hand, the $\Delta(1232)$ form-factor is decreasing relative to both the elastic as well as the second and third resonances.

4.1 $\Delta(1232)$: Transition Multipoles

Since the $\Delta(1232)$ has $J = 3/2$ there are three contributing multipoles, E_{1+}, M_{1+}, and S_{1+} whose relative contributions are model dependent. Thus this is a favorable case for studying models of baryon structure.

At low Q^2 in a pure $SU(6)$ $non-relativistic$ CQM the $N \to \Delta$ transition is purely M_{1+} in character, involving a single-quark spin-flip with $\Delta L = 0$. An E_{1+}

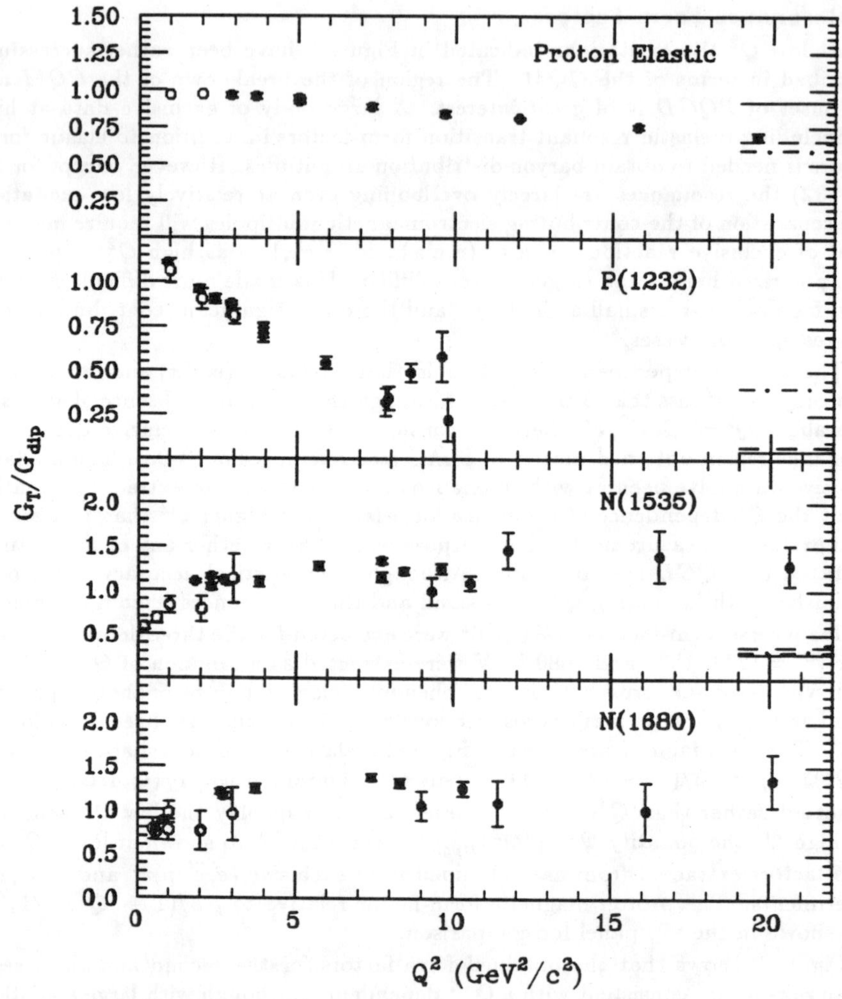

Figure 2. From St-93. (a) The absolute value of the elastic proton form-factor divided by the dipole shape G_M/G_{dipole} verses Q^2 for elastic scattering from the proton, where $G_{dipole} = \mu_P(1 + Q^2/.71)^{-2}$. The data are from Ar-86. (b-d) The quantity G_T/G_{dipole} verses Q^2 for transitions to the first, second and third resonances respectively, with $G_{dipole} = 3(1 + Q^2/.71)^{-2}$. The first resonance (b) is the $\Delta(1232)$. The second resonance (c) at lower Q^2 is due to the $S_{12}(1535)$. The third resonance at low Q^2 is dominated by the $F_{15}(1680)$. The fits for G_T were based on inclusive data referenced in St-91,93. Also shown at lower Q^2 denoted by ($\times$) are form factors derived from amplitudes obtained from exclusive $(e,e',p)\pi^\circ$ and $(e,e',p)\eta$ data.

contribution is not permitted since the Δ and N are both in $L = 0$ states which cannot be connected by an operator involving $L > 0$. The addition of a residual quark-quark color magnetic interaction adds higher L components to the Δ wave function, and thus introduces a small E_{1+} component of perhaps a few percent. The measurement of this small E_{1+}/M_{1+} ratio is one of the most interesting problems in baryon resonance physics in that it will give very powerful tests of the CQM in the low Q^2 regime. At $Q^2 = 0$ the experimental data supports the CQM prediction of M_{1+} dominance extremely well. A review by Da-90,91 finds

$$M_{1+} = 285 \pm 37 \times 10^{-3} \ GeV^{-1} \qquad E_{1+} = -4.6 \pm 2.6 \times 10^{-3} GeV^{-1}.$$

The ratio E_{1+}/M_{1+} remains very small up to Q^2 about $1 \ GeV^2/c^2$ However, at increasing Q^2, the CQM is expected to gradually become less valid and there is no reason to expect the E_{1+}/M_{1+} ratio to remain extremely small. Beyond $1 \ GeV^2/c^2$ there is very little data, however there is one experimental point at $Q^2 = 3 \ GeV^2/c^2$ (Ha-79) which has been evaluated by Bu-92 which suggests that E_{1+}/M_{1+} is still rather small, but with large errors: $Re(E_{1+}/M_{1+}) = 0.8 \pm 0.6$. This was extracted by fitting data under the assumption that E_{1+}/M_{1+} is very small so that various multipole contributions in the data are neglected. However, Wa-90 have pointed out that one can obtain an equally consistent fit to the data on which the analysis is based by assuming in fact that $E/M \sim 1$ so that we must conclude that the magnitude of E_{1+}/M_{1+} at $Q^2 = 3 \ GeV^2/c^2$ remains unknown.

What does one expect above at high Q^2? $PQCD$ predicts that only helicity conserving amplitudes should contribute at high Q^2. The multipoles may be expressed in terms of helicity conserving and non-conserving amplitudes as follows:

$$\Delta\lambda = 0 : A_{1+} = \tfrac{3}{2}M_{1+} + \tfrac{1}{2}E_{1+} \tag{3a}$$

$$\Delta\lambda = 2 : B_{1+} = E_{1+} - M_{1+} \tag{3b}$$

$$\Delta\lambda = 1 : C_{1+} = \frac{2Q^2}{p_\pi^*}S_{1+}. \tag{3c}$$

Thus, helicity conservation implies $B_{1+} = 0$, or $E_{1+} = M_{1+}$. This is quite different from the low Q^2 situation. A separation of the multipoles at high Q^2 will be extremely important.

4.2 $\Delta(1232)$: Form-Factor

The anomalously large decrease with Q^2 of the Δ form-factor seen in Figure 3b has been confirmed (see the contribution of L.M. Stuart et al. at this workshop), though one finds that uncertainties in background prescription lead to uncertainties in the rate of of falls off at higher Q^2, where the Δ amplitude is relatively smaller.

Using QCD sum rule techniques Ca-88, have calculated the distribution functions for the $\Delta(1232)$ excitation. The resulting leading order $\lambda = 1/2$ form-factor for the $P \to \Delta$ transition is very sensitive to the proton distribution functions, as shown on the extreme right in Figure 2b. The C-Z (Ch-84) and K-S (Ki-87) proton distributions are rather similar, and both yield small transition form factors,

$G_T(Q^2) \sim 0.07$ and $0.11 \ GeV^4$ respectively. This can be traced to a cancellation in the leading order term of the matrix elements connecting the symmetric $\Delta(1232)$ distribution amplitude, with the symmetric and antisymmetric proton distribution amplitude respectively, ie.

$$< \phi_\Delta | T_H | \phi_S^P > \ \text{ and } \ < \phi_\Delta | T_H | \phi_A^P >.$$

On the other hand the G-S (Ga-87) distribution does not yield this large cancellation and gives $Q^4 F_{1\Delta}(Q^2) \sim 0.61 \ GeV^4$ for large Q^2. However, the G-S distribution function has been contradicted by recent neutron form-factor measurements reported in these proceedings by P. Bosted et al.

If the leading amplitude of the P $\rightarrow \Delta$ transition is indeed small, the anomalous shape of the transition form factor might be explained as follows. At high Q^2, the leading order helicity conserving amplitude dominates over the helicity non-conserving amplitude. That is, $A_{1/2} >> A_{3/2}$.

A suppression of the $A_{1/2}$ amplitude at all Q^2 due to the cancellation of the symmetric and antisymmetric matrix elements, might then result in the dominance of the $A_{3/2}$ amplitude over a larger range of Q^2 than otherwise expected, and $Q^4 G_T(Q^2)$ would decrease as a function of Q^2. In fact the evidence that E_{1+}/M_{1+} is still small for Q^2 up to $3 \ GeV^2/c^2$, is consistent with the dominance of non-leading processes. It will be interesting in the future to determine whether $Q^4 G_T(Q^2)$ does indeed level off above $Q^2 = 10 \ GeV^2/c^2$, and where the E_{1+} amplitude becomes comparable to the M_{1+} This would confirm the PQCD description.

4.3 The Second Resonance

The the $S_{11}(1535)$ in the second resonance is one of the strongest excitations and therefore has been the subject of considerable investigation. Figure 2 shows that at high Q^2 the form-factor for the peak at W ~ 1535 MeV approaches the Q^{-4} dependence predicted by PQCD. The dominance of the $S_{11}(1535)$ over the $D_{13}(1520)$ at $Q^2 = 3 \ GeV^2/c^2$ has been demonstrated in Ha-79 The likelyhood that this peak continues to be dominated by the $S_{11}(1535)$ is good because the *apparent* width of the peak is narrow (less than 100 MeV over the entire Q^2 range covered), which would be consistent with a large η branching ratio, in which the η production threshold is close to the resonant energy. Ca-88 has calculated the $P \rightarrow S_{11}$ transition form-factor in the PQCD limit. They expected it to behave similarly to the elastic form-factor because the distribution amplitude $\Phi_{S_{11}}$ obtained is similar to that of a proton, with the antisymmetric part about half as large. The results for $Q^4 G_T(Q^2)$ obtained for the three proton distribution functions are respectively: C-Z ~ 0.60, K-S ~ 0.77 and (G-S) ~ 0.66 GeV^4, and are also plotted as bars on the extreme right in Figure 2. Although all these results are about a factor of two lower than the data, theoretical uncertainties in the distribution functions, and higher order contributions to α_s are probably great enough to encompass these discrepancies.

4.4 The Third Resonance

Figure 2 shows that at high Q^2 the form-factor for the peak near W = 1680 MeV is consistent with the predicted Q^{-4} behavior. However the errors are large, and it is not clear how many resonances are contributing to this peak.

5. Potential for New Accelerators

The region in Q^2 corresponding to the onset of PQCD in exclusive reactions is far from settled, as can be seen by the opposing views taken by C. Carlson and A. Radyuskin in this workshop. However, much of the available inclusive data appears to be consistent with PQCD scaling laws at a lower rather than higher range of Q^2 in controversy. An exception is the $\Delta(1232)$, whose form-factor falls relative to PQCD expectations up to the maximum measured Q^2. The potential for obtaining separated resonance amplitudes at high Q^2 with exclusive reactions is very good. This is particularly true since it has been demonstrated (Bl-70,St-91,Ca-90), that the non-resonant background diminishes with Q^2 at approximately the same rate as the resonancws.

This work is supported by NSF grant number PHY8905162.

References.

Ar-86 R.G. Arnold et al., Phys. Rev. Lett., **57**, 174 (1986).

Bl-70 E.D. Bloom and F.J. Gilman, Phys. Rev. Lett. **25**, 1140 (1970),
      Phys. Rev. **D4**, 2901 (1971).

Br-76 F.W. Brasse et al., Nucl. Phys. **B110**, 410 (1976).

Br-81 S.J. Brodsky and G. P. Lepage, Phys. Rev. D **24**, 2848 (1981).

Bu-92 V. Burkert, private communication, to be published.

Ca-90 C.E. Carlson and N.C. Mukhopadhyay, Phys. Rev. **D41** , 2343 (1990).

Ca-88 C.E. Carlson and J.L. Poor, Phys. Rev. **D38**, 2758 (1988).

Ch-77 V.L. Chernyak and A.R. Zhitnitsky, JETP Lett. **23**, 510 (1977)

Ch-84 V.L. Chernyak and A.R. Zhitnitsky, Phys. Rep. **112**, 173 (1984).

Da-90 R.M. Davidson and N.C. Mukhopadhyay, Phys. Rev. **D42**, 20 (1990).

Da-91 R.M. Davidson, N.C. Mukhopadhyay and R. Wittman, Phys. Rev. **D43**,
      Phys. Rev. **D43**,71 (1991).

Ga-87 M. Gari and N.G. Stefanis, Phys. Rev. **D35** 1074 (1987).

Ha-79 R. Haidan, PhD thesis, Univ. of Hamburg, unpublished,
      DESY internal report DESY F21-79/03.

Ki-87 I.D. King and C.T. Sachrajda, Nucl. Phys., **B279**, 785 (1987).

Le-80 G.P. Lepage and S.J. Brodsky, Phys. Rev. **D22**, 2180 (1980).

Li-92 Z. Li, V. Burkert, and Zh. Li, Phys. Rev. **D46**, 70 (1992)

Li-92a H-N. Li and G. Sterman, SUNY-Stony Brook report ITP-SB-92-10,
      Nucl. Phys. **B** to be published.

Li-92b H-N. Li, SUNY-Stony Brook report ITP-SB-92-25, to be published.

Sh-79 M.A. Shifman, A.I. Vainstein, and V.I. Zakharov, Nucl. Phys. **B147**
      385, 448, 519 (1979).

St-91 P. Stoler, Phys. Rev. Lett., **66**, 1003;Phys. Rev.,**D44**, 73 (1991).

St-93 P. Stoler, Physics Reports, **226**, 103 (1993)

Wa-90 G.A. Warren and C.E. Carlson, Phys. Rev. D, **42**, 3020 (1990).

The $\Delta(1232)$ Resonance Transition Form Factor

L. M. Stuart,[2,4,a] P. E. Bosted,[1] A. Lung,[1,b] L. Andivahis,[1]
J. Alster,[12] R. G. Arnold,[1] C. C. Chang,[5] F. S. Dietrich,[4] W. R. Dodge,[7,c]
R. Gearhart,[10] J. Gomez,[3] K. A. Griffioen,[8] R. S. Hicks,[6] C. E. Hyde-Wright,[13]
C. Keppel,[1] S. E. Kuhn,[11,d] J. Lichtenstadt,[12] R. A. Miskimen,[6] G. A. Peterson,[6]
G. G. Petratos,[9,a] S. E. Rock,[1] S. H. Rokni,[6,a] W. K. Sakumoto,[9]
M. Spengos,[1] K. Swartz,[13] Z. Szalata,[1] L. H. Tao[1]

[1] *The American University, Washington D.C. 20016*
[2] *University of California, Davis, California 95616*
[3] *CEBAF, Newport News, Virginia 23606*
[4] *Lawrence Livermore National Laboratory, Livermore, California 94550*
[5] *University of Maryland, College Park, Maryland 20742*
[6] *University of Massachusetts, Amherst, Massachusetts 01003*
[7] *National Institute of Standards and Technology, Gaithersburg, Maryland 20899*
[8] *University of Pennsylvania, Philadelphia, Pennsylvania 19104*
[9] *University of Rochester, Rochester, New York 14627*
[10] *Stanford Linear Accelerator Center, Stanford, California 94309*
[11] *Stanford University, Stanford, California 94305*
[12] *University of Tel-Aviv, Ramat Aviv, Tel-Aviv 69978, Israel*
[13] *University of Washington, Seattle, Washington 98195*

*Work supported in part by National Science Foundation grants PHY–87–15050 (AU), PHY–89–18491 (Maryland), PHY–88–19259 (U Penn), and PHY–86–58127 (UW); by Department of Energy contracts DE–AC03–76SF00515 (SLAC), W–7405–ENG–48 (LLNL), DE–FG02–88ER40415 (U Mass), DE–AC02–ER13065 (UR) and DE–FG06–90ER40537 (UW); and by the US–Israel Binational Science Foundation. Present addresses:

a) Stanford Linear Accelerator Center, Stanford, CA 94309

b) California Institute of Technology, Pasadena, CA 91125

c) George Washington University, Wash., D.C. 20052

d) Old Dominion University, Norfolk, VA 23529

Abstract

Old and new measurements of inclusive $e-p$ cross sections in the $\Delta(1232)$ resonance region have been combined, and a global data fit has been made. Using this fit to parameterize the nonresonant background, the transition form factors have been extracted out to a four-momentum transfer, Q^2, of 9.8 $(\text{GeV/c})^2$. The results are systematically higher than those from a previous analysis, but agree within errors. A similar analysis has been done with $e-d$ cross sections, and σ_n/σ_p in the $\Delta(1232)$ resonance region has been extracted out to a Q^2 of 7.9 $(\text{GeV/c})^2$. σ_n/σ_p for $\Delta(1232)$ production is consistent with unity, while σ_n/σ_p for the nonresonant background is constant with Q^2 at approximately 0.4.

Understanding the structure of the nucleons and their excitations is of fundamental interest. In the limit of large four-momentum transfer, leading order perturbative QCD (pQCD) is expected to be valid, but it is not clear at which Q^2 the non-leading order processes die off. The form factor analysis of Stoler[1,2] indicates that the $\Delta(1232)$ transition form factor does not exhibit the expected leading order pQCD behavior for Q^2 out to 10 $(\text{GeV/c})^2$. Instead, this form factor falls off more rapidly with Q^2 than expected. Thus, there is a need for data on baryon excitation cross sections and transition form factors at large Q^2, in order to understand this effect better and also to test form factor models.

New measurements[3] have been made by SLAC experiment NE11 of inclusive electron scattering cross sections using hydrogen and deuterium targets in the region of the $\Delta(1232)$ resonance[4] This data has been combined with low Q^2 data[5] high Q^2 data[6] from SLAC experiment E133, and high missing mass squared, W^2, data[7] A global fit has been made to this combined proton data. The components of the fit are separable into a nonresonant and three resonant components. The form of the fit functions for the nonresonant and $\Delta(1232)$ resonance are the same as those used by Stoler[1,2] while two higher resonances were modeled using a nonrelativistic Breit-Wigner form rather than a relativistic form for simplicity.

Using the global nonresonant fit, each individual cross section spectrum with $\Delta(1232)$ resonance data was refit to extract the transition form factor. The results of these fits are shown in Figure 1, where $G_D = (1.0 + Q^2/0.71)^{-1}$, μ_p is the proton magnetic moment, and F_Δ has been previously defined[1,2,4] Also shown are the diquark model fit by Kroll, *et al.*[8] the prediction from Stefanis and Bergmann[9] using their heterotic nucleon and $\Delta(1232)$ distribution amplitudes, and the asymptotic predictions (denoted by $*$) of Carlson and Poor[10] which have been evaluated at the Q^2 shown. The Kroll curve falls below the data, but the model was tuned to agree with the previous analysis[1,2] so this is not surprising. The heterotic curve lies above the data. Both curves are consistent with the data within errors. The major difference between this analysis and that of Stoler is the use of the global fit to the nonresonant component rather than fitting this component separately for each cross section spectrum. The effect on the extracted form factors is to shift them up by about one σ. The form factors, however, are still decreasing faster with

46

Figure 1. $\Delta(1232)$ transition form factors scaled by $\mu_p G_d$ extracted from fits to cross section data at each Q^2 point. Error bars are statistical only. Also shown are the diquark model fit by Kroll *et al.*[8], the heterotic prediction from Stefanis and Bergmann[9], and the asymptotic predictions (denoted by $*$) by Carlson and Poor[10], which have been evaluated at the Q^2 shown.

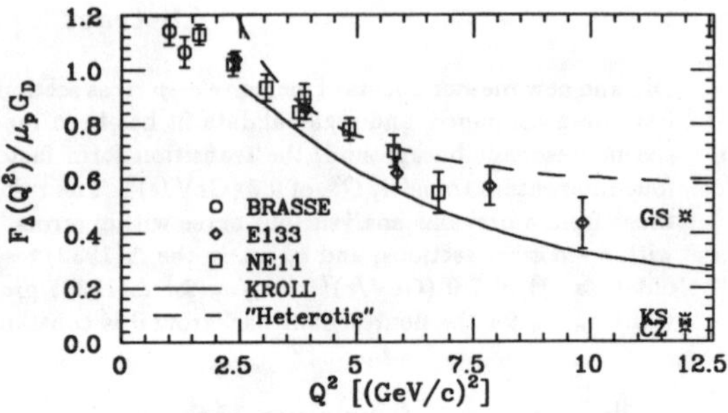

Q^2 than expected from leading order pQCD. The upward shift of the data points brings them in closer agreement with the heterotic prediction.

Using the results from the proton global fit, the deuterium data was analyzed in order to extract information on σ_n/σ_p as a function of Q^2 for the resonant and non-resonant components separately. The shape of the quasielastic peak was modeled with a non-relativistic Plane Wave Impulse Approximation (PWIA) calculation[11] using the Paris[12] deuteron wavefunction. The effects due to meson-exchange currents (MEC) were modeled using a parameterization[13] based on calculations by Laget[14] at $Q^2 = 1.75$ (GeV/c)^2. The proton resonant and nonresonant components from the proton global fit were separately Fermi smeared using the light cone smearing formalism of Sargsyan, Frankfurt, and Strikman[15]. Off-mass-shell corrections[16] were applied to the input on-mass-shell model structure functions in the smearing process. The result was smeared proton resonant and nonresonant cross sections which were fit to the deuterium data along with the quasielastic and MEC components. The fit coefficients yield information on the σ_n/σ_p ratios for the smeared components in the region of the $\Delta(1232)$ resonance as shown in Figure 2. The resonant σ_n/σ_p ratios were consistent with unity for all values of Q^2 as expected from isospin invariance. The nonresonant σ_n/σ_p ratios were roughly constant with Q^2 at a value of around 0.4. The top curve in (b) corresponds to the NMC deep inelastic model[18] evaluated at a fixed missing mass squared, $W^2 = 4.0$ GeV^2. The bottom curve is the Quark Parton Model (QPM) prediction at x = 1 ($W^2 = M_p^2$). Within errors the results are consistent with both curves.

In summary, new results on the $\Delta(1232)$ transition form factors from the combined SLAC NE11 and E133 cross sections show a small systematic upward shift from a previous analysis[1,2]. However, these results still fall faster with Q^2 than expected from leading order pQCD. Ratios of σ_n/σ_p have been extracted from deuterium cross sections using a model dependent method. The results for the $\Delta(1232)$ are consistent with unity as expected from isospin invariance, and the results for the nonresonant background are approximately 0.4 which is consistent with deep inelastic results. More details of this analysis and a model dependent study will soon be published[4].

Figure 2. σ_n/σ_p for the resonant and nonresonant cross section components in the region of the $\Delta(1232)$ resonance. The errors are statistical only. Previous data at low Q^2 from Köbberling[17] is also shown. Plot (b) also contains two curves. One is the NMC deep inelastic model evaluated at $W^2 = 4$ GeV^2, and the other is the Quark Parton Model prediction at x=1 ($W^2 = M_p^2$).

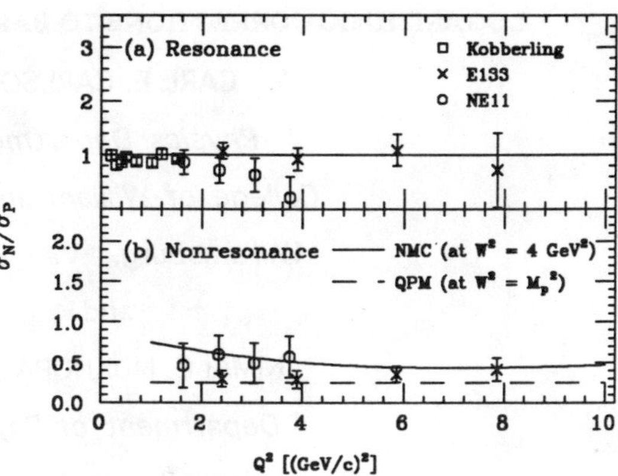

References

1 . P. Stoler, Phys. Rev. Lett. **66**, 1003 (1991).

2 . P. Stoler, Phys. Rev. D **44**, 73 (1991).

3 . P. Bosted, *et al.*, Phys. Rev. Lett. **68**, 3841 (1992), Phys. Rev. C **46**, 2505 (1992), A. Lung, *et al.*, Phys. Rev. Lett. **70**, 718 (1993).

4 . L. Stuart. *et al.*, SLAC-PUB-6305, To be submitted to Phys. Rev. C.

5 . F. W. Brasse, *et al.*, Nucl. Phys. **B110**, 413 (1976).

6 . S. Rock, *et al.*, Phys. Rev. D **46**, 24 (1992).

7 . L. W. Whitlow, SLAC-Report-357, (1990).

8 . P. Kroll, M. Schürmann, and W. Schweiger, Z. Phys. **A342**, 429. (1992).

9 . N. Stefanis and M. Bergmann, Phys. Lett. **B304**, 24 (1993).

10. C. E. Carlson and J. L. Poor, Phys. Rev. D **38**, 2758 (1988).

11. L. Durand III, Phys. Rev. **123**, 1393 (1961); I. J. McGee, Phys. Rev. **161**, 1640 (1967); **158**, 1500 (1967) as given by W. Bartel, *et al.*, Nucl. Phys. **B58**, 429 (1973).

12. M. Lacombe *et al.*, Phys. Lett. **101B**, 139 (1981).

13. L. Stuart, *et al.*, SLAC-PUB-6235, (1993).

14. J. M. Laget, Can. J. Phys. **62**, 1046 (1984); J. M. Laget, Phys. Lett. B **199**, 493 (1987).

15. M. Sargsyan, L. Frankfurt, and M. Strikman, Z. Phys. **A335**, 431 (1990).

16. D. Kusno and M. Moravczik, Phys. Rev. C, **27**, 2173 (1983).

17. M. Köbberling, *et al.*, Nucl. Phys. **B82**, 201 (1974).

18. P. Amaudruz, *et al.*, Phys. Lett. **B295**, 159 (1992).

LOGARITHMIC CORRECTIONS TO BARYON FORM FACTORS

CARL E. CARLSON

Physics Department

College of William and Mary

Williamsburg, VA 23187

and

NIMAI C. MUKHOPADHYAY

Department of Physics

Rensselaer Polytechnic Institute

Troy, NY 12180-3590

ABSTRACT

We study the logarithmic corrections to the pQCD scaling law governing the baryon form factors in the $\Delta(1232)$ resonance region as a function Q^2, . We show that these corrections are sizeable above $Q^2 \sim 6$ GeV^2. We find that the resonance–continuum duality will be violated, in general, by the logarithmic corrections at high Q^2.

1. Introduction.

In this Report we investigate logarithmic corrections to Bloom–Gilman(BD) duality, which is a relationship [1–3] between resonance physics and the physics of the deep inelastic region. Bloom and Gilman [1] observed that the ratio of a resonance bump in inelastic electron scattering to the continuum beneath the bump was generally constant with increasing Q^2 and that the smooth scaling curve seen at high Q^2 was an accurate average over the resonance bumps seen at lower Q^2, but the same Bjorken x. The first of these observations turns out not to be true [4] for the $\Delta(1232)$, although for other resonances, it is well confirmed out to high momentum transfer. The second observation appears to be true in general. In particular,for the $\Delta(1232)$, the background seems to rise [4], as the resonance falls, so that the average is constant relative to the scaling curve. Theoretically, one can understand that the Q^2 falloff of the resonance and of the scaling curve evaluated at the x value pertinent to the resonance are the same, at least as far as the powers of Q^2 are concerned. The dependence on logarithms of Q^2 have not yet been considered in this context.

We study here the logarithmic corrections to the resonance–continuum (or BD) duality discussed above. In particular, we wish to examine if the duality is violated by the logarithmic corrections, sizes of which are influenced by the structure of the hadrons. We conclude that the BD duality will be,in general, violated by logarithmic corrections at high Q^2.

Details of this work would be reported elsewhere[11].

2.Effects of logarithmic corrections on the continuum scaling function for $x \rightarrow 1$

Let us examine what effect logarithmic corrections have on the parton distribution functions, hence on the continuum scaling function. Since the resonance region draws closer to $x=1$, with increasing Q^2, we shall limit our considerations for the continuum also to $x \rightarrow 1$.

The basic tool is the Altarelli–Parisi equation,

50

$$\frac{dq(x,t)}{dt} = \frac{\alpha_s(t)}{2\pi} \int_0^1 dy\, dz\, \delta(x - yz)\, q(y,t) P_{qq}(z) \, . \tag{1}$$

Here, $t = \ln(Q^2/\Lambda^2)$, $\alpha_s(t) = 4\pi/\beta_1 t$, $\beta_1 = 11 - 2/3 n_f$, (where n_f is the number of fermion flavors), and $q(x,t)$ is a quark distribution function of a given flavor. The gluon term is omitted, as its contribution is subleading in $(1-x)$.

The splitting function is

$$P_{qq}(z) = C_F \left\{ \frac{1 + z^2}{(1 - z)_+} + \frac{3}{2} \delta(1 - z) \right\} \, , \tag{2}$$

where C_F is 4/3, and $(1-z)_+$ is defined by

$$\int_x^1 dz \frac{f(z)}{(1 - z)_+} = \int_x^1 dz \frac{f(z)}{1 - z} - \int_0^1 dz \frac{f(z)}{1 - z} \, . \tag{3}$$

We want to examine the evolution of a form like $q(x,t_0) = N_0(1-x)^b$, where b is a constant, and t_0 corresponds to some benchmark Q_0^2. Hence we use the Ansatz,

$$q(x,t) = N(x,t)(1 - x)^b \, , \tag{4}$$

in the Altarelli–Parisi equation. Systematically throwing away terms of higher order in $(1-x)$, we get

$$q(x,t) = N_0 (1 - x)^{b + \frac{4 C_F (\ln \ln Q^2)}{\beta_1}} \, , \tag{5}$$

where

$$\ln \ln Q^2 = \ln \frac{\ln Q^2 / \Lambda^2}{\ln Q_0^2 / \Lambda^2} = T(Q) \cdot \tag{6}$$

To see the size of the logarithmic correction, we examine the values of x corresponding to the peak of the Δ resonance region,

$$\frac{1}{x} = 1 + \frac{m_\Delta^2 - m_N^2}{Q^2} \cdot \tag{7}$$

We now estimate the size of the correction factor to the high-x continuum in the Δ-region for a benchmark Q^2 of $4\,\mathrm{GeV}^2$. As Q^2 varies from 4 to 21 GeV^2, the logarithmic correction to the form $(1-x)^b$ varies from 1 to 0.56.

Thus, **the logs are important!**

Let us now examine the $F_2 = \nu W_2$ data in the resonance region to the continuum scaling curve. This is shown in Fig. 1, with the log correction; the ratio $R = I/S$ is plotted versus Q^2. Here we have defined

$$I = \int_{\Delta \xi} d\zeta (F_2(\zeta, Q^2));$$

$$S = \int_{\Delta \zeta} d\zeta (F_2^{scaling}(\zeta, Q^2)) , \tag{8}$$

where ζ is the Nachtmann variable (x corrected [3,7] for the target mass effects), and $\Delta \zeta$ is a bite covering the chosen resonance region (here, the Δ). Although the error bars are large at high Q^2, one observes that the logarithmic corrections improve the constancy of the ratio R at high Q^2. Hence the inclusion of the logarithmic effects helps to make the duality idea, the low-Q^2 structure function for a given W should average to the scaling curve, appear to work better.

4. Logarithmic effects on the resonance to continuum ratio

We can derive a condition for the preservation of the BD duality in the form of the constancy of the resonance tpeak to scaling curve ratio.Writing the continuum form for $x->1$ as [6]

$$F_2^{scaling} = const. (1-x)^{b+C_1 T} \ , \qquad (9)$$

we get a condition on the baryon amplitudes, which is extremely stringent:

$$(1/2)C_1 = \frac{\sum_{ij}(E_i N_i^P N_j^R)(2 + \gamma_i^P + \gamma_j^R)}{\sum_{ij}(E_i N_i^P N_j^R)} \ . \qquad (10)$$

Since $C_1/2$ is about 1/4 and the γ_j are positive, only exceptional choices of the amplitudes N_i^P and/or N_j^P can fulfill the above equation. We find that it is not fulfilled for the cases of the Chernyak-Zhitnitsky [5] or the King-Sachrajda [8] wave function for the proton, and analogous wave functions for the $S_{11}(1535)$. It happens that every significant $E_{ij}N_i^P N_j^{S_{11}}$ is positive, violating Eq.(17). Thus,in general, the BD duality in the form of the constancy of the resonance peak to scaling curve must be logarithmically violated.

4.Conclusions

We have studied here the logarithmic corrections to the pQCD scaling laws governing the form factors at high Q^2. The region of the $\Delta(1232)$ resonance excitation has been of our special interest, in which the resonance bump has been shown to fall faster than the underlying background. We have found the logarithmic corrections to be important for $Q^2 > 6 \ GeV^2$. In general,these effects are dependent on the specific baryon wave functions. The resonance-continuum duality, a cornerstone of our understanding of the hadron structure, in general, will be violated by the logarithmic corrections at high Q^2.

5.Acknowledgments

One of us(NCM) thanks Professor Paul Stoler and his co-organizers for their kind invitation to present this paper at the Elba Workshop. He also thanks the hosts of the Workshop for their partial support. We thank Professors S. Brodsky and P. Stoler for valuable comments. The research of CEC is supported in part by the NSF (Grant#PHY-9112173), and that of NCM by the U.S. Department of Energy (Grant#DE-FG02-88ER47048-A005).

Fig. 1: The effect on the B-D scaling with logarithmic corrections included, in a manner discussed in the text.

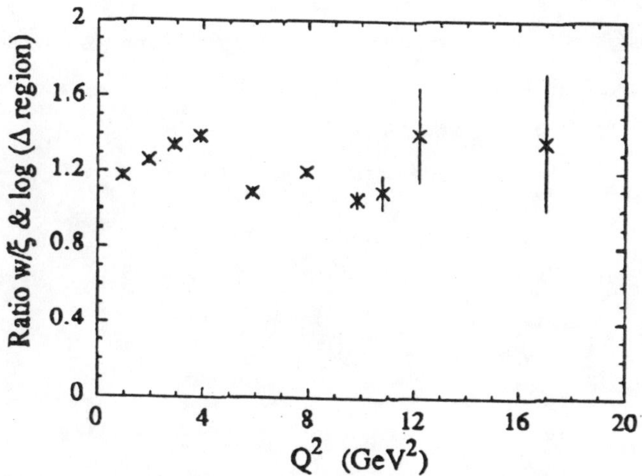

[1] E.D. Bloom and F.J. Gilman, Phys. Rev. Lett.25 (1970) 1140; Phys. Rev.D4 (1971) 2901.

[2] A. DeRu'jula, H. Georgi and H. D. Politzer, Ann. Phys.(N.Y.)103 (1977) 315.

[3] C.E. Carlson and N.C. Mukhopadhyay,Phys. Rev.D41 (1990) R2343; Phys. Rev. Lett.27 (1991) 3745..

[4] C.E. Carlson and N.C. Mukhopadhyay,Phys.Rev.D47 (1993) R1737.

[5] V.L. Chernyak and A.R. Zhitnitsky,Nucl. Phys.B246 (1984) 52.

[6] J.G. Morfin and W.K. Tung,Z.Phys.C52 (1991) 13.

[7] O. Nachtmann,Nucl.Phys.B63 (1973) 237.

[8] J.D. King and C.T. Sachrajda,Nucl. Phys.B279 (1987) 785.

[9] E.J. Eichten,I. Hinchliffe and C. Quigg,Phys. Rev.D45 (1992) 2269.

[10] P.Stoler,Phys. Rev. Lett.66 (1991) 1003; Phys. Rev.D44 (1991) 73.

[11] C.E. Carlson and N.C. Mukhopadhyay, to be published.

MESON FORM FACTORS AT LARGE Q^2

JOSEPH MILANA

Department of Physics, University of Maryland
College Park, MD 20742-4111, U.S.A.

ABSTRACT

Recent work concerning the determination of the pion's electromagnetic form factor
at both large space–like and time–like Q^2 are reviewed. In addition, new results
suggesting potentially large corrections to heavy quark symmetry are presented.

1. Introduction

I thank the organizers for their kind invitation to come to Elba. I have
been asked to discuss meson form factors, and keeping in line with the spirit of the
organizers intent, for most of the talk I will review some recent work of mine and
my collaborators concerning the determination of the pion's form factor at large Q^2.
Thanks though to the particularly vagueness of the organizers title, I will also take
this opportunity to mention some new results concerning heavy quark mesons at
large recoil. A pQCD analysis suggests large corrections to the Wisgur form factor,
some of which are *not* $O(\Lambda/M)$, and since this is a field of enormus recent interest, I
think it important the audience be at least aware of these results.

2. The space–like data revisited

I would like to raise the question, What is the Pion's Electromagnetic Form
Factor at large Q^2? I think we're all familiar with the following plot:

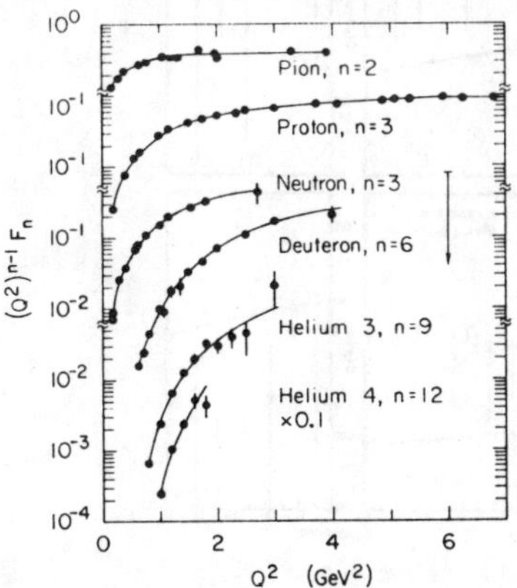

Figure 1: Hadronic electromagnetic form factors.

The proton data of course extends well out to over 30 GeV², and Paul Stoler[1] has done important work extracting baryon transition form factors. I wish to focus on the top curve. Over the years, this top curve–the purported pion's form factor–has been the source of a great deal of theoretical discussion. Is this though truly the pion's form factor? With a little reflection, one realizes that because the pion is an unstable particle, measuring it's electromagnetic form factor is a tricky business. What was actually done was to measure the electroproduction of pions from nucleons:[2]

$$e + p \rightarrow e' + \pi^+ + n. \tag{1}$$

By selecting neutrons emitted at low momentum, it was modeled that the cross-section for longitudinal polarizations of the virtual photon in a Rosenbluth decomposition is dominated by the t–channel pion pole graph, thereby relating the measured result to $g_{\pi NN}(t \approx 0)$ and $F_\pi(Q^2)$. Neglecting any possible differences between the properties of an on–shell and off–shell pion, this model has a clear prediction that:

$$\frac{d\sigma}{dt}(Q^2) \propto t^{-1}. \tag{2}$$

The curves on the right (labeled C) below shows the measured dependence on t.[2]

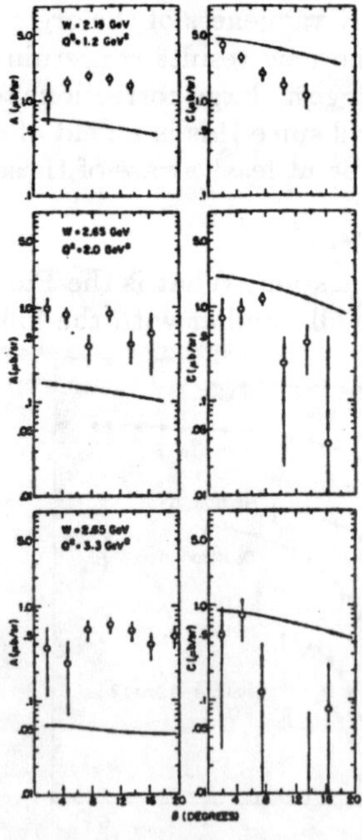

$$t \rightarrow$$

Figure 2: The measured longitudinal component of the differential cross–section.

Clearly the large Q^2 values show no convincing t–channel pole.

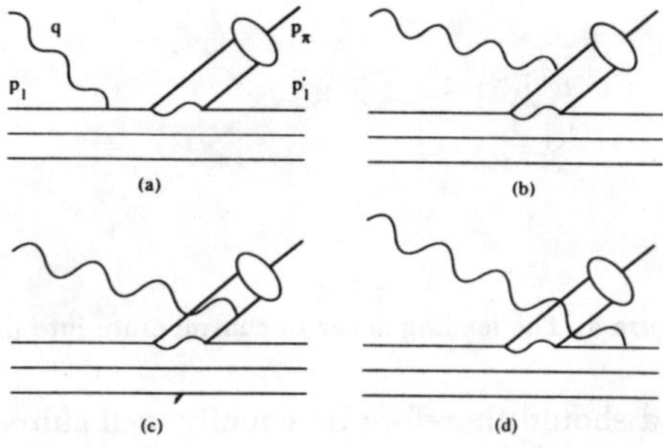

Figure 3: Background processes for pion photoproduction.

What Carl Carlson and I did[3] was to make an estimate of the background processes to the t–channel pole diagram using a PQCD formalism. We calculated the graphs shown in Fig. 3, in which a single quark from the proton interacts with the virtual photon to produce a pion at large energy. A quark is constrained to be emitted from the hard process with momentum parallel to that of the initial hadron so as to reform a neutron in the final state. Focussing on Fig. 3a, we see that in a hadronic language, this diagram corresponds to what would be called s–channel corrections. Note though that because the struck quark is way off–shell, there is effectively a sum over a large number of intermediate baryon states: the economic advantage of the pQCD formalism is obvious.

There is of course a soft physics matrix element, involving integrals over the hadron's wavefunctions. For our estimates we tried various three quark, Fock component wavefunctions on the market[4]. We found that for any of these wavefunctions, for the *exact* kinematic points used by the experimentalists to extract the pion form factor at the larger Q^2 values ($Q^2 > 1.2$ GeV^2), our background processes are of order $20 - 25\%$ corrections to the t–channel pole diagram. Since we do not know the relative sign between these two contributions, this results in an ambiguity of $O(50)\%$, for the actual value of the pion's form factor.

Part of the reason for this ambiguity is that at the larger Q^2 values t is not sufficently small to properly enhance the pole diagram. One could hope to redo the experiment and push t smaller. An absolute neccessary condition would be to demonstrate that the differential cross–section exhibits the $1/t$ behavior required by the model. Because of the kinematics involved, such an experiment would be best suited for SLAC energies. Alternatively, at the lower energies available say at CE-BAF, one could focus on π^0 production. This would provide a direct measurement of the background processes, as the t–channel diagram is absent (the electromagnetic form factor of the π^0 being identically zero). Note that such a cross–section should display the same scaling behavior with Q^2 as predicted by pQCD[13] for the

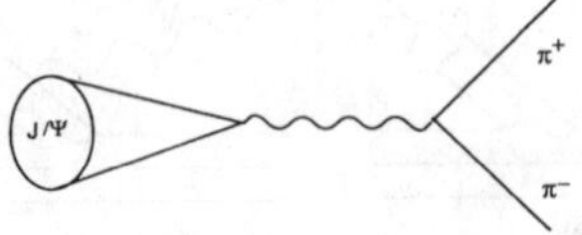

Figure 4: The leading decay of charmonium into pions.

charged pion case and should therefore be equally well suited for addressing issues concerning the validity of pQCD and of the onset of asymptopia.

3. A determination at large time–like Q^2?

Having raised doubts with the value of pion's form factor at large space–like Q^2, one is naturally lead to seek another, independent source of information for the pion form factor at large Q^2. One such source is the measured decay channels of charmonium. The leading order diagram for the decay

$$J/\psi \to \pi^+\pi^- \tag{3}$$

is shown in Fig. 4. I remind the audience that the decay of the J/ψ to two pions by the strong interaction is forbidden by G–parity. (For those like myself who are perhaps not so familiar with the rule of G–parity, it is essentially a judicious combination of parity and isospin. Assuming isospin invariance, the two pions are identical and being bosons, they must therefore be in a symmetric final state upon interchange. However by parity and angular momentum considerations [$J^P = 1^-$ for the J/ψ] they must be in a $L = 1$ state. As additional support of Fig. 4 being the leading mechanism, one should note that the decay into two π^0s has not been observed.) To extract a value for $F_\pi(M^2_{J/\psi})$, one can eliminate the dependence on the wavefunction of the J/ψ by comparing the obtained branching ratio $\Gamma^{J/\psi}_{\pi^+\pi^-}$ to the leptonic decay rate $\Gamma^{J/\psi}_{e^+e^-}$, from which one obtains that

$$\frac{\Gamma^{J/\psi}_{\pi^+\pi^-}}{\Gamma^{J/\psi}_{e^+e^-}} = \frac{F^2_\pi(M^2_{J/\psi})}{4}, \tag{4}$$

and inserting the measured branching ratios ($\Gamma^{J/\psi}_{e^+e^-} = (6.27 \pm .20)10^{-2}, \Gamma^{J/\psi}_{\pi^+\pi^-} = (1.47 \pm .23)10^{-4}$) one finds that

$$M^2_{J/\psi}F_\pi(M^2_{J/\psi}) = (.94 \pm .08)GeV^2. \tag{5}$$

Note that Eq. (5) is a factor of two or so larger than might have been expected had one extrapolated the space-like data [2] for $3.33GeV^2 \geq |Q^2| \geq 1.18GeV^2$ assuming

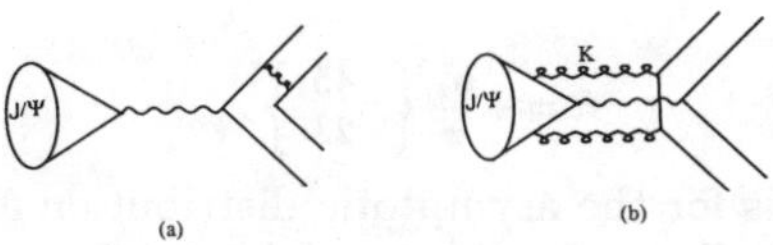

Figure 5: (a) The leading order diagram redrawn in terms of PQCD, and (b) the leading order correction that could spoil the extraction of the pion's EM form factor.

PQCD scaling. It is amusingly much closer to the largest quoted Q^2 value of [2] ($Q_o^2 F_\pi(Q_o^2) \approx .7 GeV^2$, at $Q_o^2 = 9.77 GeV^2$) although both the experimental errors and the expected background processes [3] are largest.

Considering the difficulties associated with the space-like data, one must consider the background processes that could spoil the extraction (5). One expects that the largest of these occur through strong interactions between the outgoing pions and the initial state of charmonium. The lowest order PQCD graph is shown in Fig. (5), where I have also redrawn Fig. 4 in this language. Remember that the decay via three gluons is forbidden by G-Parity. I also distinguish between diagrams such as Fig. 5b and those that are merely a single gluon added to Fig. 5a, as the latter are effectively vertex corrections to either the J/ψ wavefunction (and hence would cancel in the ratio 4) or are part of the definition of $F_\pi(M_{J/\psi}^2)$. Either way, they would not spoil the extraction (5). Unlike the background processes [3] discussed for the space–like data, the corrections considered here are higher order in $\alpha_s(Q^2)$ and thus one might expect them to be small. On the other hand there are known examples [5] in the charmonium spectrum (e.g. the decay of the η_c into hadrons) where such corrections are accompanied with an unusually large coeffecent.

To estimate the contribution to the amplitude, $\mathcal{M}_{NLO}$, of Fig. 5b and its 35 accompanying diagrams, Rick Kahler and I[6] calculated these graphs using the asymptotic[7] and the Chernyak-Zhitnitsky(CZ)[8] distribution amplitudes for the pion. Whereas neither distribution amplitude is capable of reproducing the value for $F_\pi(M_{J/\psi}^2)$ contained in Eq. (5) ($Q^2 F_\pi(Q^2) = 1.21\,\alpha_s(Q^2)$ for the CZ distribution amplitude, which is a factor 25/9 larger than the asymptotic result), we believe that between the two, considering the rather different forms for these two distribution amplitudes, we will get a good estimate for the relative size

$$R = \frac{\mathcal{M}_{NLO}}{\mathcal{M}_{LO}}. \tag{6}$$

Advoiding some technical details of the calculation which the intereseted reader can read about in our paper, (e.g. $\mathcal{M}_{NLO}$ has both a real and an imaginary part), we found that

$$R = +\frac{\alpha_s}{\pi} \left\{ \begin{array}{c} .45 \\ .23 \end{array} \right\}. \tag{7}$$

where the upper number is for the asymptotic distribution and the lower is for the CZ one. The relatively small size for (7) we understand as arising due to the difficulty of extracting additional gluons from a small color-singlet initial state and we find this result highly suggestive that charmonium provides a good laboratory[14] for extracting the pion's electromagnetic form factor at large Q^2 and that $M_J^2 F_\pi(M_J^2) \approx .9$.

4. pQCD results for Heavy Quark Mesons (briefly)

Since the work of Isgur and Wise[9], heavy quark physics and heavy quark symmetry has become a huge industry. It is important therefore to delineate the limitations of the approach. One obvious dubious application is in the limit of large recoil as occurs say in the exclusive two body decay of a B–meson. Fortunately this is precisely the region where one could expect pQCD to be a useful guide[10]. At large recoil, one finds that the Wisgur form factor[11]

$$\xi(\omega \equiv (v - v')^2) \propto \frac{1}{\omega(\omega - 2)}. \tag{8}$$

In order to obtain this univeral form, it was neccessary to take the heavy meson decay constant $f_H \propto 1/\sqrt{M_H}$[12]. Although the latter is a standard heavy quark result, it is the first time such an explicit connection between ξ and f_H has been made, and any violation in the heavy quark result in one clearly impacts the other.

Observe that (8) is a dipole in ω, which is proportional to Q^2 of the probe. It is *not* therefore the general pQCD counting rule result[13] for a meson form factor at large Q^2. Hence in the regime where $Q^2 \geq M_H^2$, such as in $e^+ + e^-$ annihilation, one expects large corrections[12].

The proportionality factor in (8) is also important and a potential source of large corrections. This factor is a sharp function of a light quark mass scale, Λ (explicitly, the dependence is Λ^{-3}). We thus see that very large corrections are formally possible when probing sectors of the heavy quark effective theory (e.g. comparing the decays, $B \to D \pi$, $B \to D^* \pi$) and where the corrections are known to be sizable on the *light quark* mass scale (e.g. the D–D* mass difference).

Acknowledgements I thank J. Napolitano for invigorating discussions concerning the pion's space–like data. This work was supported in part by the U.S. Department of Energy under grant No. DE-FG05-93ER-40762.

6. References

1. P. Stoler, Phys. Rep. **226**, 103 (1993).

2. C. J. Bebek *et al.*, Phys. Rev. D **17**, 1693 (1978).

3. C. E. Carlson and J. Milana, Phys. Rev. Lett. **65**, 1717 (1990).

4. V. L. Chernyak and A. R. Zhitnitsky, Nucl. Phys. **B246**, 52 (1984); I. D. King and C. T. Sachrajda, Nucl. Phys. **B279**, 785 (1987); M. Gari and N. G. Stefanis, Phys. Rev. D35, 1074 (1987); C. E. Carlson and F. Gross, Phys. Rev. D36, 2060 (1987).

5. R. Barbieri *et al.*, Nucl. Phys. B **154**, 535 (1979).

6. R. Kahler, and J. Milana, Phys. Rev. D47, R3690 (1993).

7. G. R. Farrar and D. R. Jackson, Phys. Rev. Lett **43**, 246 (1979); A. V. Efremov and A. V. Radyushkin, Phys. Lett. B **94**, 245 (1980).

8. V. L. Chernyak and A. R. Zhitnitsky, Phys. Rep. **112**, 173 (1984).

9. N. Isgur and M. Wise, Phys. Lett. B **232**, 113 (1989); *ibid*, **237**, 527 (1990).

10. A. Szczepaniak, E.M. Henley, and S.J. Brodsky, Phys. Lett. B **243**, 287 (1990).

11. C.E. Carlson and J. Milana, Phys. Lett. B301, 237 (1993).

12. T. D. Cohen, and J. Milana, Phys. Lett. B303, 134 (1993).

13. S. J. Brodsky and G. R. Farrar, Phys. Rev. Lett. **31**, 1153 (1973); Phys. Rev. D **11**, 1309 (1975); V. A. Matveev, R. M. Muradyan, and A. V. Tavhkhelidze, Lett. Nuovo Cimento **7**, 719 (1973); S. J. Brodsky and G. P. Lepage, Phys. Rev. D **24**, 1808 (1981).

14. For a general consideration of other background processes, see J. Milana, S. Nussinov, and M. G. Olsson, "Does $J/\psi \to \pi^+\pi^-$ fix the Electromagnetic Form Factor $F_\pi(t)$ at $t = M^2_{J/\psi}$?", U. of MD PP #94-001, (1993).

Measuring the Charged Pion Form Factor at CEBAF

D. J. Mack

Physics Division, CEBAF

Newport News, VA 23606

U.S.A.

Abstract

The need for improved measurements of F_π at high Q^2 is reviewed. Improved data up to $Q^2 = 1.5\ (GeV/c)^2$ will be obtained in an initial series of measurements in endstation C at CEBAF. Upgrade of the CEBAF beam energy to 6 GeV will permit high quality data up to $Q^2 = 5\ (GeV/c)^2$.

1. Introduction

The charged pion form factor, F_π, is used to conveniently summarize information about the (non-pointlike) distribution of charge in the pion. It is particularly interesting for tests of QCD because the pion wave function is relatively simple when compared to that of the nucleon. The pion wave function is also normalized in the asymptotic limit by the pion decay constant. Another important feature is that the PQCD diagram for elastic electron-pion scattering involves the exchange of only a single gluon between the two quarks, whereas two gluons must be exchanged between the three quarks in the nucleon case. This lowers the Q^2 threshold at which the scattering amplitude is expected to become dominated by PQCD. High quality data above $Q^2 = 1\ (GeV/c)^2$ would allow one to test models in the transition from non-perturbative to perturbative QCD.

The most direct means of measuring F_π is to scatter pions from atomic electrons. At very low Q^2, such experiments have shown[1] that F_π is given to a good approximation by the vector meson dominance prediction

$$F_\pi = \frac{1}{1 + Q^2/M_\rho^2}.$$ (1)

However, even with 300 GeV pion beams, the maximum Q^2 measured with this technique is only $.28\ (GeV/c)^2$.

At $Q^2 \gg 1\ (GeV/c)^2$, the PQCD prediction for F_π is[2]

$$F_\pi = \frac{-2f_\pi^2}{b\,ln(Q^2)}\frac{1}{Q^2}$$ (2)

where f_π is the pion decay constant. If QCD is the correct theory of the strong interactions, then the character of F_π must change considerably between these two extremes of Q^2. But how is one to explore this interesting transition region?

Following an early suggestion by Frazer[3], F_π can be determined at higher Q^2 by pion electroproduction. In this case the target is the cloud of virtual pions surrounding the proton, with the likelihood of finding a pion of a given momentum proportional to $g_{\pi NN}(t)$. For sufficiently small values of $-t$, the longitudinal response σ_L is dominated by the contribution from the pion pole

$$\sigma_L \propto \frac{-t g^2(t)}{(t - m_\pi^2)^2} F_\pi^2(Q^2) \tag{3}$$

With properly chosen kinematics the longitudinal response can be much larger than the transverse response. In practice, one does not use this simple expression but rather the best available model of pion electroproduction treating F_π as a free parameter.

2. Previous Experimental Work

A large body of experimental work has established the validity of extracting F_π via pion electroproduction. Both monopole and dipole form factor fits to the elastic scattering data are, when extrapolated, in good agreement with the low Q^2 electroproduction data. This suggests that a number of possible difficulties with the technique, such as non-negligible physics backgrounds in the longitudinal response, uncertainty in the $g_{\pi NN}$ form factor, and pion rescattering do not cause significant problems at low Q^2.

Carlson and Milana[4] have pointed out that physics backgrounds and the $g_{\pi NN}$ form factor uncertainties may become significant as one increases Q^2 and moves away from the pion pole. To some extent these issues were addressed long ago by Bebek et al.[5]. In Figure 16 of that work (not reproduced here), experimentally equivalent values of F_π were extracted from data covering a wide range of distances to the pion pole. Barring a fortuitous series of cancelling effects, it appears that even at $Q^2 = 2 \ (GeV/c)^2$ these effects were not significant. These effects may become important, however, in a measurement of higher precision.

The CEA-CSA electroproduction data base[6] is shown in Figure 1. There are inconsistencies near $Q^2 = 1 \ (GeV/c)^2$ which need to be cleaned up. Above $Q^2 = 1\text{-}2$ $(GeV/c)^2$, the data have very large statistical errors. Systematic errors were large since for some of the data points no backward angle data exist, and for others the forward and backward angle measurements were from different experiments. From top to bottom in the same figure the curves are the extrapolated monopole form factor[1], an estimate of soft contributions[7], a Bethe-Salpeter model[8], and the asymptotic QCD prediction[2]. Clearly the existing data above $Q^2 = 2 \ (GeV/c)^2$ have little power to differentiate between any of these models.

3. Proposed CEBAF Measurements

The anticipated high duty factor and luminosity of the CEBAF accelerator and endstation C will permit much smaller statistical errors and much larger reals to

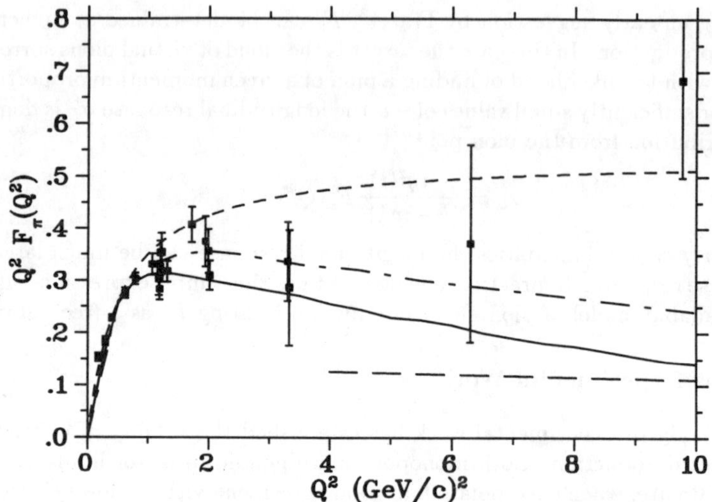

Figure 1: Existing data for $F_\pi Q^2$ from electroproduction at large Q^2.

randoms ratios. Because the time needed to acquire data will be relatively short, we will also be able to spend more time checking for systematic errors.

The proposal submitted by our collaboration[9] to the CEBAF Program Advisory Committee is in two phases. Phase I covers the Q^2 range of .5-1.5 $(GeV/c)^2$, and is well suited to an accelerator with maximum energy of 4 GeV. We will make L-T separations in a single experiment with the same apparatus. Although improved measurements are most badly needed at higher Q^2, the various models begin to diverge above $Q^2 = 1$ $(GeV/c)^2$, so precise measurements at $Q^2 = 1$ and 1.5 $(GeV/c)^2$ will still be useful. We also plan to do a better search of physics backgrounds than has been achieved previously. Finally, we will be able to tightly constrain the t dependence of the $g_{\pi NN}$ form factor. Phase I of our proposal is approved.

Phase II covers the Q^2 range of 2-5 $(GeV/c)^2$ and requires beam energies as high as 6 GeV. In this region existing data are of poor quality, so the interest in an improved series of measurements is high. Unfortunately, CEBAF has not yet approved this or any other measurement which requires beam energies greater than 4 GeV. There appear to be no weak links in the accelerator[10], however, which would prevent achieving beam energies of 5.5 GeV. Commissioning of the endstation C beamline is scheduled for mid-1994. If all goes well, the Phase I measurement described here will take place in 1996, after the first series of shakedown experiments.

Anticipated errors for Phases I and II are shown in Figure 2. Statistical and systematic errors have been combined in quadrature, and the Bethe-Salpeter model of Jacob and Kisslinger[8] was assumed. Data of this quality would be extremely valuable for testing any model of the transition from non-perturbative to perturbative QCD.

4. References

1. S.R. Amendolia *et al.*, Phys. Lett. **146B**, 116 (1984), and Nucl. Phys. **B277**, 168 (1986).

2. G.R. Farrar and D.R. Jackson, Phys. Rev. Lett. **43**, 246 (1979).

3. W.R. Frazer, Phys. Rev. **115**, 1763 (1959).

4. C.E. Carlson and J. Milana, Phys. Rev. Lett. **65**, 1717 (1990).

5. C.J. Bebek *et al.*, Phys. Rev. D **13**, 25 (1976).

6. C.J. Bebek *et al.*, Phys. Rev. Lett. **37**, 1326 (1976); Phys. Rev. D **17**, 1693 (1978).

7. N. Isgur and C.H. Llewellyn Smith, Phys. Rev. Lett. **52**, 1080 (1984).

8. O.C. Jacob and L.S. Kisslinger, Phys. Lett. B **243**, 323 (1990).

9. D.J. Mack *et al.*, CEBAF proposal PR-093-021.

10. A. Hutton, private communication.

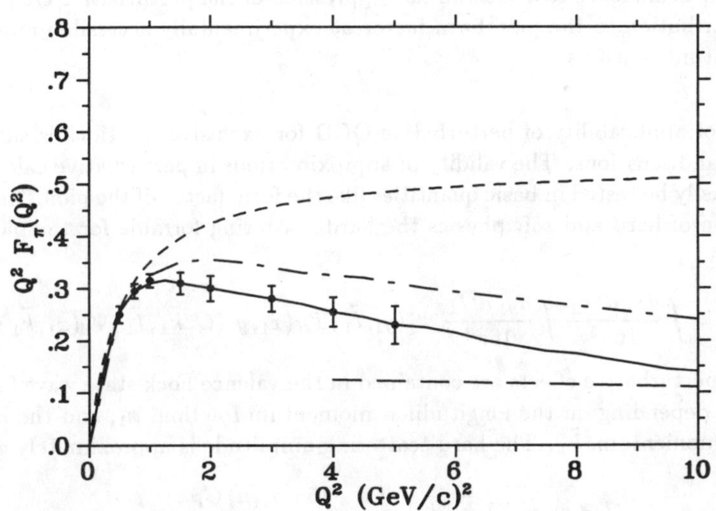

Figure 2: Errors from the CEBAF proposal assuming the Bethe-Salpeter model of Jacob and Kisslinger.

THE PION FORM FACTOR:
ROLE OF INTRINSIC TRANSVERSE MOMENTUM

R. Jakob *

Fachbereich Physik, Universität Wuppertal,
42097 Wuppertal, Germany

and

P. Kroll **

Theory Division, CERN, CH-1211 Geneva, Switzerland

ABSTRACT

The effect of the intrinsic transverse momentum dependence of the pion
wave function in the calculation of the pion form factor is compared
with Sudakov corrections. It turns out that the intrinsic transverse
momentum leads to a substantial suppression of the perturbative QCD
contribution to the pion form factor at experimentally accessible momentum transfers.

The range of applicability of perturbative QCD for exclusive reactions is subject to
controversial discussions. The validity of approximations in perturbative calculations
can most easily be tested in basic quantities like the form factor of the pion. Assuming
factorization of hard and soft physics the hard-scattering formula for the pion form
factor reads

$$F_\pi(Q^2) = \int \frac{dx_1 \, d^2 k_\perp}{16\pi^3} \int \frac{dy_1 \, d^2 l_\perp}{16\pi^3} \, \Psi_0^*(y_1, \vec{l}_\perp) \, T_H(x_1, y_1, Q, \vec{k}_\perp, \vec{l}_\perp) \, \Psi_0(x_1, \vec{k}_\perp). \quad (1)$$

where non-perturbative effects are contained in the valence Fock state wave function,
$\Psi_0(x_1, \vec{k}_\perp)$, depending on the longitudinal momentum fraction, x_1, and the intrinsic
transverse momentum, $\vec{k}_\perp$. The hard-scattering amplitude is approximately given as

$$T_H(x_1, y_1, Q, \vec{k}_\perp, \vec{l}_\perp) = \frac{16\pi \, \alpha_s(\mu) \, C_F}{x_1 y_1 Q^2 + (\vec{k}_\perp + \vec{l}_\perp)^2}, \quad (2)$$

where the denominator is the squared momentum of the virtual gluon exchanged
between the constituents to share the momentum transfer.

From the work of Li and Sterman[2] we learned that the explicit $k_\perp$- and $l_\perp$-dependence

---

*Supported by the Deutsche Forschungsgemeinschaft
**On leave of absence from Fachbereich Physik, Universität Wuppertal, Germany

in the hard scattering amplitude should be retained in order to render the perturbative calculation self-consistent. Moreover, Sudakov corrections arising from vertex-like and self-energy corrections (in axial gauge) should be taken into account as well. In order to include this corrections, it is advantageous to reexpress Eq. (1) in terms of the Fourier transform variable $\vec{b}$, associated with the transverse separation of the constituents

$$F_\pi^{pert}(Q^2) \;=\; \int_0^1 \frac{dx_1\,dy_1}{(4\pi)^2} \int_{-\infty}^\infty d^2b\; \hat{\Psi}_0^*(y_1,\vec{b})\,\hat{T}_H(x_1,y_1,Q,b,t)\,\hat{\Psi}_0(x_1,-\vec{b})$$
$$\times \exp\left[-S(x_1,y_1,b,Q)\right] \tag{3}$$

where $\hat{\Psi}$ and $\hat{T}$ are the Fourier transforms of Ψ and T. The Sudakov suppression function, $S(x,y,b,Q)$ (given explicitly by Li and Sterman[2]), tends to suppress contributions from large b at large momentum transfer, Q. Therefore Li and Sterman use the approximation

$$\hat{\Psi}_0(x_1,\vec{b}) \to \hat{\Psi}_0(x_1,\vec{b}=0) = 4\pi\,\frac{f_\pi}{2\sqrt{6}}\,\phi(x_1), \tag{4}$$

which is exact at asymptotically large Q. The distribution amplitude, $\phi(x)$, is defined and normalized in the usual way as

$$\frac{f_\pi}{2\sqrt{6}}\,\phi(x_1,\mu) = \int \frac{d^2k_\perp}{16\pi^3}\,\Psi_0(x_1,\vec{k}_\perp) \qquad \int_0^1 dx_1\,\phi(x_1,\mu) = 1 \tag{5}$$

where $f_\pi(=133\,\text{MeV})$ is the pion decay constant.

But is the approximation (4) valid at Q^2 of the order of a few GeV ?

In order to check this approximation we make a phenomenological ansatz for the $k_\perp$-dependence of the wave function

$$\Psi_0(x_1,\vec{k}_\perp) = \frac{f_\pi}{2\sqrt{6}}\,\phi(x_1)\,\Sigma(x_1,\vec{k}_\perp) \tag{6}$$

where

$$\Sigma(x_1,\vec{k}_\perp) = 16\pi^2\beta^2\,g(x_1)\,\exp\left(-g(x_1)\beta^2 k_\perp^2\right), \tag{7}$$

$g(x_1)$ can be either 1 or $1/x_1x_2$, the latter choice is motivated by harmonic oscillator wave functions[3]. $f_\pi(=133\,\text{MeV})$ is the usual pion decay constant. The parameter β is determined by requiring for the root mean square transverse momentum, $\langle k_\perp{}^2\rangle^{1/2}$, a value of $350\,\text{MeV}$ in accordance with the pion charge radius. The value for β is justified by a constraint derived from $\pi^0 \to \gamma\gamma$ [3]. With this ansatz the Fourier transform of the wave function reads

$$\hat{\Psi}_0(x_1,\vec{b}) = \hat{\Psi}_0(x_1,\vec{b}=0)\,\exp\left(-\frac{b^2}{4g(x_1)\beta^2}\right). \tag{8}$$

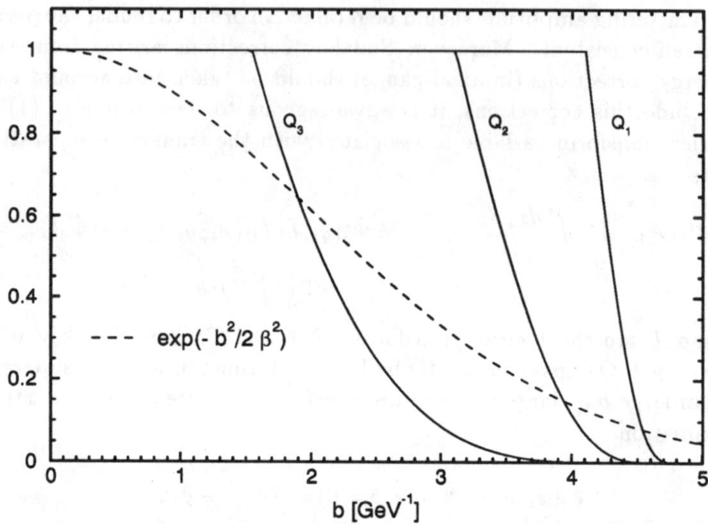

Figure 1: The Sudakov factor, evaluated at $x_1 = y_1 = 1/2$, and the Gaussian $\exp\left(-b^2/2\beta^2\right)$ as functions of the transverse separation b. The Gaussian is shown for a r.m.s. transverse momentum of $350\,\mathrm{MeV}$ (dashed line). The Sudakov factor is evaluated at $Q_1 = 2\,\mathrm{GeV}$, $Q_2 = 5\,\mathrm{GeV}$ and $Q_3 = 20\,\mathrm{GeV}$ with $\Lambda_{QCD} = 200\,\mathrm{MeV}$ (solid lines).

Setting the exponential to unity corresponds to the aproximation (4). To examine the validity of the approximation one has to compare the exponential in Eq. (7) with the Sudakov factor in Eq. (3). In figure 1 this is done for a specific configuration, $x_1 = y_1 = 0.5$, where incoming and outgoing wave function contain the same exponential. Clearly the Gaussian is not negligible compared to the Sudakov factor and plays the major rôle in the behaviour of the integrand in Eq. (3) for momentum transfer, Q, less than $20\,\mathrm{GeV}$.

Numerical evaluations of the pion form factor confirm the observations made in figure 1. The $\vec{k}_\perp$-dependence of the wave function, i.e. the Gaussian in the Fourier transform, $\hat{\Psi}(x,\vec{b})$, provides considerable suppression of the perturbative QCD contribution to the pion form factor. This is shown in figure 2 for the distribution amplitude [4] $\phi(x) = 30x(1-x)(2x-1)^2$ (with $g = 1$). As compared with the standard perturbative calculation, neglecting the $k_\perp$-dependence of the hard scattering amplitude, the Sudakov corrections lead to some suppressions, which, however, is augmented when the intrinsic $k_\perp$-dependence of the pion wave function is taken into account as well. Other examples of wave functions are discussed elsewhere[5]. The results are similar in trend.

We conclude on the basis of our phenomenological studies that intrinsic transverse momenta have to be taken into account for a reliable quantitative estimate of the

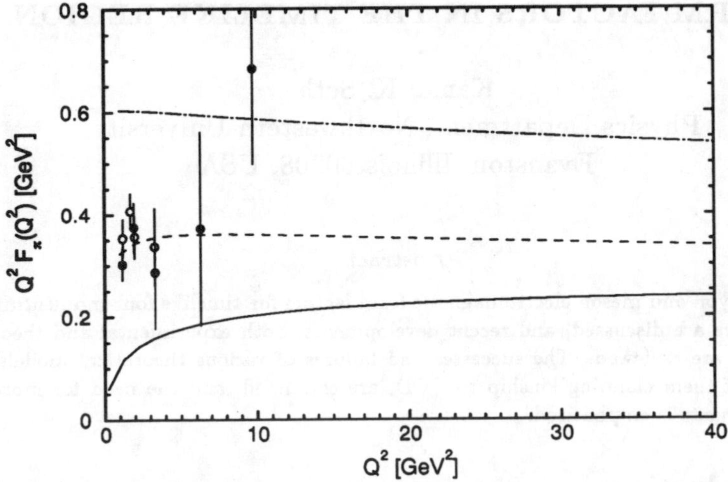

Figure 2: The pion form factor as a function of Q^2 evaluated with $\phi(x) = 30x(1-x)(2x-1)^2$, $g = 1$ and $\Lambda_{QCD} = 200\,\mathrm{MeV}$. The dash-dotted line is obtained with an α_s cut-off at 0.5 neglecting transverse momenta and the dashed line including Sudakov effects but ignoring the intrinsic $k_\perp$-dependence in the wave function. The solid line represents the complete result obtained from the same formula but taking into account both the Sudakov factor and the intrinsic $k_\perp$-dependence for $\langle k_\perp^2 \rangle = 350\,\mathrm{MeV}$. Data are taken from Bebek et al.[6] (o 1976, • 1978).

perturbative QCD contribution to the pion form factor. It turns out that the perturbative QCD contribution is presumably too small as compared to data. Other contributions (higher twist) may play an important rôle in the few GeV region.

References

1. G. P. Lepage and S. J. Brodsky, *Phys. Rev.***D22** (1980) 2157.

2. H. Li and G. Sterman, *Nucl. Phys.***B381** (1992) 129.

3. G. P. Lepage, S. J. Brodsky, T. Huang and P. B. Mackenzie, *Banff Summer Institute, Particles and Fields 2*, eds. A. Z. Capri and A. N. Kamal (Plenum Press, New York 1983), p. 83.

4. V. L. Chernyak and A. R. Zhitnitsky, *Nucl. Phys.***B201** (1982) 492.

5. R. Jakob and P. Kroll, *preprint* CERN-TH.6900/93

6. C. J. Bebek et al., *Phys. Rev.***D13** (1976) 25 and **D17** (1978) 1693

FORM FACTORS IN THE TIMELIKE REGION

Kamal K. Seth

Physics Department, Northwestern University

Evanston, Illinois 60208, USA

Abstract

Baryon and meson electromagnetic form factors for timelike four-momentum transfers are discussed, and recent developments, both experimental and theoretical, are reviewed. The successes and failures of various theoretical models, most of them claiming kinship to QCD, are examined, and the need for more experiments is emphasized.

1 Introduction

While it is true that we now bestow the aura of being 'fundamental' to point-like particles only, the real world remains dominated by particles which are composite and have structure. The study of these structures is the study of form factors. In principle one can be interested in the structure or distribution of any kind of charges and currents, electromagnetic, color, or flavor in the composite object. However, in practice, we generally mean the structure of electromagnetic charges and currents. In this talk I will confine myself to electromagnetic form factors of the first member of each of the two families of hadrons we have, the nucleon as the quintessential baryon, and the pion as the first among mesons.

In nonrelativistic quantum mechanics (first order perturbation theory, one photon approximation) the form factor of an extended object is the Fourier transform of its density distribution, $\rho(r)$

$$F(q^2) = \int d^3 r \rho(r) e^{i\vec{q}\cdot\vec{r}/\hbar} \tag{1}$$

At $q = 0$ this gives the total 'charge'. At small momentum transfers it leads to the rms radius of the density distribution

$$F(q^2) \approx 1 - \frac{q^2}{6\hbar^2} < r^2 > \tag{2}$$

At large momentum transfers it leads to a more detailed knowledge of $\rho(r)$. The form factor can be interpreted as a measure of the probability that the composite 'object' can take a large momentum transfer and still hang together, i.e. it provides a map of the large momentum components in the wave function of the composite object.

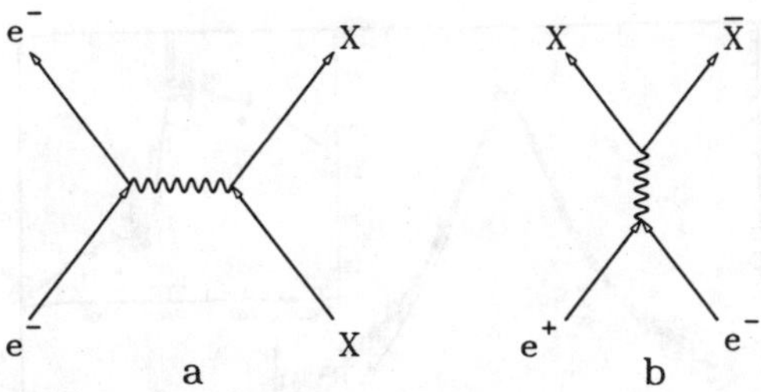

Figure 1: (a) Spacelike momentum transfer in the elastic scattering of electrons by a charged hadron X. (b) Timelike momentum transfer in the pair creation of X and $\bar{X}$ in the annihilation of e^+ and e^-.

Properly speaking, the momentum transfer q in equations 1 and 2 should be the four-momentum transfer Q. But there are two regimes of four-momentum transfer,

$$\text{spacelike, or } Q^2 \geq 0, \tag{3}$$
$$\text{and timelike, or } Q^2 < 0 \tag{4}$$

$$\text{where, } Q^2(\text{4-momentum})^2 \equiv |\vec{q}|^2(\text{3-momentum})^2 - E^2 \tag{5}$$

in our adopted metric. What we said above in Eqs. 1 and 2 about Fourier transforms and density distributions applies only to spacelike momentum transfers. These momentum transfers occur in electron elastic scattering (purely 3-momentum transfer) from a charged hadron X, as illustrated in Fig. 1(a).

$$e^- + X \to e^- + X \tag{6}$$

Timelike momentum transfers occur in pair-creation processes (purely energy transfer), as illustrated in Fig. 1(b) for the production of hadron X and its antiparticle $\bar{X}$,

$$e^+ + e^- \to X + \bar{X}. \tag{7}$$

The form factors for this process tell us something different. Since negative four-momentum corresponds to the mass of a real particle, this process puts the virtual photon almost on the mass-shell. It tells us about the cloud of 'mesons' which may surround the 'bare' hadron. There is one more difference. The physically accessible region of timelike four-momentum transfer is only $-Q^2 = s > (2m_X)^2$.

2 Form Factors in Theory

2.1 The Vector Dominance Model

Even before the discovery of the first vector meson, Nambu [1] had conjectured that the virtual photon might interact with hadrons not just electromagnetically, but also

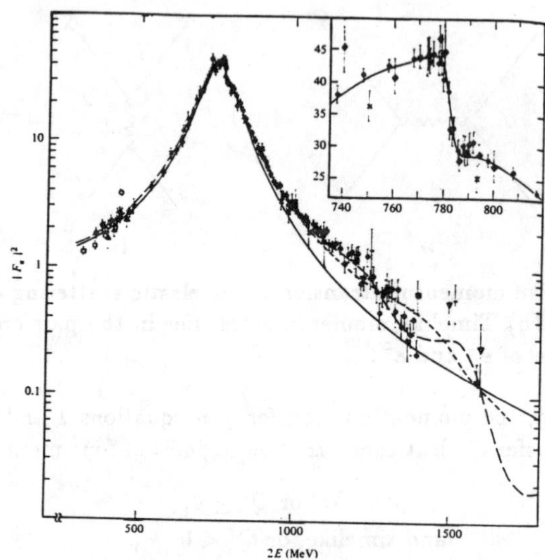

Figure 2: Pion form factor $|F_\pi|^2$ for timelike momentum transfers versus $2E = \sqrt{s} = \sqrt{-Q^2}$. The inset shows the region of $\rho - \omega$ interference in detail. The solid curves show VDM fits with only ρ and ω mesons. Other curves [15] show VDM fits including higher mass mesons.

strongly via vector mesons. These intermediary vector mesons, ρ, ω, ϕ, ... were later discovered, and the cult of vector dominance was born [2]. According to this model, the vector meson intermediaries should show up as 'resonances' in the form factors for timelike momentum transfers equal to their masses, if these masses occur in the physically accessible region. A remarkable confirmation of this idea is seen in the pion form factor in the timelike region. As shown in Fig. 2, the ρ and the ω mesons (and the interference between them) are clearly visible in the pion form factor. The model has been improved and polished, so that in the hands of its afficianados it now includes the known vector mesons and their known and unknown recurrences [3, 4] and involves as many as 20 free parameters. Admittedly, with so many parameters the attractiveness of this model is reduced. As we shall see later, the model has other difficulties also.

2.2 QCD-based Models

With the advent of QCD, the baseline model for form factors changes. Hadronic form factors become intimately related to the quark wave functions of hadrons and to the general problem of exclusive reactions in QCD.

Brodsky and Farrar [5] have shown that the amplitude for exclusive processes in QCD is the convolution of two parts:

- a wave function part, $\phi(x_i, Q^2)$, which contains the nonperturbative aspects of

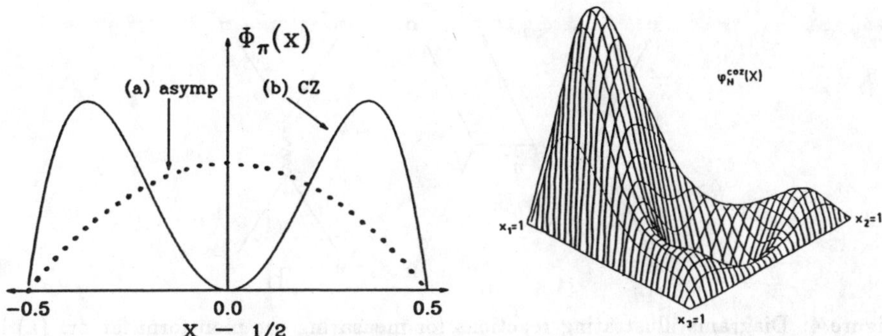

Figure 3: (Left) Quark distribution amplitudes for the pion: (a) the asymptotic amplitude $\propto x(1-x)$, (b) the amplitude proposed by Chernyak and Zhitnitsky $\propto x(1-x)(1-2x)^2$ (adapted from [12]). (Right) Quark distribution amplitude for the proton as proposed by Chernyak, Oglobin and Zhitnitsky [7].

the hadrons' constituents, the quark distribution amplitudes, and

- a <u>hard</u> scattering part, T_H, which can be calculated in perturbative QCD, because it consists of a series of short distance interactions between collinear quarks with longitudinal momentum fractions, x_i.

This description leads to 'QCD counting rules', one of which is that

$$F_n(Q^2) \sim (1/Q^2)^{n-1} \tag{8}$$

or, $F_n(Q^2) \sim 1/Q^2$ for mesons ($n=2$) and $F_n(Q^2) \sim 1/Q^4$ for baryons ($n=3$).

There is general agreement that the above must be true asymptotically, but a royal battle rages about when asymptopia is approached. Lepage, Brodsky and coworkers [6] and Chernyak and Zhitnitsky [7] believe that, precocious or not, it begins already at $Q^2 \approx 5$ GeV^2. Isgur and Llwellyn Smith [8] and Radyushkin [9] believe that Q^2 larger by an order of magnitude or two are needed to justify the hard scattering approximation. Some relief from the objections raised against the use of PQCD in exclusive processes was provided by the work of Sterman and collaborators [10] who showed that the objectionable large contributions from the soft end-point regions are much reduced when transverse momenta of the quarks and the Sudakov suppression are taken into account. Nevertheless, these developments have failed to quench the criticism and the controversy continues unabated [11, 12]. At least part of the problem lies with the rather counter-intuitive, lumpy, aesthetically non-pleasing (the politically correct equivalent of 'ugly') wave functions required for PQCD to work (see Fig. 3). I must confess to sharing the distaste for these wave functions and subscribing to Einstein's faith that 'subtle He is; malicious He is not'. Except for expressing this philosophical bias, I will not even attempt to enter this debate. I will however note that since after 350 years even Fermat's last theorem has been solved (today's (Jan.

74

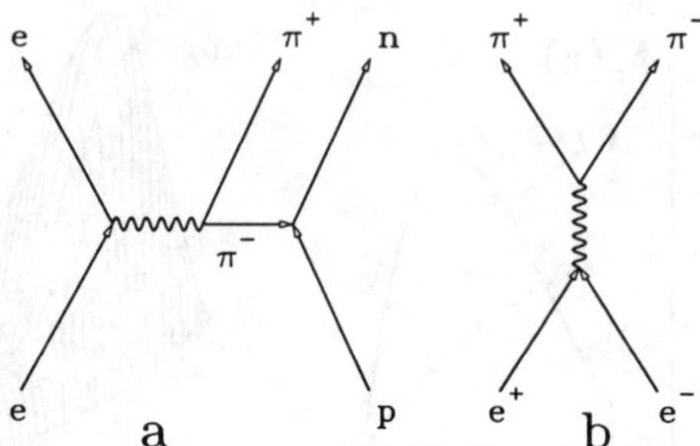

Figure 4: Diagrams illustrating reactions for measuring the pion form factor: (a) in the spacelike region, (b) in the timelike region.

25, 1993) Int. Herald Tribune), one can hope that given sufficient time even this controversy will be settled!

3 Pion Form Factors

Because the pion spin is zero, only one form factor is needed for pions. This simplifies life a lot. However, because one does not have a pion target, measurement of pion form factors for spacelike Q^2 are difficult.

Pion form factors for spacelike momentum transfers can be measured in two ways. The first method is to scatter pions from electrons bound in atoms. This transfers very little momentum even with the highest energy pion beams. Therefore, this technique has been used only to determine the pion radius. In the best of these experiments [13], done with a 300 GeV π^- beam, form factors were measured in the range $0.014 \leq Q^2 \leq 0.122$ GeV^2 and the radius of the pion was measured to be $< r^2 >^{1/2} = 0.657(12)$ fm.

For large spacelike momentum transfers the only reactions available are

$$e^- + p \rightarrow e^- + \pi^+ + n, \tag{9}$$
$$\text{and} \quad e^- + n \rightarrow e^- + \pi^- + p \tag{10}$$

The diagram illustrating these pion production reactions is shown in Fig. 4(a). The extraction of the pion form factor from data for these reactions is model-dependent because of the presence of the strongly interacting nucleon in the final state. The existing data are confined to $Q^2 \leq 10$ GeV^2 and date back to 1978. They all come from essentially one source [14]. These data are shown in Fig. 5 where $Q^2 F_\pi(Q^2)$ is plotted against Q^2. We note that the precision of the data for $Q^2 \geq 3$ GeV^2 is quite poor. Let me remind you that these are the data which are at the heart of the PQCD controversy mentioned earlier. Needless to say, it is absolutely essential that precision measurements of pion form factors should be made at CEBAF up to the highest momentum transfer possible. We should, however, keep in mind the

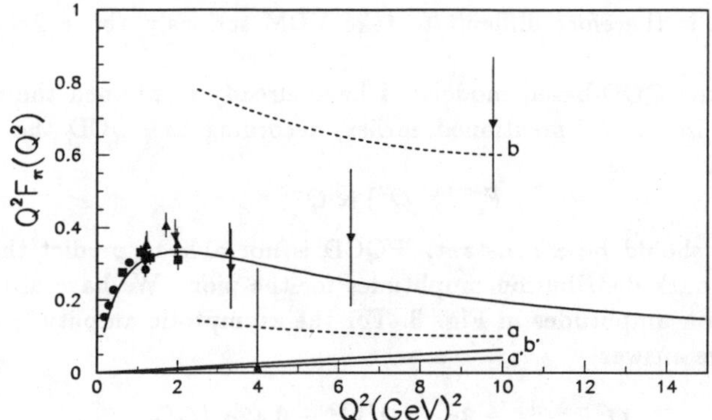

Figure 5: Pion form factors in the spacelike region. The data are from [14]. the solid curve which nearly goes through the data points is the QCD sum-rule prediction of [19]. The curves marked a and a' are the 'naive' and the 'legal' predictions using the asymptotic distribution amplitudes; curves marked b and b' are the 'naive' and 'legal' predictions using Chernyak and Zhitnitsky's amplitudes. Curves a,a',b and b' are based on Ref. [12].

objections raised by Carlson and Milana [16] against the assumptions made in the extraction of the pion form factor at large Q^2 by this electroproduction technique.

Pion form factors for timelike momentum transfers invariably come from the reaction (Fig. 4(b))

$$e^+ + e^- \to \pi^+ + \pi^- \tag{11}$$

Two rather extensive sets of these measurements come from Novosibirsk and Orsay. These are shown in Fig. 2. Barkov et al [15] have reported $|F_\pi|^2$ for timelike momentum transfers in the range $0.36 \le \sqrt{s} \le 1.40$ GeV, or $0.130 \le |Q^2| \le 1.952$ GeV^2. The errors in these data range from $\sim 3\%$ for low Q^2 to 35% at the highest Q^2. For $|Q^2| > 1.96$ GeV^2 the existing data are extremely sparse and are generally of poor quality. (The only exception is the datum point for $e^+e^- \to J/\psi \to \pi^+\pi^-$.) One would very much like to improve on the large Q^2 data. Unfortunately, CEBAF does not have colliding beams (at least not yet), and the Tau-Charm Factory and the Beauty factories are still in the distant future.

Many attempts have been made to explain the form factor of the pion for timelike Q^2 in terms of the VDM. In Fig. 2 some of these are shown. We draw attention to only two features of the fits shown there. The solid lines show the VDM prediction with only ρ and ω, and the interference between them (see inset). It is seen that in the region of the ρ and ω the VDM fit is quite satisfactory. However, it becomes quite poor for larger Q^2. This is interpreted by the VDM specialists to indicate the necessity to put in other vector mesons, in particular the ρ and ω recurrences. Unfortunately, while better fits are indeed obtained with the increased number of resonances and parameters [4], the recurrences employed do not seem to correspond to any known particles [17]. It is therefore difficult to take VDM seriously above $2E = \sqrt{s} = 1$ GeV.

What about the QCD-based models? I have already mentioned the raging controversy on this issue. As mentioned earlier, according to PQCD the asymptotic prediction is that

$$F_\pi^{\text{asymp}}(Q^2) \propto Q^{-2} \tag{12}$$

Thus, $Q^2 F_\pi(Q^2)$ should be a constant. PQCD is not able to predict this constant unless we have quark distribution amplitudes for the pion. We have already shown several distribution amplitudes in Fig. 3. For the asymptotic amplitude $\propto x(1 - x)$ one gets the naive answer

$$Q^2 F_\pi^{\text{asymp}} = 8\pi\alpha_s(Q^2)f_\pi^2 = 0.42\alpha_s(Q^2) \tag{13}$$

using $f_\pi = 0.13$ GeV. The result is shown in Fig. 5 for typical values of the running coupling constant $\alpha_s(Q^2)$. Quite apart from the controversy about what fraction of this result is 'legal' (from 1/2 to 1/10 according to Isgur and Llwellyn Smith [12]) the result is already a factor 2 to 3 smaller than the experimental values shown in the figure. We also have the result using Chernyak and Zhitnitsky's [7] amplitudes $\propto x(1 - x)(1 - 2x)^2$. To begin with, these predict a factor ~ 2 larger values of $Q^2 F_\pi$ than the experimental results. As shown in Fig. 5, these too are whittled down to 1/10 to 1/100 of the experimental results when 'legal' intervention is made [12]. Li and Sterman [10] have shown that when Sudakov suppression is properly taken into account the situation improves considerably, but not enough. The 'saturated' value of $Q^2 F_\pi(Q^2)$ is found to be ~ 0.1 GeV^2 for the asymptotic distribution amplitudes, and ~ 0.2 GeV^2 for the CZ amplitudes.

In a recent paper Jakob and Kroll [18] have argued that the PQCD prediction for $F_\pi(Q^2)$ is further suppressed upon the inclusion of the intrinsic transverse momentum dependence of the quarks in the pion, in addition to the Sudakov corrections. They show that with this inclusion the predicted form factor using CZ amplitudes falls to nearly half the experimental value.

From QCD sum rules and local quark-hadron duality we have the prediction of Nesterenko and Radyushkin [19] that

$$F_\pi(Q^2) = 1 - \frac{1 + (6s_0/Q^2)}{(1 + 4s_0/Q^2)^{3/2}} \tag{14}$$

With $s_0 = 4\pi^2 f_\pi^2 \approx 0.7$, one obtains the results which are also shown in Fig. 5. These agree reasonably with the data. Since the asymptotic rate of fall-off of $F_\pi(Q^2)$ is Q^{-4} in the relation above, we need good data at large Q^2 to really tell how good this prediction is.

As far as the pion form factor in the timelike region is concerned, we are only aware of VMD-based predictions. Some of these are shown in Fig. 2. In addition, we have the prediction by Dubnicka and coworkers [4] which is illustrated in Fig. 6.

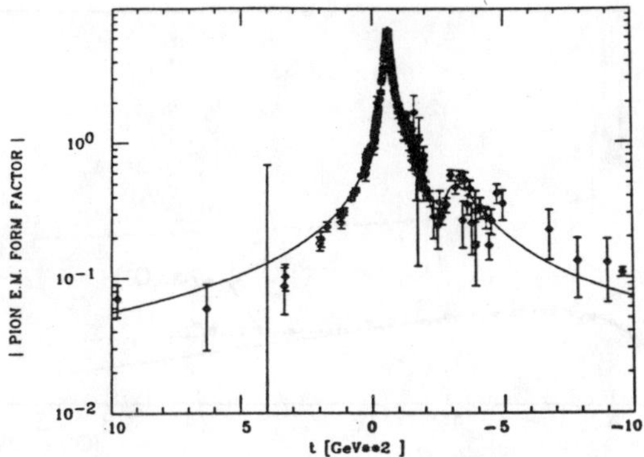

Figure 6: Pion form factors and the extended VDM fit due to Dubnicka et al [4].

As mentioned earlier, the quality of the experimental data for $-Q^2 > 2.6$ GeV^2 is generally very poor, and it is difficult to assess how good the VMD predictions are, even with their 20 parameters. The form factor is apparently dominated by resonances in the timelike region. However, we may compare the experimental timelike and spacelike form factors at the highest Q^2. We find that $F_\pi(Q^2 = 9.8 \text{ GeV}^2) = 0.070(19)$ and $F_\pi(-Q^2 = (m^2_{J/\psi}) = 9.6 \text{ GeV}^2) = 0.098(8)$ [20]; the spacelike and timelike values are essentially in agreement. At this workshop we have a contribution by J. Milana on this subject [21].

As far as the experimental prospects for the future are concerned, there is no reason why high intensity e^+e^- colliders, currently under consideration for the Tau-Charm Factory and Beauty factories, cannot measure pion form factors in the timelike region up to much higher energies and with much higher precision than has been possible so far. The measurements are of fundamental importance to the understanding of the domain of validity of PQCD.

4   Proton Form Factors

In the spacelike momentum transfer region great progress has been made in the last ten years. Precision measurements now exist for the proton magnetic form factors $G_M(p)$ up to $Q^2 = 31.3$ GeV^2 [22, 23], proton electric form factors $G_E(p)$ up to $Q^2 = 8.83$ GeV^2, and magnetic and electric form factors of the neutron up to $Q^2 = 4$ GeV^2 [24]. At this workshop we will hear about these measurements in detail from Peter Bosted [25]. I am going to confine my remarks to form factor measurements in the timelike momentum transfer region, and only mention those results from the spacelike region which are of direct relevance to my discussion.

The measurements in the spacelike region have shown that:

$$G_m(p)/\mu_p = G_E(p) \quad , \text{ to within } \pm 15\% \text{ for } Q^2 < 8.83 GeV^2 \tag{15}$$

$$F_2(Q^2) = F_1(Q^2)/Q^2 \quad , \text{ to within } \pm 15\% \text{ for } Q^2 \geq 2.5 GeV^2 \tag{16}$$

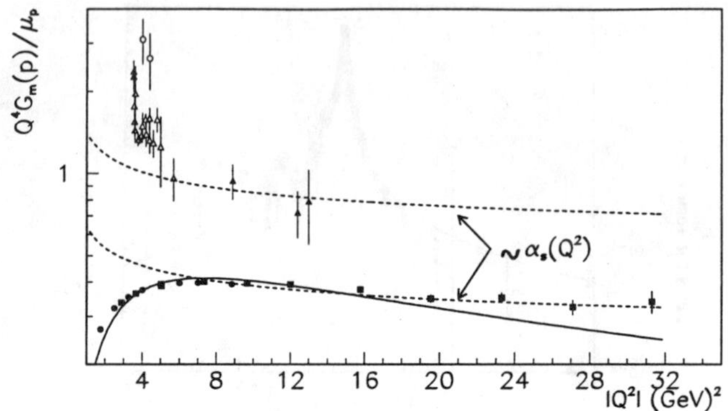

Figure 7: Proton form factors $|G_m(p)| \times (Q^4/\mu_p)$ as a function of Q^2. Results for both the spacelike and timelike region are shown. Two data points for neutrons have also been plotted by multiplying $G_m(n)$ by (μ_p/μ_n). The dashed curves are arbitrarily normalized fits to $\alpha_s^2(Q^2)$. The solid curve is from [19].

$$G_M(p)/\mu_p = G_m(n)/\mu_n \quad \text{, to within} \pm 15\% \text{ for } Q^2 \le 4.0 GeV^2 \tag{17}$$

$$Q^4 G_m(p) \propto \alpha_s(Q^2) \qquad \text{, for } Q^2 \ge 5.0 GeV^2 \tag{18}$$

For timelike momentum transfers the physically accessible region begins at $Q^2 = 4m_p^2 = 3.52$ GeV^2. Form factor measurements have been made with the reactions:

$$e^+ + e^- \to \bar{p} + p \quad [26, \ 27, \ 28, \ 29] \tag{19}$$

$$e^+ + e^- \to \bar{n} + n \qquad [28, \ 29] \tag{20}$$

$$\bar{p} + p \to e^+ + e^- \qquad [30, \ 31, \ 32] \tag{21}$$

These data are shown in Fig. 7. The data of Refs. [26]-[31] are confined to a small region around the threshold, $3.52 \le Q^2 \le 5.69$ GeV^2. The only data at large Q^2 are due to Fermilab experiment E760 [32], which is designed to study charmonium spectroscopy in $p\bar{p}$ annihilation:

$$\bar{p} + p \to [c\bar{c}]_R \to \text{electromagnetic final states} \tag{22}$$

The vector states of charmonium $(J/\psi, \psi')$ decay into e^+e^-, but other states, $\eta_c(^1S_0)$, $\chi_J(^3P_J)$ and $^1P_1$ can not decay into e^+e^-. Electron pairs observed at these resonances are almost entirely form factor events. The cross sections for $\bar{p}p \to e^+e^-$ at these energies are extremely small. For example, at $s = 13$ GeV^2 (searching for η_c') the measured cross section is ~ 1 picobarn out of a total annihilation cross section of ~ 70 milibarn. Nevertheless, because of excellent identification of electrons, E760 has succeeded in measuring $|G_m(p)|$ for timelike momentum transfers of 8.9 GeV^2 (η_c), 12.4 GeV^2 ($^1P_1$) and 13.0 GeV^2 (η_c' search) [32]. The results are shown in Fig. 7. The dashed curve through the data shows the arbitrarily normalized prediction of PQCD, i.e.,

$$Q^4 G_m(Q^2) \propto \alpha_s^2(Q^2) \tag{23}$$

where α_s is the running coupling constant of QCD. It is seen that the curve fits the data very well for $Q^2 \geq 5$ GeV^2. We notice that this is very reminiscent of the behavior of $Q^4 G_m(Q^2)$ for spacelike momentum transfers ≥ 5 GeV^2 (Eq. 15). However, we also notice that the results for the timelike momentum transfers are nearly a factor two higher than the corresponding ones for spacelike momentum transfers. It is our understanding that this difference can not be understood in the context of the conventional PQCD, according to which at large Q^2 the form factors for spacelike and timelike momentum transfers should be identical [33].

There is one other discrepancy to note in Fig. 7. In the figure we have also plotted the results of a very recent first measurement at Frascati of the magnetic form factor of the neutron in the timelike region [29] by means of the reaction $e^+ + e^- \rightarrow n + \bar{n}$. We note the normalized form factors for the neutron $|G_m|/\mu_n$ are nearly a factor two larger than the corresponding form factors $|G_m(p)|/\mu_p$ for the proton. This is at variance with the experimental observation (Eq. 17) in the spacelike region where $|G_m(p)|/\mu_p \approx |G_m(n)|/\mu_n$.

I will let Peter Bosted tell you all about what theory has to say about proton and neutron form factors for spacelike momentum transfers. Let us look into what the theory has to say about the nucleon form factors in the timelike region.

First, the VDM predictions. Körner and Kuroda have calculated cross sections for the reaction $e^+e^- \rightarrow p\bar{p}$ as a function of Q^2 in the timelike region [3]. Their results for the proton form factor are a factor ~ 3 lower than the E760 results at both $-Q^2 = 8.9$ and 13.0 GeV^2. The extended VDM calculations of Dubnicka [4] were made to fit the full set of experimental results in the spacelike region and the near-threshold data in the timelike region (Fig. 8). The author did not have the E760 results at large Q^2 available. The first thing to note about Dubnicka's predictions is that for a given value of $|Q^2|$ the form factors in the timelike region are nearly a factor two higher than those in the spacelike region. Nevertheless, at $-Q^2 = 8.9$ GeV^2 her predicted form factor $|G_m(p)|$ is 0.024, i.e., about 40% lower than the E760 result. This, by itself is perhaps not too bad. However, Dubnicka [4] also makes predictions for the neutron form factor in the timelike region. Her predictions at $-Q^2 = 4.0$ and 4.4 GeV^2 are $|G_m(n)| = 1.5$ and 1.0 respectively. Both are a factor 4 larger than the experimental results. The fact that Dubnicka's predictions for timelike Q^2 for $|G_m(p)|$ are a factor two lower and those for $|G_m(n)|$ are a factor four higher can be understood by the observations that the isoscalar and isovector parts of the Dirac and Pauli form factors $F_1(Q^2)$ and $F_2(Q^2)$ enter with different signs for the proton and neutron in the definition of $|G_m|$.

As far as the QCD based predictions are concerned, Peter Bosted will tell you about them for the spacelike region. He will probably tell you that nothing works over the entire range of momentum transfers and that the PQCD controversy continues. Just

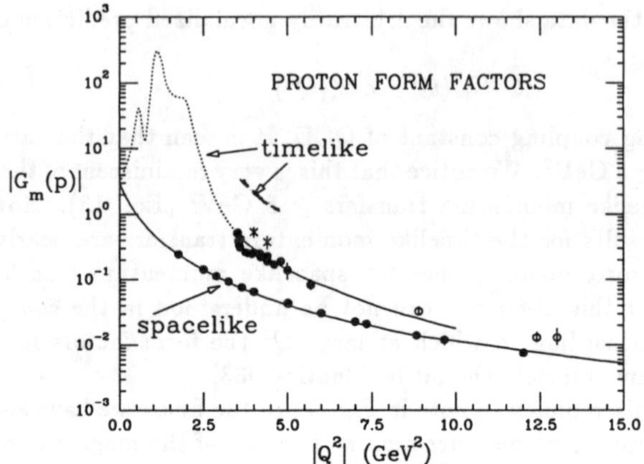

Figure 8: Proton magnetic form factors according to the extended VDM calculations of Dubnicka [4]. The data are from [22] - [32]. The two asterisked data points and the dashed curve are $G_m(n)\mu_p/\mu_n$.

for fun in Fig. 7 I once again show the predictions of Nesterenko and Radyushkin based on the QCD sum-rule methods [19].

There are not many QCD based predictions for $|G_m(p)|$ in the timelike region. As mentioned earlier, conventional PQCD predicts that $|G_m(p)|$ in the spacelike and timelike regions should be identical, at least in the large Q^2 limit. The data in Fig. 7, however, show that this is not so. For timelike Q^2 the form factor is a factor two larger.

Hayer [34] has calculated the proton magnetic form factor in the timelike region using several 'candidate' proton distribution amplitudes and the PQCD hard scattering approximation with Sudakov suppression a la Sterman [10]. He obtains nearly constant values of $Q^4 F_1(p) \approx Q^4 G_m(p)/(1 + \frac{1}{Q^2})$ for $s = 4 - 64$ GeV^2. His results for several different distribution amplitudes based on QCD sum-rules are the same within $\pm 15\%$. For the Chernyak-Oglobin-Zhitnitsky amplitudes the predictions are between 60% and 30% smaller than the E760 experimental results. A much worse disagreement occurs for the neutron form factor. $F_1(n)/F_1(p)$ is predicted to be $\sim$ 1/4 whereas the experimental data at $-Q^2 = 4$ GeV^2 suggest $F_1(n)/F_1(p) \approx 2$.

The only other calculation of the proton form factor in the timelike region is that due to Kroll et al [35]. Dr. Schürmann will talk about this in detail at this workshop [36]. Let me describe it only briefly. In this calculation a diquark-quark model is assumed for the proton. Form factors for spin-isospin scalar and vector diquarks are introduced phenomenologically. A diquark form factor cut-off parameter is also used. Using the QCD hard scattering formalism with Sudakov suppression both the Dirac (F_1) and the Pauli (F_2) form factors are calculated. To go from the spacelike to the

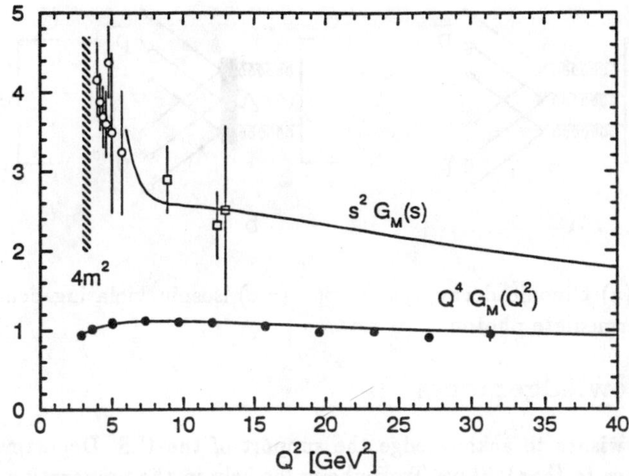

Figure 9: $Q^4 G_m(p)$ for spacelike momentum transfers (curve with solid data points) and timelike momentum transfers (curve with open data points) as predicted by the diquark model of Kroll et al [35].

timelike region only the sign of Q^2 is changed. A good fit to $G_m(p)$ in both the spacelike and timelike regions is obtained (see Fig. 9). Diquark models have always been a bit controversial. So, I will avoid getting into the defense argument. I will leave all that in the able hands of Manfred Schürmann [36].

Let me end this talk with a plea for more precision data on the pion and nucleon form factors, especially in the timelike region. We have already seen how these data directly adress questions relating to the validity of PQCD and to the structure of nucleons which are our world. What could be more *fundamental* than that?

Let me also give two examples of how *useful* the measurement of $G_m(p)$ in the timelike region is. The E760 measurements provide a peek into a very interesting aspect of hadronization, still a rather mysterious subject. From the knowledge of $G_m(p)$ one obtains the result that

$$\frac{\sigma_T(R_c \to ggg \to p\bar{p})}{\sigma_T(R_c \to ggg \to \text{hadrons})} \approx 15(5) \times \frac{\sigma_T(e^+e^- \to \gamma^* \to p\bar{p})}{\sigma_T(e^+e^- \to \gamma^* \to \text{hadrons})}, \qquad (24)$$

i.e., hadronization into $p\bar{p}$ is preferred by an order of magnitude via gluons than via a virtual photon. This also means that photons prefer other hadrons. Vector mesons?

A related question on which our measurements shed light is the magnitude of isospin violating component of $J/\psi \to p\bar{p}$ form factor (see Fig. 10). Using the formalism of Claudson, Glashow and Wise [37], we find that this amplitude is $24\pm6\%$ of the total:

$$E_{em}(J/\psi \to p\bar{p})/E(J/\psi \to p\bar{p}) = 0.24 \pm 6 \qquad (25)$$

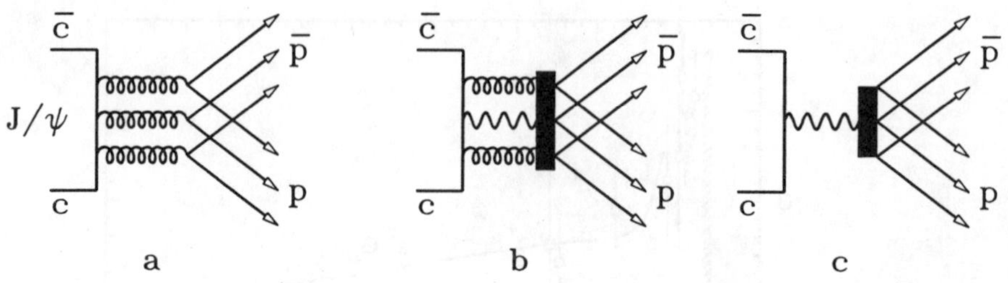

Figure 10: (a) Gluonic decay $J/\psi \to p\bar{p}$. (b,c) isospin violating decays $J/\psi \to p\bar{p}$ with an intermediate photon.

5 Acknowledgements

The author wishes to acknowledge the support of the U.S. Department of Energy. He also wishes to thank Steve Trokenheim for help in the preparation of the written version of this talk.

References

[1] Y. Nambu, *Phys Rev.* **106** (1957) 1366; W.R. Frazer and J.R. Fulco, *Phys. Rev.* **117** (1960) 1609.

[2] J.J. Sakurai, *Ann. Phys. (N.Y.)* **11** (1960) 1609; M. Gell-Mann and F. Zachariasen, *Phys. Rev.* **124** (1961) 953.

[3] J.G. Korner and M. Kuroda, *Phys. Rev.* **D16** (1977) 2165.

[4] S. Dubnicka, *Nuovo Cimento* **A100** (1988) 1, **A104** (1991) 363.

[5] S.J. Brodsky and G.R. Farrar, *Phys. Rev. Lett.* **31** (1973) 1153; similar ideas were independently developed by V.A. Matveev et al, *Lett. Nuovo Cimento* **7** (1973) 719.

[6] G.P. Lepage and S.J. Brodsky, *Phys. Rev.* **D22** (1980) 2157; G.P. Lepage et al, in *Banff Summer Institute proceedings - Particle and Fields 2*, ed. A.Z. Capri and A.N. Kamal (Plenum, New York, 1983) p. 83, *ibid* p. 143.

[7] V.L. Chernyak and I.R. Zhitnitsky, *Phys. Reports* **112** (1984) 173; *Nucl Phys.* **B246** (1984) 52.; V.L. Chernyak, A.A. Ogloblin and I.R. Zhitnitsky, *Z. Phys.* **C42** (1989) 569.

[8] N. Isgur and C.H. Llewellyn Smith, *Phys. Rev. Lett.* **52** (1984) 1080.

[9] A.V. Radyushkin, *Acta Physica Polonica* **B15** (1984) 403.

[10] J. Botts and G. Sterman, *Nucl. Phys.* **B325** (1989) 62; H. Li and G. Sterman, *Nucl. Phys.* **B381** (1992) 129.

[11] A.V. Radyushkin, *Nucl. Phys.* **A532** (1991) 141c.

[12] N. Isgur and C.H. Llewelyn Smith, *Phys. Lett.* **B217** (1989) 535; also *Nucl. Phys.* **B317** (1989) 526.

[13] S.R. Amendolia et al, *Phys. Lett.* **B146** (1984) 116.

[14] C.J. Bebek et al, *Phys Rev.* **D17** (1978) 1693, and references therein.

[15] L.M. Barkov et al, *Nucl. Phys.* **B256** (1985) 365, and references therein.

[16] C.E. Carlson and J. Milana, *Phys. Rev. Lett.* **65** (1990) 1717.

[17] Review of Particle Properties, *Phys. Rev.* **D45** (1992) S1.

[18] R. Jakob and P. Kroll, *Phys. Lett.* **B315** (1993) 463.

[19] V.A. Nesterenko and A.V. Radyushkin, *Phys. Lett.* **B115** (1982) 410; *Phys. Lett.* **B128** (1983) 439.

[20] J. Milana, S. Nussinov and M.G. Olsson, *Phys. Rev. Lett.* **71** (1993) 2533.

[21] J. Milana, elsewhere in these proceedings.

[22] R.G. Arnold et al, *Phys Rev. Lett.* **57**(1986) 174.

[23] P.E. Bosted et al, *Phys. Rev. Lett.* **68**(1992) 3841.

[24] A. Lung et al, *Phys. Rev. Lett.* **70** (1993) 718.

[25] P. Bosted, elsewhere in these proceedings.

[26] B. Delcourt et al, *Phys. Lett.* **B86** (1979) 395.

[27] D. Bisello et al, *Nucl. Phys.* **B224** (1983) 379.

[28] D. Bisello et al, *Z. Phys.* **C48** (1990) 23.

[29] A. Antonelli et al, *Phys. Lett.* **B313** (1993) 283.

[30] G. Bassompiere et al, *Phys. Lett.* **B68** (1977) 477; also *Nuovo Cimento* **73A** (1983) 347.

[31] G. Bardin et al, *Phys. Lett.* **B255** (1991) 149; also *Phys. Lett.* **B257** (1991) 514.

[32] T. Armstrong et al, *Phys. Rev. Lett.* **70** (1993) 1212.

[33] V.L. Chernyak, priv. comm.

[34] T. Hayer, *Phys. Rev.* **D47** (1993) 3875.

[35] P. Kroll, T. Pilsner, M. Schürmann and W. Schweiger, *Phys. Lett.* **B316** (1993) 546.

[36] M. Schürmann, elsewhere in these proceedings.

[37] M. Claudson, S.L. Glashow and M.B. Wise, *Phys. Rev.* **D25** (1982) 1345.

On the Proton Form Factors in the Time-like Region

S.M. Bilenky[a,b,c], C. Giunti[b,c] and V. Wataghin[b,c]

(a) Joint Institute of Nuclear Research, Dubna, Russia

(b) INFN Torino, Via P. Giuria 1, I–10125 Torino, Italy

(c) Dipartimento di Fisica Teorica, Università di Torino

Abstract

The process $\bar{p}p \to e^- e^+$ is considered in the general case of polarized initial particles. A relation between the difference of the phases of the electromagnetic form factors G_M and G_E in the time-like region and measurable asymmetries is derived. It is shown that the moduli of the form factors can be determined from measurements of the total unpolarized cross section and of the integral asymmetry for longitudinally polarized (or transversely polarized) $\bar{p}$ and p. The behaviour of the proton form factors at high q^2 in the time-like region is also discussed. From the Phragmén-Lindelöf's theorem it follows that the asymptotical behaviour of the form factors in the space-like and time-like regions must be the same. An analysis of experimental data in both regions based on perturbative QCD is presented.

The charge and magnetic form factors of the nucleon, $G_E(q^2)$ and $G_M(q^2)$, are classical objects of investigation. For a long time these fundamental quantities have been investigated in the region of space-like q^2. Starting from the seventies some informations on the electromagnetic form factors of the proton in the time-like region have been obtained. Recently rather accurate measurements of the proton form factors in the time-like region, from $q^2 = 4M^2$ up to $q^2 = 4.2\,GeV^2$, have been done at LEAR [1]. Some informations on the electromagnetic form factor of the proton at high time-like q^2 were also obtained at Fermilab [2].

There exist several QCD calculations of the electromagnetic form factors in the space-like region [3]. According to our knowledge there are no QCD-based calculations of the form factors in the time-like region. The phenomenological models which try to describe the behaviour of the form factors in both space-like and time-like regions are based on the vector meson dominance models [4]. However, even taking into account all known meson resonances it is not possible to obtain a statistically acceptable description of all the existing experimental data.

The understanding of the behaviour of nucleon electromagnetic form factors still remains a challenge for the theory. It is clear that any additional information about the form factors which could be obtained from experiment is very important.

Taking into account possible future developments of the experiments at LEAR [5], we have analyzed [6] which additional informations on the proton form factors in the time-like region can be obtained from the investigation of the process

$$\bar{p}p \to e^- e^+ \tag{1}$$

with a polarized proton target and/or a polarized antiproton beam.

The nucleon form factors in the time-like region are complex. In the case of unpolarized initial particles the cross section depends only on the squared moduli $|G_M|^2$ and $|G_E|^2$. The study of process (1) with polarized initial particles could allow to obtain informations also about the phase difference $\chi = \chi_M - \chi_E$, where $\chi_M = Arg\, G_M$ and $\chi_E = Arg\, G_E$, which is an important characteristic of the form factors in the time-like region.

In ref.[6] we gave the differential cross section of process (1) in the general case of polarized initial particles and we analyzed differential and integral asymmetries. Here we present some results for the integral asymmetries.

The value of $\sin \chi$ can be obtained from measurements of the cross section of process (1) with an unpolarized antiproton beam and a polarized proton target (or a polarized antiproton beam and an unpolarized proton target). If the target polarization $\vec{P}_\perp$ is orthogonal to the beam direction, for the asymmetry integrated over the angle φ between $\vec{P}_\perp$ and $\vec{n}$ (the unit vector orthogonal to the reaction plane) from $-\pi/2$ to $\pi/2$ and over ϑ (the angle between the momenta of the antiproton and the electron in the c.m.s.) from 0 to $\pi/2$ we have

$$A_\perp = \frac{\dfrac{4M}{\pi\sqrt{q^2}}|G_M||G_E|\sin\chi}{2|G_M|^2 + \dfrac{4M^2}{q^2}|G_E|^2}. \tag{2}$$

Information about $\cos \chi$ can be obtained from measurements of the cross section of process (1) with a transversely polarized beam and a longitudinally polarized target (or a longitudinally polarized beam and a transversely polarized target). For the asymmetry integrated over the angle φ' between $\vec{P}'_\perp$ (the antiproton polarization vector) and $\vec{n}$ from 0 to π and over the angle ϑ from 0 to $\pi/2$ we have

$$A_{\perp;\|} = -\frac{\dfrac{4M}{\pi\sqrt{q^2}}|G_M||G_E|\cos\chi}{2|G_M|^2 + \dfrac{4M^2}{q^2}|G_E|^2}. \tag{3}$$

From Eqs.(2) and (3) we obtain the following relation between the phase difference χ and the integral asymmetries

$$\tan\chi = -\frac{A_\perp}{A_{\perp;\|}}. \tag{4}$$

Thus measurements of the integral asymmetries $A_\perp$ and $A_{\perp;\|}$ would allow us to determine the phase difference χ directly from experimental data. Notice that the phase difference χ can be determined unambiguously with this method (in addition to Eq.(4) it is necessary to take into account the sign of $A_\perp$ or $A_{\perp;\|}$). Let us stress that the knowledge of the moduli $|G_M|$ and $|G_E|$ is not necessary in order to obtain χ with the help of Eq.(4).

Since $G_E(4M^2) = G_M(4M^2)$, it is clear that the asymmetry $A_\perp$ vanishes at the threshold. It is possible to show that the asymmetry $A_\perp$ goes to zero at $q^2 \to \infty$. In fact the electromagnetic form factors $G_{E,M}(q^2)$ are limiting values of the functions $G_{E,M}(z)$, $G_{E,M}(q^2) = \lim_{\epsilon \to 0^+} G_{E,M}(q^2 + i\epsilon)$, which are analytical in the upper half of the complex z plane and increase at infinity not faster than a power of z. We can apply [7] to the form factors the Phragmén-Lindelöf's theorem [8]. From this theorem it follows that the form factors have the same asymptotical behaviour in the space-like and time-like regions. In the space-like region the form factors are real. This means that the form factors are real in the time-like region at asymptotically high q^2 and from Eq.(2) it follows that $A_\perp \to 0$ at $q^2 \to \infty$.

In conclusion, let us discuss in some more detail the asymptotical behaviour of the electromagnetic form factors of the nucleon in the time-like region. In accordance with the quark counting rule [9], at high $|q^2|$ the form factors of the nucleon behave as $G_M(q^2) \sim F_1(q^2) \sim 1/q^4$. The quark-gluon interaction leads to violation of scaling and additional logarithmic q^2 dependence of the form factors. Let us write in the space-like region ($q^2 < 0$)

$$\frac{G_M(q^2)}{\mu_p} \underset{q^2 \to -\infty}{\sim} \frac{C_s}{q^4}\, \Phi(q^2)\,, \tag{5}$$

where $\mu_p = 2.79$ is the proton magnetic moment in nuclear magnetons. From the leading order perturbative QCD it follows that $\Phi(q^2)$ is given by [10]

$$\Phi(q^2) = \alpha_s^2(-q^2)\left[\ln\left(-\frac{q^2}{\Lambda^2}\right)\right]^{-4/3\beta}, \qquad \text{with} \qquad \alpha_s(-q^2) = \frac{4\pi}{\beta\ln\left(-\dfrac{q^2}{\Lambda^2}\right)}, \tag{6}$$

where $\beta = 11 - 2n_f/3$, n_f is the number of flavours and Λ is the QCD scale parameter. The value of the constant C_s is determined by the wave function of the nucleon [10,11]. The Phragmén-Lindelöf's theorem implies that in the time-like region ($q^2 > 0$) the form factors have the following asymptotical behaviour

$$\frac{G_M(q^2)}{\mu_p} \underset{q^2 \to \infty}{\sim} \frac{C_t}{q^4}\, \Phi(q^2) \tag{7}$$

with $C_t = C_s$.

Let us compare the behaviour of the form factors at large space-like and time-like momentum transfer. In a recent SLAC experiment [12] the elastic electron-proton cross section was measured in a wide range of momentum transfer, from $-q^2 = 2.9\,GeV^2$ to $-q^2 = 31.3\,GeV^2$. From these measurements the values of the form factor $G_M(q^2)$ at high $-q^2$ can be extracted (the contribution to the cross section of the form factor $G_E(q^2)$ at high $-q^2$ is small and can be neglected). With $\Lambda = 100\,MeV$ we obtain $C_s = 12.3 \pm 0.2\,GeV^4$ ($\chi^2/NDF = 4.6/5$) and with $\Lambda = 200\,MeV$ we get $C_s = 7.8 \pm 0.1\,GeV^4$ ($\chi^2/NDF = $

10.2/5). Let us notice that the quality of the fit depends rather strongly on the value of Λ (smaller values of Λ are preferable).

The cross section of the process $\bar{p}p \to e^-e^+$ at high q^2 ($q^2 = 8.9, 12.4, 13.0\,GeV^2$) was measured recently in a Fermilab experiment [2]. We made a fit of these data using Eq.(7). For $\Lambda = 100\,MeV$ we obtained $C_t = 30.9^{+4.1}_{-4.8}\,GeV^4$ ($\chi 2/NDF = 0.29/1$) and for $\Lambda = 200\,MeV$ we obtained $C_t = 21.2^{+2.8}_{-3.3}\,GeV^4$ ($\chi^2/NDF = 0.29/1$). Thus the experimental data in the space-like as well as in the time-like regions of q^2 are described by expressions (5) and (7), respectively. The accuracy of the data in the time-like region is much worse than that in the space-like region. As a consequence, the corresponding accuracy of the determination of the constant C_t in the time-like region is much worse than that of the constant C_s in the space-like region. However, the average values of C_t and C_s are so different that, even with such a low accuracy in the determination of C_t, we can conclude that these constants are different (C_t is more than 3σ higher than C_s). From our point of view this difference means that the range of high q^2 values investigated in present experiments is not asymptotic. Nonperturbative effects [13], nonleading log corrections and other effects could be important in this region.

References

[1] G. Bardin et al., Phys. Lett. B 255 (1991) 149; Phys. Lett. B 257 (1991) 514.

[2] T. Armstrong et al., Phys. Rev. Lett. 70 (1993) 1212.

[3] For a recent review see: P. Kroll, Invited talk given at the IVth International Symposium on Pion-Nucleon Physics and the Structure of the Nucleon, Bad Honnef, September 1991.

[4] V. Wataghin, Nucl. Phys. B 10 (1969) 107. P. Ceselli, et al., Workshop on Physics at LEAR with Low-Energy Cooled Antiprotons, Erice 1982, p. 365. V. Bardek, et al., Nuovo Cimento A 75 (1983) 368. S. Dubnička, Nuovo Cimento A 100 (1988) 1; 103 (1990) 469; 103 (1990) 1417; 104 (1991) 1075. S.I. Bilenkaya, et al., Nuovo Cimento A 105 (1992) 1421.

[5] See Proceedings of the Second Biennial Conference on Low-Energy Antiproton Physics LEAP '92, Courmayeur, Italy, September 1992.

[6] S.M. Bilenky, C. Giunti and V. Wataghin, DFTT 13/93, to be published in Z. Phys. C.

[7] A.A. Logunov, N. van Hieu and I.T. Todorov, Annals of Physics 31 (1965) 203.

[8] E.C. Titchmarsh, "The theory of functions", Oxford University Press, London 1939. M. Sugawara and A. Kanazawa, Phys. Rev. 123 (1961) 1895. N.N. Meĭman, Sov. Phys. JETP 16 (1963) 1609.

[9] V.A. Matveev, R.M. Muradyan and A.N. Tavkhelidze, Lettere al Nuovo Cimento 7 (1973) 719. S.J. Brodsky and G.R. Farrar, Phys. Rev. Lett. 31 (1973) 1153.

[10] S.J. Brodsky and G.P. Lepage, Phys. Rev. Lett. 43 (1979) 545, 1625; Phys. Rev. D 22 (1980) 2157; Physica Scripta 23 (1981) 945.

[11] V.L. Chernyak and I.R. Zhitnitsky, Nucl. Phys. B246 (1984) 52. M.Gari and N.G. Stefanis, Phys. Lett. B 175 (1986) 462; Phys. Rev. D 35 (1987) 1074.

[12] R.G. Arnold et al., Phys. Rev. Lett. 57 (1986) 174; A.F. Sill et al., SLAC-PUB-4395, October 1992.

[13] N. Isgur and C.H. Llewellyn Smith, Phys. Rev. Lett 52 (1984) 1080; Phys. Lett. B 217 (1989) 535. A.V. Radyushkin, Acta Phys. Pol. B15 (1984) 403.

ON EXCLUSIVE REACTION IN THE TIME-LIKE REGION

M. Schürmann[1]
Fachbereich Physik, Universität Wuppertal,
42097 Wuppertal, Germany

ABSTRACT

The electromagnetic form factors of the proton in the time-like region and two-photon annihilations into proton-antiproton are calculated at moderately large s with a variant of the Brodsky-Lepage hard-scattering formalism where diquarks are considered as quasi-elementary constituents of baryons. The proton wave function and the parameters controlling the diquark contributions are determined from fits to space-like data.

We are going to report here on a recent investigation on the magnetic form factor G_M^p of the proton in the time-like region and two-photon annihilations into proton-antiproton[1]. To calculate the processes in question a variant of the Brodsky-Lepage picture[2] for exclusive reactions is used in which a baryon is assumed to consist of a quark and a diquark. The diquark, being a cluster of two valence quarks and a certain amount of glue and sea quark pairs, is regarded as a quasi-elementary constituent, which partly survives medium hard collisions.

The quark-diquark model of baryons has turned out to work rather well for exclusive reactions in the space-like region[3,4]. It is particularly well suited for moderately large momentum transfer since the diquark picture models non-perturbative effects, in fact correlations in the baryon wave functions, which are known to play an important role in that kinematical region. With a common set of parameters specifying the diquarks and process independent distribution amplitudes (DAs) for the involved hadrons a good description of a large number of exclusive reactions in the space-like region has been accomplished by now. Among these reactions are the electromagnetic form factors of baryons, Compton scattering off protons and photoproduction of mesons.

In the hard-scattering model a form factor or a scattering amplitude is expressed by a convolution of DAs with hard-scattering amplitudes calculated in collinear approximation within perturbative QCD. In a collinear situation in which intrinsic transverse momenta are neglected and all constituents of a hadron have momenta parallel to each other and parallel to the momentum of the parent hadron, one can write the valence Fock state of the proton in a covariant fashion (omitting colour indices for

[1]Supported in part by the Bundesministerium für Forschung und Technologie, FRG, under contract number 06 Wu 765

convenience)

$$|p, \lambda\rangle = f_S \Phi_S(x_1) B_S u(p, \lambda) + f_V \Phi_V(x_1) B_V (\gamma^\alpha + p^\alpha/m) \gamma_5 u(p, \lambda)/\sqrt{3} \qquad (1)$$

u is the spinor of the proton, p and m its momentum and mass, respectively. The two terms in (1) represent configurations consisting of a quark and either a spin-isospin zero (S) or a spin-isospin one (V) diquark, respectively. The couplings of the diquarks with the quarks in an isospin 1/2 baryon lead to the flavour functions

$$B_S = u\, S_{[u,d]} \qquad\qquad B_V = [u V_{\{u,d\}} - \sqrt{2} d\, V_{\{u,u\}}]/\sqrt{3}. \qquad (2)$$

The DA $\Phi_{S(V)}(x_1)$, where x_1 is the momentum fraction carried by the quark, represents a light-cone wave function integrated over transverse momentum. The constant $f_{S(V)}$ is the configuration space wave function at the origin.

In applications of the diquark model, Feynman diagrams contributing to the hard-scattering amplitudes for the processes of interest, are calculated for point-like particles. The explicit form of the different couplings can be found elsewhere[3,4]. However, in order to take into account the composite nature of the diquarks phenomenological vertex functions have to be introduced. Advice for the parametrizations of the 3-point functions, ordinary diquark form factors, is obtained from the requirement that asymptotically the diquark model evolves into the pure quark model of Brodsky and Lepage. In view of this the diquark form factors in the space-like region are parametrized as

$$F_S(Q^2) = F_S^{(3)}(Q^2) = 1/(1+Q^2/Q_S^2) \qquad F_V(Q^2) = F_V^{(3)}(Q^2) = 1/(1+Q^2/Q_V^2)^2. \quad (3)$$

In accordance with the required asymptotic behaviour the n-point functions $(n \geq 4)$ are parametrized as

$$F_S^{(n)}(Q^2) = a_S F_S(Q^2) \qquad F_V^{(n)}(Q^2) = a_V F_V(Q^2)/(1 + Q^2/Q_V^2)^{(n-3)}. \qquad (4)$$

The constants $a_{S,V}$ are strength parameters, which take into account absorption due to diquark excitation and break-up. The relations (4) represent an effective parametrization valid at large space-like Q^2. It is not possible to continue these parametrizations in a unique way to the time-like region since we do not know the exact dynamics of the diquark system. We define a continuation to the time-like region as follows: Q^2 is replaced by $-s$ in eq. (4). The correct asymptotic behaviour is so guaranteed. This simple replacement however leads to poles in the diquark form factors which have no real physical meaning. In order to avoid these unphysical poles we keep the diquark form factors constant once they have reached a certain value, say c_0. The results presented here are not very sensitive to the value of c_0 since the cut-off of the diquark form factors is only effective in the end-point regions. The role of this parameter is expected to diminish further if Sudakov form factors are taken into account. For the numerical studies the following proton DAs

$$\Phi_S(x_1) = 25.97 x_1 x_2^3 \exp\left[-b^2(m_q^2/x_1 + m_S^2/x_2)\right]$$
$$\Phi_V(x_1) = 22.29 x_1 x_2^3 (1 + 5.8 x_1 - 12.5 x_1^2) \exp\left[-b^2(m_q^2/x_1 + m_V^2/x_2)\right] \tag{5}$$

and the set of parameters

$$\begin{aligned} f_S &= 73.85\,\text{MeV}, \quad Q_S^2 = 3.22\,\text{GeV}^2, \quad a_S = 0.15, \\ f_V &= 127.7\,\text{MeV}, \quad Q_V^2 = 1.50\,\text{GeV}^2, \quad a_V = 0.05, \quad \kappa_V = 1.39; \end{aligned} \tag{6}$$

is used. The DAs and the parameters have been determined by fits to the space-like data of the nucleon form factor[3,4]. The masses in the exponentials are constituent masses ($m_{\text{quark}} = 330\,\text{MeV}, m_{\text{diquark}} = 580\,\text{MeV}$) since they enter through a rest frame wave function. The oscillator parameter b^2 is taken to be $0.248\,\text{GeV}^{-1}$.

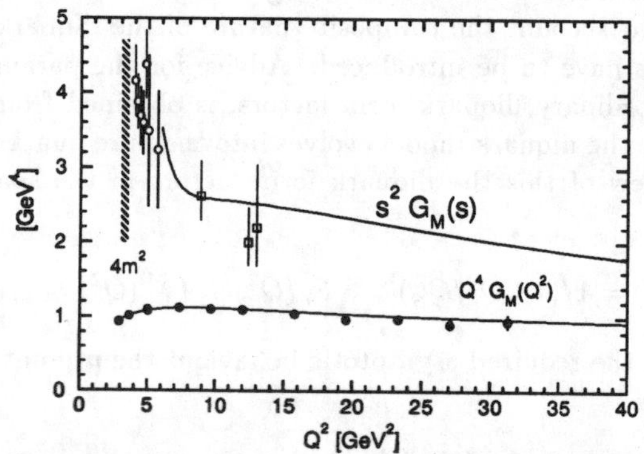

Fig. 1 The magnetic form factor of the proton in the time-like and space-like (at $Q^2 = -s$) regions. The solid lines represent the predictions of the diquark model. The time-like data ($\circ$, $\square$) are taken from Ref.[5,6], the space like-data ($\bullet$) from Ref.[7].

In Fig. 1 the predictions for the magnetic form factor of the proton are presented. Good agreement is obtained with the data of the E760 collaboration[5] for $c_0 = 1.3$. G_M does not yet behave as $1/s^2$. It falls down somewhat faster, approaching slowly the value of the space-like form factor. Typical vector meson dominance models cannot explain the large difference between the values of the time-like and space-like form factors in the $6 - 15\,\text{GeV}$ region. The electric form factor G_E behaves in a way similar to G_M, but has a value about 15% larger than that of G_M for s between 6 and $15\,\text{GeV}$. Armstrong et al.[5] assumed $|G_E| = |G_M|$ in the analysis of their $e^+e^- \to \bar{p}p$ data. The diquark model provides justification for this assumption.

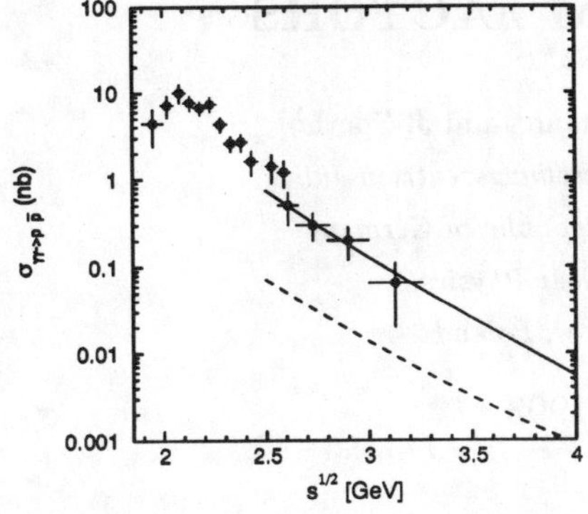

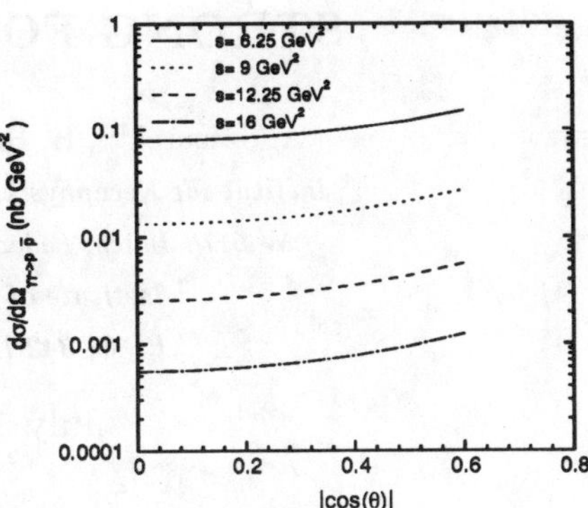

Fig. 2 solid line: The integrated cross-section ($|\cos\theta| \leq 0.6$) for $\gamma\gamma \to p\bar{p}$ as a function of $\sqrt{s}$. dash line: Result from Ref.[9] of the pure quark picture using the CZ-DA.

Fig. 3 The differential cross-section for $\gamma\gamma \to p\bar{p}$ as a function of ($|\cos\theta|$).

In Fig. 2 the integrated cross-section ($|\cos\theta| \leq 0.6$) for two-photon annihilations into $p\bar{p}$ is shown as a function of $\sqrt{s}$. For large values of $\sqrt{s}$ ($\geq 2.5\,\mathrm{GeV}$) good agreement with the CLEO data[8] is found, whereas the pure quark model[9] fails by about a factor of 8. The differential cross-section Fig. 3 is found to increase smoothly from 90^0 downwards smaller angles.

References

1. P. Kroll, Th. Pilsner, M. Schürmann and W. Schweiger, CERN-TH.6869/93

2. G. P. Lepage and S .J. Brodsky, *Phys. Rev.* **D22** (1980) 2157

3. P. Kroll, Proc. of the Adriatic Research Conference on Spin and Polarization Dynamics in Nucl. and Particle Physics, Trieste (1988); P. Kroll, W. Schweiger and M. Schürmann, *Z. Phys.* **A338** (1991) 339 and **A342** (1992) 429

4. P. Kroll, W. Schweiger and M. Schürmann, *Int. Jour. Mod. Phys.* **A6** (1991) 4107

5. T. A. Armstrong et al. (E760 collaboration), *Phys. Rev. Lett.* **70** (1993) 1212

6. D. Bisello et al. (DM2 collaboration), *Z. Phys.* **C48** (1990) 23

7. R. G. Arnold et al., *Phys. Rev. Lett.* **57** (1986) 174

8. B. Ong (representing the CLEO Collaboration), The Proc. of the 7th Meeting of the American Physical Society, Division of Particles and Fields (1992)

9. G. R. Farrar, E. Maina and F. Neri, *Nucl. Phys.* **B259** (1985) 702, **B263** (1986) 746

STRONG FORM FACTORS

A. Szczurek[1,2], H. Holtmann[1] and J. Speth[1]

[1] *Institut für Kernphysik, Forschungszentrum Jülich,*

W-5170 Jülich, Federal Republic of Germany

[2] *Institute of Nuclear Physics,*

PL-31-342 Krakow, Poland

July 15, 1993

Abstract

Mesonic models predict violation of SU(2) symmetry in the nucleon sea which seems to be necessary to explain the violation of the Gottfried Sum Rule. Careful analysis of the Drell-Yan processes in p-p and p-n collisions should provide complementary data to the Gottfried Sum Rule on the $\overline{d}/\overline{u}$ asymmetry. We use high energy experimental data for neutron, Λ and Δ^{++} production to restrict unknown vertex form factors. Large one-loop corrections in semileptonic hyperon decays can be compensated by a tiny shift in the values of D and F. The meson cloud is an important ingredient in understanding the spin structure of the nucleon.

1　Introduction

It has been customarily assumed that the nucleon sea is flavor symmetric ($\overline{d}_p(x) = \overline{u}_p(x)$). There is no general principle that forces one to this hypothesis other than the fact that it appears as a natural consequence of a perturbative approach to the nucleon's parton distributions. There is no justification for such a restricted point of view to proton structure and there is strong indication that nonperturbative physics is crucial. Specifically, the observed violation of the Gottfried Sum Rule (GSR)[1] provides experimental evidence that the nucleon sea is not flavor symmetric. There remains no quantitatively compelling explanation for the effect.

The meson cloud model provides a natural explanation for the excess of $\overline{d}$ over $\overline{u}$ quarks already in its simplest form in which the proton contains components of a bare proton and π^0 and a bare neutron and π^+. In this presentation we review a generalization of this simple idea, developed recently in Jülich. We discuss the role of the meson cloud in deep-inelastic lepton scattering, especially in connection with the GSR and present predictions for the nucleon-nucleon induced dilepton production at high energies. In addition, we discuss the one-loop meson corrections in the semileptonic hyperon decays and for the spin structure of the nucleon.

2 Gottfried Sum Rule

The GSR addresses the value of the integral over x of the difference of the $F_2(x)$ structure function of the proton (p) and neutron (n). It is written[1] as

$$\int_0^1 [F_2^p(x) - F_2^n(x)] \, \frac{dx}{x} \; = \; \frac{1}{3} \int_0^1 \left(\left[u_p^v(x) - d_p^v(x) \right] + 2 \left[\overline{u}_p(x) - \overline{d}_p(x) \right] \right) dx, \quad (1)$$

where $u_p^v(x) \equiv u_p(x) - \overline{u}_p(x)$, etc., and charge symmetry has been assumed, i.e., $u_p(x) = d_n(x)$, etc. If one further makes the customary assumption that $\overline{u}_p(x) = \overline{d}_p(x)$, then

$$\int_0^1 [F_2^p(x) - F_2^n(x)] \, \frac{dx}{x} \; = \; \frac{1}{3}. \quad (2)$$

Equation (2) is referred to as the GSR. However the most recent measurement[2] of the relevant structure functions over the interval $0.004 \leq x \leq 0.8$ yields, when extrapolated to $0 \leq x \leq 1$,

$$\int_0^1 [F_2^p(x) - F_2^n(x)] \, \frac{dx}{x} = 0.24 \pm 0.016 \quad (3)$$

at $Q^2 = 5$ GeV^2.

Taken at face value, a comparison of Eqs. (1), (2), and (3) implies

$$\int_0^1 \left[\overline{d}_p(x) - \overline{u}_p(x) \right] dx = 0.135 \pm 0.024, \quad (4)$$

in marked disagreement with the customary assumption. It appears impossible[3] to generate such a large difference in the $\overline{d}$ and $\overline{u}$ distributions from perturbative processes; hence, the answer likely lies with more complicated nonperturbative physics. For example, there have been a few[4, 5, 6, 7, 8, 9, 10] attempts to calculate the difference due to virtual meson emission. In the absence of such calculations, all that can be done is to reparametrize the $\overline{u}_p(x)$ and $\overline{d}_p(x)$ distributions so that they agree with the observed violation of the GSR.

Typically one defines

$$\overline{d}_p(x) = \overline{q}(x) + \frac{\Delta(x)}{2} \quad (5)$$

and

$$\overline{u}_p(x) = \overline{q}(x) - \frac{\Delta(x)}{2} \quad (6)$$

where

$$\int_0^1 \Delta(x)dx = 0.135 \pm 0.024. \quad (7)$$

The initial reparametrization [11] used $\Delta(x) = A(1 - x)^k$, which placed the $\overline{d}$-$\overline{u}$ difference at large x $(x > 0.05)$ and led to very large values for $\overline{d}_p(x)/\overline{u}_p(x)$ for $x \geq 0.1$. These large ratios have been ruled out by a recent reanalysis of earlier Drell-Yan data. More recent[12] parametrizations have a form $\Delta(x) = B(1 - x)^\ell/x^{0.5}$, which places the bulk of the difference at smaller x.

---

[1] The structure functions $F_2(x)$ are functions of Q^2, as are the quark distribution functions $q(x)$. The Q^2 dependence is suppressed to keep the expressions from being too cumbersome.

3 Hybrid meson-baryon model of the nucleon

In this section we briefly review a recent hybrid meson-baryon model of the nucleon[10] and present its prediction for the asymmetry of the light sea antiquarks. In this model the nucleon is viewed as a quark core, termed a bare nucleon, surrounded by the mesonic cloud. The nucleon wave function can be schematically (we neglect the isospin degrees of freedom for simplicity) written as a superposition of a few principle Fock components

$$|N\rangle_{dressed} = Z^{1/2}\left[\,|N\rangle_{bare} + \alpha\,|N\pi\rangle + \beta\,|\Delta\pi\rangle + (...)\,\right]. \tag{8}$$

The factor Z measures the probability that the physical nucleon contains a bare nucleon. The model of Ref.[10] includes all the mesons required in the description of the low energy nucleon-nucleon and hyperon-nucleon scattering. Furthermore it ensures charge conservation, number and momentum sum rules.

The x-dependence of the structure functions in the meson cloud model can be written as a sum of components corresponding to the expansion given by Eq.(8).

$$F_2^N(x) = Z\left[F_{2,core}^N(x) + \sum_{MB}\left(\delta^{(M)}F_2(x) + \delta^{(B)}F_2(x)\right)\right]. \tag{9}$$

The contributions from the virtual mesons and baryons can be written as a convolution of the meson (baryon) structure functions and its longitudinal momentum distribution in the nucleon

$$\delta^{(M)}F_2(x) = \int_x^1 dy\, f_M(y) F_2^M(x\!\!\!\diagup y). \tag{10}$$

Equation (10) can be written in an equivalent form in terms of the quark distribution functions

$$x\delta^{(M)}q_f(x) = \int_x^1 dy\, f_M(y)\frac{x}{y}q_f^M(\frac{x}{y}). \tag{11}$$

The longitudinal momentum distributions of virtual mesons (or baryons) can be calculated assuming a model of the vertex, and depends on the coupling constants and form factors. For the dominant contributions ($\pi N, \pi\Delta$) they are given by simple formulas

$$f_{\pi(N)}(y) = \frac{3g_{N\pi N}^2}{16\pi^2}\, y \int_{-\infty}^{t^{max}} dt \frac{[G(t)]^2}{(t-m_\pi^2)^2}\,(-t), \tag{12}$$

$$f_{\pi(\Delta)}(y) = \frac{2g_{N\pi\Delta}^2}{24\pi^2}\, y \int_{-\infty}^{t^{max}} dt \frac{[G(t)]^2}{(t-m_\pi^2)^2}\,[(m_N+m_\Delta)-t]\left[\frac{(m_N^2-m_\Delta-t)^2}{4m_\Delta^2}-t\right]. \tag{13}$$

The main ingredients of the model are the vertex coupling constants, the parton distribution functions for the virtual mesons and baryons, and the vertex form factors which account for the extended nature of the hadrons. The coupling constants are assumed to be related via SU(3) symmetry. The measured free-hadron parton distributions are used. In Ref.[10] an overall cut-off parameter has been used for simplicity.

This led to the value of 0.28 for the GSR, about 5/9 of the original NMC effect. In principle the cut-off parameter can be slightly different for different components. In our model the GSR is a delicate interplay between πN and $\pi \Delta$ components and depends on their probabilities in the nucleon wave function. By lowering the probability of the $\pi \Delta$ component one could achieve a better agreement with the NMC result[2]. This is a crucial point in order to understand the NMC result and requires a more detailed analysis. Phenomenologically the $N\pi\Delta$ vertex can be determined[13] by comparison of the one pion exchange model (OPEM) with high energy data

$$(a) \quad p(p, \Delta^{++})X,$$
$$(b) \quad p(\overline{p}, \Delta^{++})X. \tag{14}$$

The cross section for the p(h,B)X reaction in the OPEM can be expressed as

$$\frac{d\sigma}{dx_B dp_\perp} = \tilde{f}(x_B, p_\perp)\sigma_{h\pi}^{tot}, \tag{15}$$

where $\tilde{f}(x_B, p_\perp)$ is a function which involves the $N\pi B$ form factor and $\sigma_{h\pi}^{tot}$ is the total cross section for $h\pi$ collision. While reaction (14a) involves $\sigma_{\pi^- p}^{tot}$ which is well known experimentally, reaction (14b) involves $\sigma_{\pi^- \overline{p}}^{tot}$, which can be easily related to $\sigma_{\pi^+ p}^{tot}$ – also well known. Experimentally these cross sections approximately scale in a broad range of energies, making application of OPEM very simple.

In Fig.1 we show prediction of the OPEM with the experimental data for $p(p, \Delta^{++})X$ [14] and $p(\overline{p}, \Delta^{++})X$ [15] reactions for different values of the cut-off parameter. The data for both reactions prefer a cut-off of about 1.0 GeV in a dipole parametrization. We note that the slight energy dependence of cross sections has been disregarded here and we have taken 24 mb for the $\sigma_{\pi p}^{tot}$ cross section.

In the following we will use cut-off parameter of 1.2 GeV for the components with octet baryons and of 1.0 GeV for the components with decuplet baryons. The softer form factor for the $N\pi\Delta$ vertex than for $N\pi N$ or $NK\Lambda$ vertices seems be a universal feature of high energy scattering[16, 17].

In Fig.2 we present the difference $x(\overline{d} - \overline{u})$ obtained in the framework of the meson cloud model and for various parametrizations with asymmetric sea-quark distributions. It is interesting to notice that the mesonic model is qualitatively similar both in shape and magnitude to the result of a very recent global fit of Martin, Stirling and Roberts (MSR) to the world deep-inelastic and Drell-Yan data[12]. For instance, the GSR obtained from our model is 0.255, with 0.235 in the experimentally measured region. This can be compared with the value of 0.26 obtained from the recent MSR fit. The Eichten-Hinchliffe-Quigg[8] and Ellis-Stirling [11] parametrizations give larger differences; however, they have been adjusted to reproduce the NMC value[2] of the GSR.

In Table 1 we present the contributions of each of the components in Eq.(8). One notes that only the processes involving virtual π and ρ mesons contribute to the asymmetry. The ratio $\frac{\overline{d}(x)-\overline{u}(x)}{\overline{d}(x)+\overline{u}(x)}$ turns out to be rather insensitive to the choice of parton distributions in mesons and in the bare baryons.

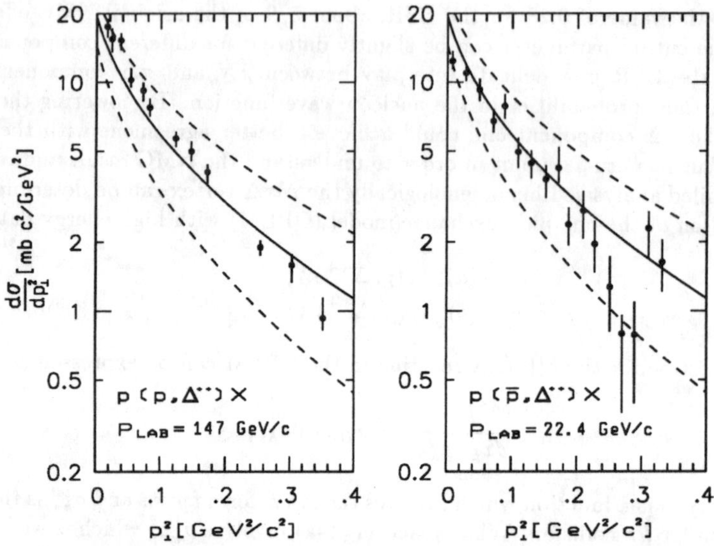

Fig. 1: *Differential cross section $d\sigma/dp_T^2$ as a function of p_T^2 for Δ^{++} production in (a) pp collison[14] measured at Fermilab and (b) $\bar{p}p$ collision[15] measured at Serpukhov. The solid line is for cut-off of 1.0 GeV, whereas the dotted lines are for 0.8 GeV and 1.2 GeV.*

Table 1. *A list of the processes contributing to the violation of the $\bar{u}$-$\bar{d}$ symmetry in the nucleon sea. In addition we show also probabilities (in %) of the individual Fock components in the expansion of the nucleon wave function given by Eq.(8) and numbers of antiquarks contained in the proton wave function.*

component	probability	$\bar{u}$	$\bar{d}$	$\bar{d} - \bar{u}$
$bare N$	59.38	-	-	-
$\pi + N$	21.15	0.0353	0.1763	0.1410
$\rho + N$	1.73	0.0029	0.0144	0.0115
$\pi + \Delta$	10.69	0.0712	0.0356	-0.0356
$\rho + \Delta$	0.30	0.0020	0.0010	-0.0010
$symm.cont.$	6.75	0.0256	0.0256	-
sum	100.00	0.1370	0.2529	0.1159

4 Drell-Yan processes

The Drell-Yan (DY) process with incident nucleons can be made extremely sensitive to the $\bar{d}_p(x)/\bar{u}_p(x)$ distribution in the target[11].

The Drell-Yan process[18] involves the electromagnetic annihilation of a quark

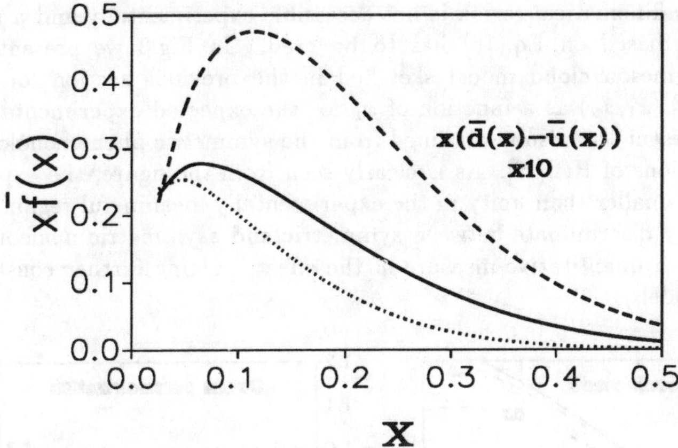

Fig. 2: The difference $x\overline{d}(x) - x\overline{u}(x)$ magnified by the factor 10. Result of the meson cloud model is shown by the solid line. We present also the difference for asymmetric phenomenological parametrizations: Eichten-Hinchliffe-Quigg[8] (dashed curve) and Martin-Stirling-Roberts[12] (dotted curve).

(antiquark) from the incident hadron A with an antiquark (quark) in the target hadron B. The resultant virtual photon materializes as a dilepton pair $(\ell^+\ell^-)$. The cross section for the DY process can be written as

$$\frac{d\sigma^{AB}}{dx_1 dx_2} = \frac{4\pi\alpha^2}{9s x_1 x_2} K(x_1, x_2) \sum_f e_f^2 \left[q_A^f(x_1)\overline{q}_B^f(x_2) + \overline{q}_A^f(x_1) q_B^f(x_2) \right], \qquad (16)$$

where s is the square of the center-of-mass energy and x_1 and x_2 are the longitudinal momentum fractions carried by the quarks of flavor f. The $q_A^f(x_1)$ and $q_B^f(x_2)$ are the quark distribution functions of the beam and target, respectively. The factor $K(x_1, x_2)$ accounts for the higher-order QCD corrections that enter the process.

The valence quark distribution in the nucleon falls off as a much smaller power of $(1 - x)$ than does its sea quark distribution, so that as $x \to 1$, there are only valence quarks and no sea quarks. Indeed, if x_1 is selected such that $x_1 = x_2 + 0.3$, the first term in Eq. (16) dominates the second term by a factor of more than 10. Thus, forming the ratio of the Drell-Yan yields for pp to pn, we have

$$\frac{d\sigma_{\rm DY}^{pp}}{d\sigma_{\rm DY}^{pn}}\bigg|_{x_F > 0.3} \simeq \frac{4u_p(x_1)\overline{u}_p(x_2) + d_p(x_1)\overline{d}_p(x_2)}{4u_p(x_1)\overline{d}_p(x_2) + d_p(x_1)\overline{u}_p(x_2)}, \qquad (17)$$

where it is assumed that $K_p(x_1, x_2) = K_n(x_1, x_2)$ and $u_p(x) = d_n(x)$, etc. In the limit $x_1 \to 1$, $\frac{d_p(x_1)}{u_p(x_1)} \to 0$, and in that limit one has

$$\frac{d\sigma_{\rm DY}^{pp}}{d\sigma_{\rm DY}^{pn}}\bigg|_{\substack{x_F > 0.3 \\ x_1 \to 1}} = \frac{\overline{u}_p(x_2)}{\overline{d}_p(x_2)}. \qquad (18)$$

In practice this kinematical region is not accessible experimentally and a more general expression based on Eq.(16) has to be used. In Fig.3 we present the prediction of the meson cloud model sketched in the previous section for the ratio $\sigma_{DY}^{pn}(x_1, x_2)/\sigma_{DY}^{pp}(x_1, x_2)$ as a function of x_2 for the expected experimental range of x_1. We also present the results obtained from the symmetric phenomenological sea-quark distributions of Ref.[19]. As is clearly seen from the figure, the experimental ratio, larger or smaller than unity in the experimentally meaningful region $x_2 < 0.3$, should definitely discriminate between symmetric and asymmetric nucleon seas. It would also give a quantitative measure of the effect, putting further constraints on the mesonic models.

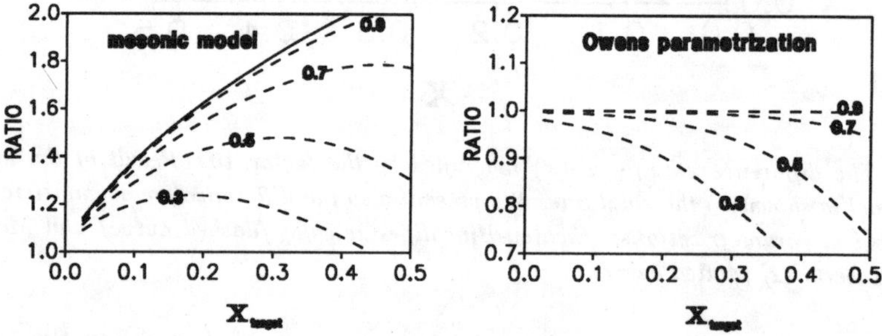

Fig. 3: The ratio $\sigma_{DY}^{pn}/\sigma_{DY}^{pp}$ as a function of the target x_2 for selected values of the beam x_1. Predictions of the meson cloud model are shown on the left side, while the result of a phenomenological parametrization [19] with symmetric antiquark distributions is presented on the right side.

The ratio of the DY yield from a nucleus with $N \neq Z$ to that from an isoscalar target such as deuterium, is sensitive to the $\overline{d}_p(x) - \overline{u}_p(x)$ difference. These ratios have been measured by the E772 Collaboration at FNAL[20] for carbon, calcium, iron and tungsten targets. Elementary algebra leads to the following result:

$$\frac{2d^2\sigma_{DY}^{pA}}{Ad^2\sigma_{DY}^{pd}} = \frac{2Z}{A} + \frac{N-Z}{A}\frac{2\sigma_{DY}^{pn}(x_1, x_2)}{(\sigma_{DY}^{pp}(x_1, x_2) + \sigma_{DY}^{pn}(x_1, x_2))}, \tag{19}$$

where Z, N, A are the number of protons, neutrons and the atomic number, respectively. In Fig.4 we present the ratios for iron and natural tungsten targets as a function of x_2 (target) for fixed value $x_1 = 0.3$, which correspond roughly to the experimental situation. For comparison we present also the experimental data of the E772 Collaboration. In addition we show the results with the flavor symmetric sea distributions of Ref.[19]. No large difference between the result of the mesonic model and the symmetric parametrization can be seen. In our opinion the data do not select in a convincing way the symmetric vs. asymmetric distributions. We present also in Fig.4 similar results with the flavor asymmetric sea-quark distributions of Refs.[22, 11, 8]. As can be seen from the figure, the data exclude such asymmetric parametrizations.

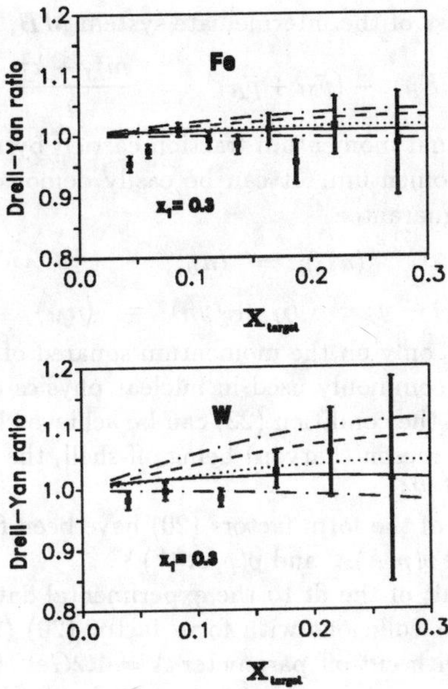

Fig. 4: The ratio $2\sigma_{DY}^{pA}/A\sigma_{DY}^{pd}$ as a function of the target x_2 for iron and natural tungsten. The experimental data are taken from Ref.[20] and Ref.[21]. Predictions of the meson-cloud model are shown by the solid line and the result with symmetric sea quark distribution by the long dashed line. Results obtained from the phenomenological parametrizations with the asymmetry built in are shown for comparison: Field-Feynman[22] (dotted curve), Eichten-Hinchliffe-Quigg[8] (dashed curve), Ellis-Stirling[11] (dash-dotted curve). Note that no nuclear effects have been included.

5 Strong form factors

In the following two sections we shall discuss evaluation of the axial-vector current matrix elements, taking into account meson radiation corrections. In contrast to the problems discussed in the previous sections, the axial-vector current matrix elements are very sensitive to the details of the model. This requires very careful analysis of the model parameters. Here the form factors of the vertices are especially important.

In the following instead of the covariant method of Ref.[10] we use the light cone method proposed recently [16, 23]. The advantage of the method is that it guarantees local gauge invariance and energy-momentum sum rules automatically. The essence of the method is a use of form factors which fulfill certain symmetries. Following the conventional normalization at the pole, the form factors are

$$G_{MB/N} = \exp\left[\frac{m_N^2 - m_{MB}^2}{2\Lambda^2}\right],\tag{20}$$

where m_{MB} is invariant mass of the intermediate system MB,

$$m_{MB} = (E_M + E_B)^2 - (\vec{p}_M + \vec{p}_B)^2 = \frac{m_M^2 + k_\perp^2}{y} + \frac{m_B^2 + k_\perp^2}{1 - y}, \qquad (21)$$

with y beeing the longitudinal momentum fraction carried by meson M and $k_\perp$ the respective perpendicular momentum. It can be easily demonstrated that the form factors (20) automatically guarantee

$$\langle n_M \rangle = \langle n_B \rangle, \qquad (22)$$

$$f_M(y) = f_B(1 - y) \ , \quad \langle y_M \rangle + \langle y_B \rangle = \langle n_M \rangle = \langle n_B \rangle. \qquad (23)$$

The form factors dependent only on the momentum squared of the virtual meson or baryon (monopole, dipole) commonly used in nuclear physics do not guarantee the required symmetries. While the condition (22) can be achieved by a correlated choice of cut-off parameter for the meson (baryon) being off-shell, the momentum sum rule (23) is unavoidably violated [24].

The cut-off parameters of the form factors (20) have been fixed by fitting to the high energy data: $p(p, n)X$, $p(p, \Lambda)X$ and $p(p, \Delta^{++})X$.

In Fig.5 we show a result of the fit to the experimental data[25, 26] for neutron production in proton-proton collisions with form factor (20) (left panel) compared with a dipole form factor with cut-off parameter $\Lambda = 1.2 GeV$ [10] (right panel). In this fit, assuming a one meson exchange mechanism, both π^+ and ρ^+ meson exchange have been included. The two-parameter fit with form factor (20) gives resonable agreement with the experimental data for all $k_\perp$. The fit is not ambiguous since the π^+ exchange dominates at small $k_\perp$ and the ρ^+ exchange becomes important at larger $k_\perp$. We find the cut-off parameter $\Lambda = 1.08 \pm 0.05$ in formula (20) for both π and ρ exchange. Almost exactly the same cut-off parameter has been found for the $p(p, \Lambda)X$ reaction for both K and K^* exchange (details will be presented in Ref.[17]). It is interesting to notice that the position of the peak in the cross section as a function of x is correctly reproduced. The discrepancy at small momentum fraction x carried by the neutron should not be taken seriously as here multi-meson emission and target fragmentation are expected to play the dominant role.

The traditional dipole form factor produces a fit of good quality at small $k_\perp$. It does not predict, however, correct $k_\perp$-evolution of the maximum of the cross section, and fails at larger $k_\perp$'s.

The cut-off parameters for the decuplet baryons have been fixed by fitting to the $p(p, \Delta^{++})X$ and $p(\bar{p}, \Delta^{++})X$ experimental data. Both π and ρ exchanges have been included. Since in this case the fit was slightly ambiguous, the same cut-off parameter has been assumed for the π and ρ exchange. The cut-off parameter found from the fit is 0.98 ± 0.05 – slightly less then for the case of octet baryons. This small difference is, however, important for the Gottfried Sum Rule (details will be discussed in Ref.[17]).

6   Semileptonic decays of the octet baryons

The matrix elements of the current operators for the semileptonic decays of the baryons belonging to the octet can be parametrized in terms of the q^2-dependent form

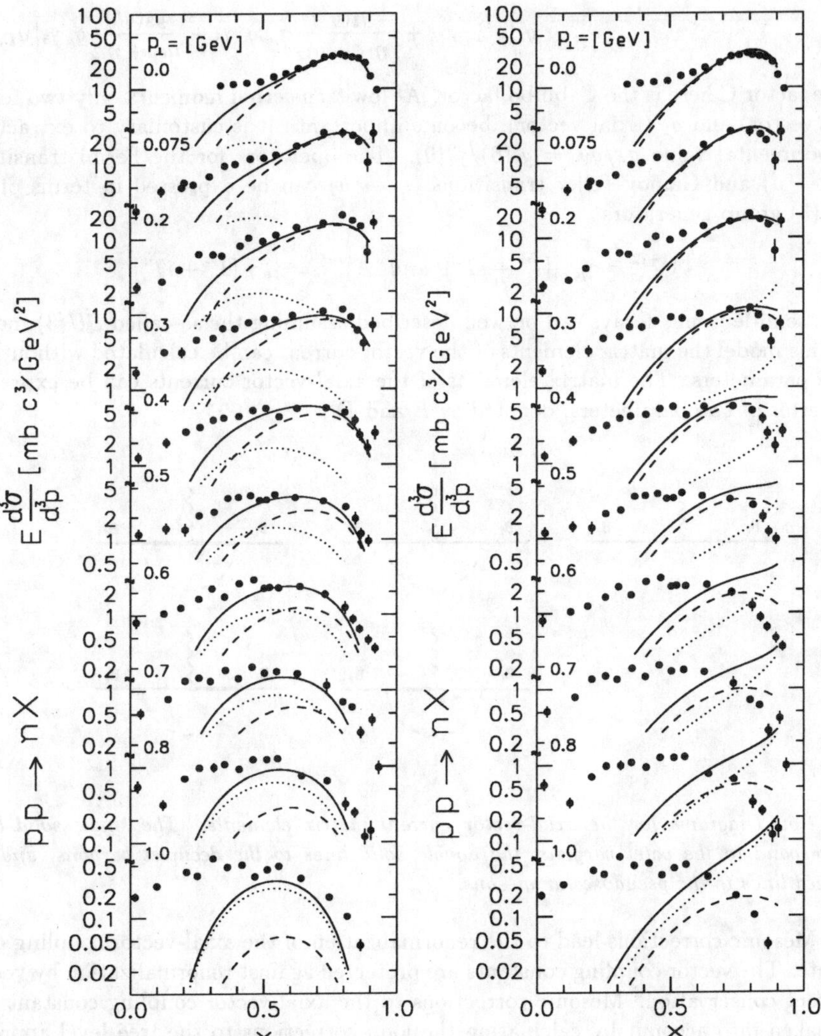

Fig. 5: Differential cross section for the neutron production in the $p(p, n)X$ reaction as a function of the longitudinal momentum fraction of the outgoing neutron for different values of $k_\perp$. The results obtained with the form factors (20) (left panel) are compared with those obtained with traditional dipole form factors (right panel).

factors,

$$\langle B_f | V_\mu + A_\mu | B_i \rangle = C u_{B_f}(p') \left[f_1(q^2)\gamma_\mu + i \frac{f_2(q^2)}{m_1 + m_2}\sigma_{\mu\nu}q^\nu + \frac{f_3(q^2)}{m_1 + m_2}q_\mu + \right. \tag{24}$$

$$g_1(q^2)\gamma_\mu\gamma_5 + i\frac{g_2(q^2)}{m_1 + m_2}\sigma_{\mu\nu}q^\nu\gamma_5 + \frac{g_3(q^2)}{m_1 + m_2}q_\mu\gamma_5\Big]u_{B_i}(p).$$

The factor C here is the Cabbibo factor. At low transferred momenta only two terms, f_1 (vector) and g_1 (axial vector), become important. It is customary to extract the experimental value $g_A/g_V = g_1(0)/f_1(0)$. The operators for the Fermi transitions $(d \to u)$ and Gamov-Teller transitions $(s \to u)$ can be expressed in terms of the $SU(3)$ group generators

$$A_\mu^{1+i2} = \gamma_\mu\gamma_5[T^1 + iT^2] \quad and \quad A_\mu^{4+i5} = \gamma_\mu\gamma_5[T^4 + iT^5]. \tag{25}$$

The semi-leptonic decays can be well described assuming the so-called $SU(3)$ model. In this model the matrix elements of the vector current can be calculated without any free parameters. The matrix elements of the axial-vector currents can be expressed in terms of two parameters, denoted as F and D.

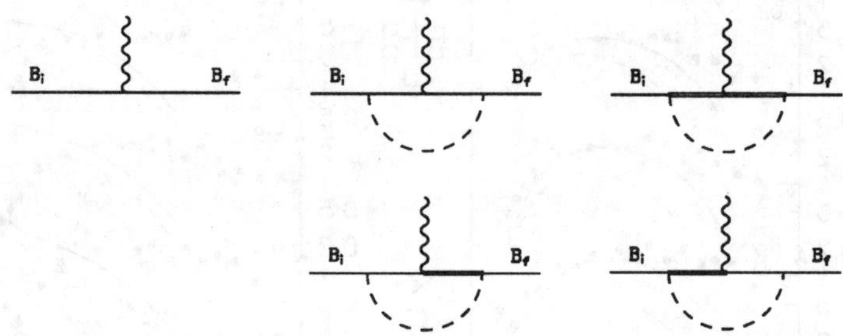

Fig. 6: Diagrams for the axial-vector current matrix elements. The single solid lines correspond to the octet baryons, the double solid lines to the decuplet baryons, and the dashed lines to the pseudoscalar mesons.

Mesonic corrections lead to the renormalization of the axial-vector coupling constants. The vector coupling constants are protected against renormalization by vector current conservation. Mesonic corrections to the axial-vector coupling constant can be taken into account by calculating the loop corrections to the tree level approximation (see Fig.6). Here we report on the results with the inclusion of intermediate pseudoscalar mesons and associated octet and decuplet baryons only (the calculations with inclusion of vector mesons are in progress and will be presented elsewhere [17]).

Including the diagrams shown in Fig.6 and assuming the $SU(3)$ symmetry for axial-vector currents, there exist 4 independent coupling constants: F and D for the transitions between the octet baryons, g_{10-10} for the transitions within the decuplet of baryons, and $g_I = g_{8-10} = g_{10-8}$ for the mixed transitions (interference diagrams). In order to reduce number of free parameters g_{10-10} and g_I have been fixed to their $SU(6)$ values and, as in the case of the simple $SU(3)$ fit, only F and D have been fitted to the experimental data on g_A/g_V. The obtained values of F and D are not far from the $SU(6)$ limit ($F = 0.47, D = 0.91$ compared with $F = 0.44$, $D = 0.82$ in the pure $SU(3)$ fit). The resulting values of g_A/g_V are compared in Table 2 with the $SU(6)$ tree result and that obtained in the $SU(3)$ tree model with parameters fitted to the semileptonic decay data. The χ^2 values presented in Table 2 give an idea of the fit quality. The quality of the fit within our model is very similar to that obtained within the traditional $SU(3)$ fit. It would even improve with allowance for the variation of g_{10-10} and g_I. A completely unrestricted fit could result, however, in unphysical values of parameters due to very limited number of experimental data.

Table 2. The value of g_A/g_V in different models in comparison with the experimental data.

decay	$SU(6)$	$SU(3)$	our model	experiment
$1.\, n \to p$	1.67	1.26	1.25	1.2573 ± 0.0028
$2.\, \Sigma^- \to \Lambda$	0.82	0.67	0.72	0.60 ± 0.03
$3.\, \Lambda \to p$	-1.22	-0.87	-0.88	-0.857 ± 0.018
$4.\, \Xi^- \to \Lambda$	0.41	0.20	0.20	0.31 ± 0.06
$5.\, \Sigma^- \to n$	0.33	0.38	0.42	0.34 ± 0.05
χ^2/N	4369	2.0	4.9	

7 Spin structure of the nucleon

In recent years there has been much excitement about polarised deep-inelastic scattering [27, 28]. After a Regge extrapolation of the deep-inelastic structure function $g_1^p(x)$ to $x = 0$, EMC found

$$S_{EJ}^p = \int_0^1 g_1^p(x) \; dx \; = \; 0.126 \pm 0.010(stat) \pm 0.015(syst). \tag{26}$$

In the parton model of Feynman g_1^p can be expressed as

$$g_1^p(x) = \frac{1}{2} \left[\frac{4}{9} \Delta u(x) + \frac{1}{9} \Delta d(x) + \frac{1}{9} \Delta s(x) \right], \tag{27}$$

where $\Delta q_f(x) = q_f^\uparrow(x) - q_f^\downarrow(x)$. The quantity

$$g_A^0 = \Delta u + \Delta d + \Delta s \tag{28}$$

determines the fraction of the proton spin which is carried by its quarks. The result (26) combined with semileptonic decay data yields a most surprising result

$$g_A^0 = 0.120 \pm 0.094(stat) \pm 0.138(syst), \tag{29}$$

which is consistent with zero. This result is often called the proton spin crisis. The experimental result has inspired one of the most serious debates in the last decade about the theoretical interpretation of the spin and its connections to QCD. Here we concentrate only on the classical part of the problem in the framework of our meson cloud model of the nucleon.

The parameters fixed in the previous section can be used to estimate the effect of the meson cloud on the spin structure of the nucleon. The matrix elements of the flavor singlet axial-vector current can be calculated analogously to those for the semileptonic decays.

Table 3. Quark polarizations, the Ellis-Jaffe sum rules for proton and neutron and the Bjorken sum rule in the $SU(6)$ model and in our meson cloud model (MCM).

model	Δu	Δd	Δs	g_A^0	S_{EJ}^p	S_{EJ}^n	S_B
$SU(6)$	4/3	-1/3	0.0	1.0	5/18	0.0	5/18
$MCM(SU(6))$	1.128	-0.326	0.002	0.804	0.233	-0.009	0.242
$MCM(SU(3))$	0.858	-0.399	0.001	0.461	0.169	-0.041	0.210

In Table 3 we compare the results of our model with those in the classical $SU(6)$ model, so succesfull in the description of the neutron-to-proton magnetic moment ratio. As seen from the table, the meson cloud model predicts a strong reduction of the spin carried by quarks in comparison to the naive $SU(6)$ model. In addition to S_{EJ}^p we present the Ellis-Jaffe sum rule for the neutron S_{EJ}^n and the Bjorken sum rule S_B. Although we get a strong reduction from the 5/18 of the $SU(6)$ model for the S_{EJ}^p, the meson cloud model alone cannot account for the experimentally measured value. Obviously other effects, such as those discussed in Ref.[29], must play an important role. The effect of the meson cloud cannot, however, be neglected in the total balance of the proton spin. As far as S_{EJ}^n is considered, the experiments at CERN and at SLAC are under way. Recently some preliminary results have been announced [30]. The value of S_{EJ}^n and S_B obtained in our model are consistent with those obtained by SMC [30]:

$$S_{EJ}^n = -0.08 \pm 0.04(stat) \pm 0.04(syst), \tag{30}$$

$$S_B = 0.20 \pm 0.05(stat) \pm 0.04(syst). \tag{31}$$

8 Conclusions

In the light of recent experiments on the deep-inelastic muon scattering by nucleons, the understanding of the nucleon structure has become one of the most intriguing problems of particle and nuclear physics. The recent observation of the Gottfried Sum Rule violation suggests a flavor asymmetry of the light sea quarks in the nucleon. Although the asymmetry does not contradict any fundamental principles, the flavor symmetry of the nucleon sea had become a part of the folklore in the particle physics community.

An asymmetry occurs in a natural way within certain non—perturbative models of the nucleon, built with valence quarks surrounded by a meson cloud. There exist two classes of models of this type. In the chiral quark model[8] the mesons are directly coupled to quarks. In the model discussed in the present paper, the nucleon is regarded as a quark core surrounded by the meson cloud. This picture of the nucleon provides a good description of nucleon electric polarizabilities[31]. It also has a close connection to models of low-energy hadron-hadron scattering[32, 33]. The value for the Gottfried Sum Rule is almost in agreement with that obtained by NMC.

In contrast to phenomenological parametrizations of the $\bar{d}$-$\bar{u}$ asymmetry, the meson cloud model discussed here predicts asymmetry concentrated at small x, quite similar to a recent fit[12] to the world data for DIS and Drell-Yan processes. As a consequence, the model predicts only 10%-20% deviations to be observed in a Fermilab experiment[34] measuring the relative dilepton yield in proton-proton and proton-deuteron Drell-Yan processes.

The mesonic corrections lead to the renormalization of the axial-vector current matrix elements, which seems to be unwanted in the light of the success of the simple Cabibbo theory. Large one-loop corrections to g_A, which explicitely violate SU(3) symmetry, can be compensated by a shift in the values of the symmetric (D) and antisymmetric (F) axial coupling constants. It appears accidental that values for D and F, consistent with experimental data, can be obtained in both the SU(3) fit and by the inclusion of the meson cloud.

We get a significant flow of the nucleon spin to the angular momentum of the mesonic cloud. Although the meson cloud model does not reproduce completely the EMC result for the Ellis-Jaffe sum rule, the contribution of the meson cloud cannot be neglected in the total balance of the proton spin.

We wish to thank J. Durso, G.T. Garvey and N.N. Nikolaev for valuable discussions. This work was supported in part by the Polish KBN grant 2 2409 9102.

References

[1] K. Gottfried, *Phys.Rev.Lett.* **18**(1967)1174.

[2] P. Amaudruz et al, *Phys.Rev.Lett.* **66**(1991)2712.

[3] D.A. Ross and C.T.Sachrajda, *Nucl.Phys.* **B149**(1979)497.

[4] E.M. Henley and G.A. Miller, *Phys. Lett.* **B251**(1990)453.

[5] A. Signal, A.W. Schreiber, and A.W. Thomas, *Mod.Phys.Lett.* **A6**(1991)271.

[6] W. Melnitchouk, A.W. Thomas, and A.I. Signal, *Z.Phys.* **A340**(1991)85.

[7] S. Kumano and J.T. Londergan, *Phys. Rev.* **D44**(1991)717.

[8] E. Eichten, I. Hinchliffe, and C. Quigg, *Phys.Rev.* **D45**(1992)2269.

[9] W-Y.P. Hwang, J. Speth, and G.E. Brown, *Z.Phys.* **A339**(1991)383.

[10] A. Szczurek and J. Speth, *Nucl.Phys.* **A555**(1993)249.

[11] S.D. Ellis and W.J. Stirling, *Phys.Lett.* **B256**(1991)258.

[12] A.D. Martin, W.J. Stirling, and R.G. Roberts, *Phys.Rev.* **D47**(1993)867.

[13] G.G. Arakelyan and A.A. Grigoryan, *Sov.J.Nucl.Phys.* **34**(1981)745.

[14] D. Brick et al, *Phys. Rev.* **D21**(1980)632.

[15] E.G. Boos et al, *Nucl.Phys.* **B151**(1979)193.

[16] V.R. Zoller, *Z.Phys.* **C53**(1992)443.

[17] H. Holtmann, A. Szczurek, and J. Speth, *paper in preparation.*

[18] S.D. Drell and T.M. Yan, *Ann. Phys.* **66**(1971)578.

[19] J.F. Owens, *Phys. Lett.* **B266**(1991)126.

[20] D.M. Alde et al, *Phys.Rev.Lett.* **64**(1990)2479.

[21] P.L. McGaughey et al, *Phys.Rev.Lett.* **69**(1992)1726.

[22] R.D. Field and R.P. Feynman, *Phys. Rev.* **D15**(1977)2590.

[23] V.R. Zoller, *ITEP preprint* ITEP-80-92 (1992).

[24] P.J. Mulders, A.W. Schreiber, and H. Meyer, *Nucl.Phys.* **A549**(1992)498.

[25] W. Flauger and F. Monning, *Nucl.Phys.* **B109**(1976)347.

[26] V. Blobel et al, *Nucl.Phys.* **B135**(1978)379.

[27] J. Ashman et al, *Phys.Lett.* **B206**(1988)364.

[28] J. Ashman et al, *Nucl.Phys.* **B328**(1989)1.

[29] R.L. Jaffe and A. Manohar, *Nucl.Phys.* **B337**(1990)509.

[30] B. Adeva et al. (SMC). *Phys. Lett.* **B302**(1993)533.

[31] V. Bernard, N. Kaiser, and U.G. Meissner, *Nucl.Phys.* **B373**(1992)346.

[32] R. Machleidt, K. Holinde, and Ch. Elster, *Phys.Rep.* **149**(1987)1.

[33] B. Holzenkamp, K. Holinde, and J. Speth, *Nucl.Phys.* **A500**(1989)485.

[34] G.T. Garvey et al, *FNAL proposal, P866* (1992).

Photon absorption on complex nuclei

N. Bianchi, <u>V. Muccifora</u>, A. Deppman, E. De Sanctis,
A. Ebolese, A. Fantoni, P. Levi Sandri, V. Lucherini,
E. Polli, A.R. Reolon, P.Rossi.
INFN - Laboratori Nazionali di Frascati,
P.O. 13, I-00044 Frascati, Italy

M. Anghinolfi, P. Corvisiero, L. Mazzaschi, V. Mokeev, G. Ricco,
M. Ripani, M. Sanzone, M. Taiuti, A.Zucchiatti.
INFN - Sezione di Genova e Dipartimento di Fisica Universitá,
Via Dodecaneso Genova,Italy

G.Gervino
Dipartimento di Fisica dell'Universitá di Torino
Via P. Giuria, I-10125 Torino, Italy.

Abstract

We present the results of the measurements of the total photoabsorption
cross section performed at Frascati in the higher nucleon resonance region. In
the Δ region our results agree with existing data, while at higher energies they
show a damping in the excitation of the higher nucleon resonances, clearly seen
in the photon absorption on proton and deuteron.

The total photonuclear absorption cross section has been measured over a
wide range in mass number and photon energy, yielding information on the influence
of the nuclear medium on the intrinsic properties of nucleons in nuclei, as well as
on the hadronic nature of the photon. In the Δ-resonance region, [1,2] the data for
the total cross section per nucleon, measured for various nuclei from ^{6}Li to ^{238}U,
show that the resonance shape and strength is nearly universal, thus indicating an
incoherent volume-photoabsorption mechanism (see Fig.1a). The response of bound
nucleons differs from that for the free nucleon, mainly because of Fermi motion, Pauli
blocking and the propagation and interaction of the Δ in the nucleus; however, the
total strength of the interaction in this region is conserved. Above about 2 GeV, on
the contrary, the total cross section shows some coherent behaviour; its A dependence

can be written as $\sigma_{\gamma A} = A^{\alpha}\sigma_{\gamma N}$, where A is the mass number and $\sigma_{\gamma N}$ is the cross section for a free nucleon.

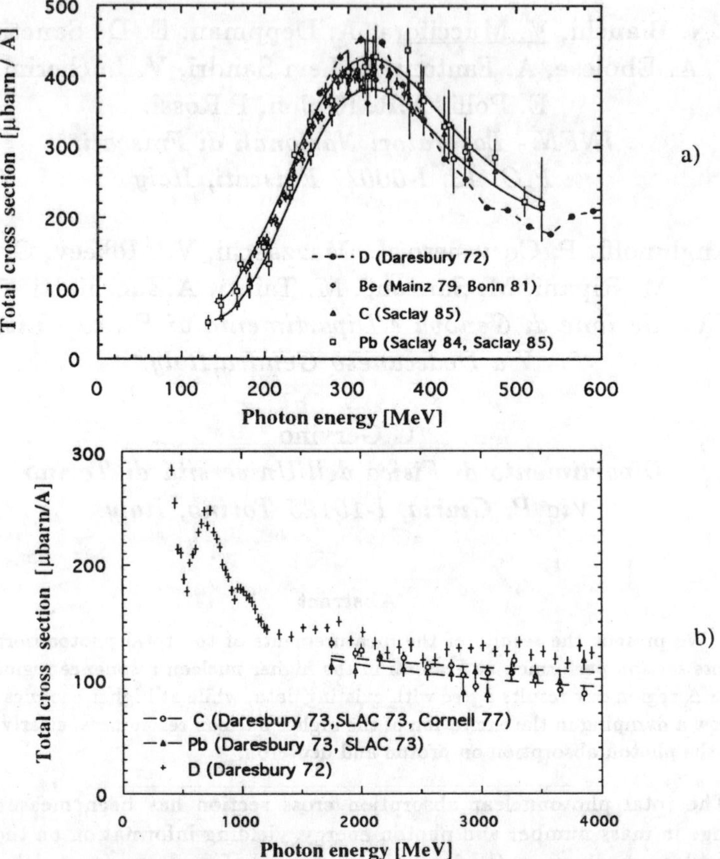

Figure 1: a) Some experimental values of cross section over the Δ energy region; the region between solid curves describe the universal behavior. b) Experimental values of cross section on Deuteron (D), C, Pb in the few GeV region and the Regge form fit to experimental data $(\sigma(k) = a + b \cdot k^{-1/2})$.

The present experimental data at very high energies suggest a value $\alpha \approx 0.9$, intermediate between a volume absorption, $\alpha = 1$, typical of a bare electromagnetic probe, and a surface absorption, $\alpha = 2/3$, typical of a hadronic probe. This is the well known shadowing effect, that is usually explained in the framework of the Vector-Meson-Dominance (VMD) Model. The available experimental data show a shadowing-like behavior at energy as low as 1.8 GeV; however, the amount of shadowing found in various studies is contradictory and does not agree quantitatively with theoretical predictions.[3] The energy dependence of the total cross section, in the few-GeV region, can be described well by a Regge behavior of the form $\sigma_{\gamma A}(k) = a + b \cdot k^{-1/2}$, where k is the photon energy (see Fig.1b).

Until a few years ago, in the region between 0.5 GeV and 1.8 GeV data were scarce and inaccurate. There were precise data only for the proton [4] and deuteron [5]. There also were some low-precision data for Be, C ,Cu and U obtained at Erevan [6,7] sparsely spread in photon energy and fluctuacting considerably more than the quoted experimental errors . Very recently, three measurements, using three different experimental apparata and techniques have been performed at Frascati in the energy range between 0.2 and 1.2 GeV:

a) simultaneous measurements of the photofission cross sections for ^{238}U and ^{232}Th using Parralel-Plate Avalance Detectors [8] (PPADs).

These detectors, acting at the same time as nuclear fissile targets and fission products detectors, consisted of 58 aluminum plates (thickness 50μm and 65 mm diameter) coated on one side with a uniform deposition of fissile material, for a total thickness of about 120 mg/cm^2. The plates were alternately connected to ground or to a potential of 430.0 ± 0.1 V and the chamber was filled with isobutane maintained at a pressure of 7.60 ± 0.01 mbar. Both voltage and pressure, whose changes would affect the detector gain and efficiency, were continuously controlled during data collecting. A special calibration detector, put inside the same chamber of the main detector, was used to measure precisely the effective thickness of the main PPAD.

b) total cross-section measurements on Li, C, Al , Cu, Sn and Pb with the photohadronic method, using a NaI crystal as the hadronic detector and a lead-glass counter as the electromagnetic detector[9]. The photo-hadronic method consists in measuring the photoproduction rate of hadronic events rejecting the wastly preponderant electromagnetic events by an angular separation. This method, applied succesfully in all the measurements of total cross section above the Δ resonance, provides a direct and absolute measurement of the total nuclear photoabsorption cross section. The photon beam interacted in a $0.1X_0$ thick target; a NaI crystal (which consisted of three cylinder sectors each 32 cm long and 15 cm thick) surrounding the target detected the charged particles and the neutral pions produced by the photonuclear interaction in the target, while the electromagnetic events close to the photon direction were vetoed by the lead-glass detector, positioned 70 cm downstream. Hadronic absorption of a photon of given energy was then indicated by a coincidence between signals from one tagging channel and the NaI detector, without a coincidence pulse in the shower detector above a fixed threshold.

c) total cross-section measurements on Be and C with the transmission technique, using a BGO crystal as the photon spectrometer [10]. This method consists in measuring the total attenuation cross section and subtracting the atomic absorption cross section computable for light nuclei with high accuracy.

In Fig.2, Frascati results on light nuclei obtained with the transmission tecnique c) [10] and also with the photo-hadronic method b) [9], are shown together with the more recent data obtained from other laboratories and also with the universal nuclear behaviour in the Δ region.

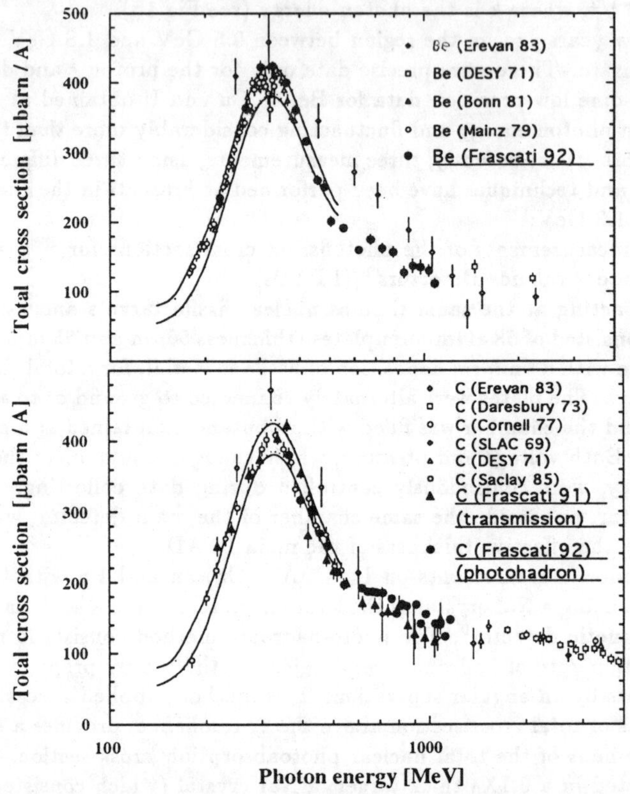

Figure 2: The Frascati data on Be and C in the intermediate energy region, compared with the most recent data from other laboratories and with the Δ universal behavior.

In Fig.3 we present the Frascati results of the photofission measurements a) both on ^{238}U [8] and on ^{232}Th [11]. The ^{232}Th photofissility was calculated from the simultaneous measurement of photofission cross section of ^{238}U in the same energy range and assuming a photofissility equal to one for the ^{238}U. In the energy range 250-1200 MeV the fissility values of ^{232}Th lie approximately between 0.6 and 0.8, showing a weak but clear increase with the energy, demostrating that for this nucleus the saturation values is not completely reached. Moreover in this figure is shown a very simple fit in the form $lnW_f(k) = A - Bk^{-1/2}$ that seems to reproduce well the energy behaviour of our data and to connect properly to the available low energy data, and also the prediction of a Monte Carlo based on an intranuclear cascade and evaporation calculation (INC).

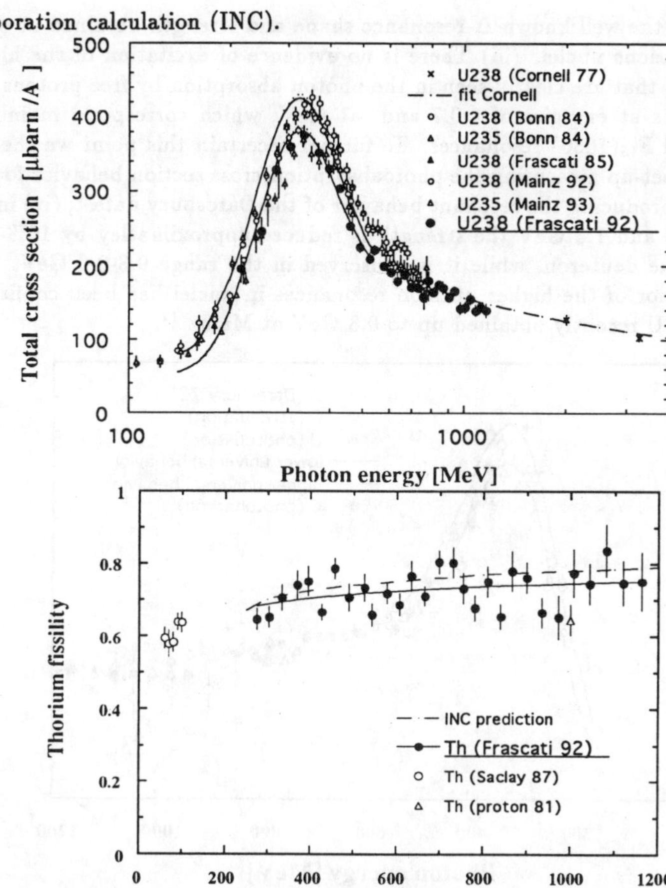

Figure 3: The Frascati data on ^{238}U and the photofissility values for ^{232}Th obtained as described in the text, solid line is a fit in the form $lnW_f(k) = A - Bk^{-1/2}$, dashed line is the result of a INC calculation.

In Fig. 4, Frascati results on C, obtained with the photo-hadronic method and with the transmission method, together with those on ^{238}U, obtained with photofission, are compared with the available deuteron[5] data over the entire energy range and with the universal nuclear behavior in the Δ region. From these comparisons we conclude that: (i) The results of the three measurements are in excellent agreement with each other, showing good control of systematic errors and confirming that the fission probability of ^{238}U is very close to one up to 1.2 GeV. (ii) In the energy region below 0.5 GeV, the values of the cross section per nucleon, $\sigma_{\gamma A}/A$, reproduce, within the

112

systematic errors, the well known Δ-resonance shape and strength obtained at other laboratories on various nuclei. (iii) There is no evidence of excitation of the higher baryon resonances that are clearly seen in the photon absorption by free protons and deuterons as peaks at energies of ≈ 0.7 and ≈ 1 GeV, which correspond mainly to the $D_{13}(1520)$ and $F_{15}(1680)$ resonances. To further ascertain this point we checked our experimental set-up measuring the photoabsorption cross section behavior for the deuteron [9] and reproducing the resonant behavior of the Daresbury data [5]; (iv) in the range between 0.6 and 1.2 GeV the strength is reduced approximatley by 10%-20% with respect to the deuteron, while it is conserved in the range 0.3-0.6 GeV. This unexpected behavior of the higher nucleon resonances in nuclei has been confirmed by the data on ^{238}U recently obtained up to 0.8 GeV at Mainz [12].

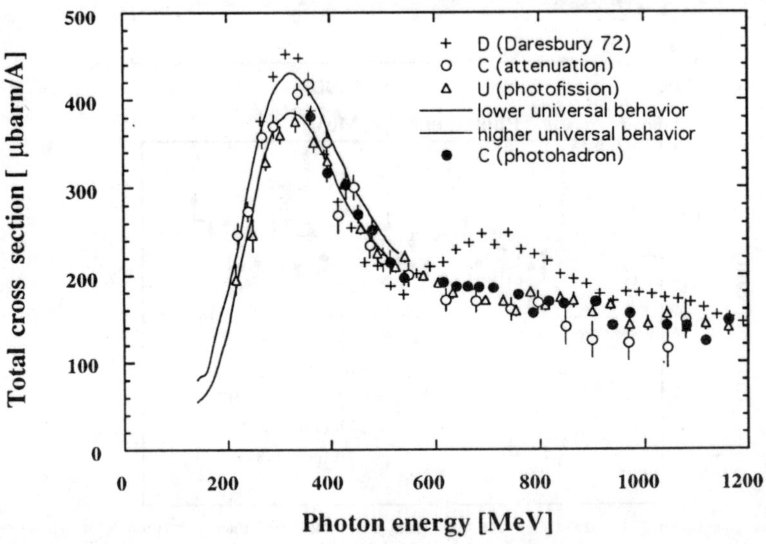

Figure 4: The Frascati data on C, obtained both with photoadronic method (close circles) and with transmission method (open circles), and on ^{238}U (open triangles), compared with the deuterium data and with the universal nuclear behavior in the Δ region.

In order to give a physical interpretation of these results, in particular those of points (iii) and (iv), we have developed a model [13] in which, beginning with the excitation of nucleon resonances deduced from experimental results on the proton and deuteron, we attempt to understand these processes inside nuclei. Within the framework of this model we account for: a) the Fermi motion described by a Fermi gas model with $P_F=280$ MeV, b) the Pauli blocking factor, c) the interaction of the produced N* resonances with the surrounding nucleons. Concerning point c), the N*N$\rightarrow$NN cross sections were used as the free parameters necessary to reproduce the experimental results on ^{238}U. As shown in Fig. 5 this model is able to reproduce fairly well the non resonant behavior of the Frascati data, assuming an interaction cross section of about 20 mb for the Δ (which is in good agreement with the experimental values inferred from inverse reactions) and one of about 80 mb for the D_{13} and F_{15} resonances. Furthermore, the model clearly overstimates both the Frascati results near 1 GeV and the Regge behavior of the Daresbury data above 1.8 GeV.

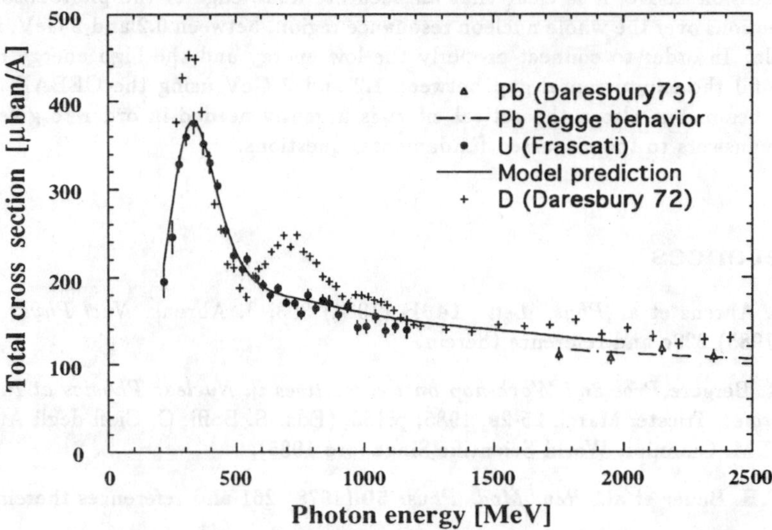

Figure 5: Comparison between the Frascati data on ^{238}U, the Daresbury data on deuteron and lead, and the prediction of a model based on standard nuclear effects.

Therefore, to account for this reduction of the strength, one has to resort to other causes, such as the onset of the shadowing effect or perhaps the damping of the excitation of higher resonances in the nuclear medium. This damping might be more effective for deformed resonances like $D_{13}(1520)$ and $F_{15}(1680)$ which correspond to state of nonzero orbital angular momentum, than for spherical ones like the Δ [$P_{33}(1232)$], which is excited by a simple quark spin flip. According to the VMD model, the propagation lenght λ of the hadronic fluctuation of the real photon is proportional to the photon energy k and inversely proportional to the square of the hadronic mass μ :$\lambda = 2k\ /\ \mu^2$. Hence, for example, a ρ-meson intermediate state of 10 GeV has a coherence lenght of about 7 fm, comparable to the size of a nucleus and therefore large enough to induce shadowing; whereas a shadowing-like effect at energy as low as 1 GeV cannot be attributed to vector mesons, but perhaps only to a lower mass nonresonant hadronic state made up of two pions with the same quantum numbers as the photon. Moreover, the low-and high energy regions are connected by sum rules and this connection has been used to establish constraints for the integrated total photonuclear cross section in the nucleon resonance region by taking account of the behavior of photon interactions at asymptotic energies. In particular, Weise[14] has shown that one can reconcile the data for the enhancement factor below pion threshold with those for the shadowing effect at high energies with a dispersion-relation approach, and has proposed that some non- negligible shadowing effects in nuclei should manifest themselves in the nucleon-resonance region. From the discussion above, it is clear that an accurate knowledge of the photoabsorption cross sections over the whole nucleon resonance region, between 0.2 and 2 GeV, is very desirable. In order to connect properly the low energy and the high energy region, we will fill the experimental gap between 1.2 and 2 GeV using the CEBAF tagged photon beam, but also a theoretical effort is urgently needed in order to give more definite answers to the discussed fundamental questions.

References

1. J. Ahrens et al.,*Phys. Lett.* **146B** (1984) 303; J. Ahrens, *Nucl Phys.* **A466** (1985) 229c and reference therein.

2. R. Bergère,*Proc.2nd Workshop on Perspectives in Nuclear Physics at Int. Energies*, Trieste, March 25-29, 1985, p.153, (Eds. S. Boffi, C. Ciofi degli Atti and M.M. Giannini, World Scientific,Singapore 1985).

3. T.H. Bauer et al., *Rev. Mod. Phys.* **50** (1978) 261 and references therein.

4. T.A. Armstrong et al.,*Phys. Rev.* **D5** (1972) 1640.

5. T.A. Armstrong et al.,*Nucl. Phys.* **B41** (1972) 445

6. E.A. Arakelyan et al., *Sov. J. Nucl. Phys.* **38** (1983) 589.

7. E.A. Arakelyan et al., *Sov. J. Nucl. Phys.* **52** (1990) 878.

8. N. Bianchi et al.,*Phys. Letters* **B299** (1993) 219 .

9. N. Bianchi et al., *Phys. Letters* **B309** (1993) 5.

10. M. Anghinolfi et al., *Phys. Rev.* **C47** (1993) R922.

11. N. Bianchi et al., submitted to *Phys. Rev. C.*

12. T. Frommhold et al., *Phys. Lett.* **B295** (1992) 28.

13. L. Kondratyuk et al., in preparation.

14. W. Weise, *Phys.Rev. Lett.* **31** (1973) 773 and *Phys.Rep.* **2** (1974) 53.

Inclusive electron scattering from an oxygen and argon jet target

M.Anghinolfi, P.Corvisiero, A.Longhi, L.Mazzaschi, V.Mokeev,

G.Ricco, M.Ripani, M.Taiuti, A.Teglia, A. Zucchiatti

INFN Genova

Via Dodecaneso 33 -16148 Genova,Italy

N.Bianchi, E.DeSanctis, A.Fantoni, V.Lucherini, V.Muccifora,

E.Polli, A. Reolon, P. Rossi

INFN Frascati

Via E. Fermi, 00044 -Frascati,Italy

Inclusive electron scattering experiments on complex nuclei give informations on general properties of the nucleus and on possible modifications of the baryon structure in the nuclear environment. Data collected in different experiments with different nuclei and transverse-longitudinal separation performed in the quasi-elastic and dip regions up to 550 MeV/c momentum transfer[1] indicate that final state interactions, nuclear correlations and meson exchange currents must be considered in order to reproduce the observed absorption strength. In this work we report on an inclusive electron scattering experiment using an oxygen and argon jet target placed in the ADONE storage ring at Frascati. Spectra have been collected at different angles and energies for momentum transfer up to 800 MeV/c and excitation energies beyond the Δ resonance. The preliminary results are presented and a special analysis of the experimental data has been developed to single out nuclear matter effects in the high energy transfer region.

The experiment was performed in ADONE, the Frascati storage ring, using electrons from .5 up to 1.5 GeV scattered from a jet target placed in a straight section[2]. Radiative corrections depending on the square of thickness were negligible due to the extremely low values of the jet density, multiple scattering of the electrons in the target was absent as well and, since the target was operating under the vacuum of the machine, no window background subtraction was needed. The scattered electrons were measured by a scintillation detector composed by a front part which allowed mass separation and angular definition, and by a rear part consisting of a basket of 20 BGO crystals to measure the energy of the scattered electrons and to improve their separation from proton and pion background. This calorimeter , described in detail in ref.3 has a moderate energy resolution but still sufficient to measure quasi-elastic and Δ peaks; since no magnetic field was present, the scattered electron spectrum was simultaneously detected from threshold to quasi-elastic peak, minimizing systematic errors.

Luminosity was determined using a very simple system to detect Moller electrons scattered at 30^0 with respect to the incoming beam : the cross section of this process is perfectly known, it has a large value so that a good monitoring counting rate was easily achieved. This system has been found to be a reliable relative monitor.

Radiative corrections have been applied to all data: a standard procedure has been used for the emission of soft photons or hard photons before scattering (Schwinger, elastic and continuum)[4],while a specific code was instead developed to calculate the correction relative to the emission of hard photons after scattering[5].

Inclusive spectra have been collected at different initial energies and scattering angles. Measurements have been performed at forward angle (32^0.) from .5 GeV up to 1.5 GeV incident energy , at 37^0 for calibration purposes and at backward angles where the transverse component of the absorption dominates. We did not perform longitudinal-transverse separation; however we chose energies and angles in order to have, in some cases, the same three-momentum transfer in the quasi-elastic peak both at forward and backward angles. In this approach one is able to compare some models to the experimental results, looking first at the dominant transverse component at the backward angle and then at the complete nuclear response in the forward direction where both longitudinal and transverse strengths are present.

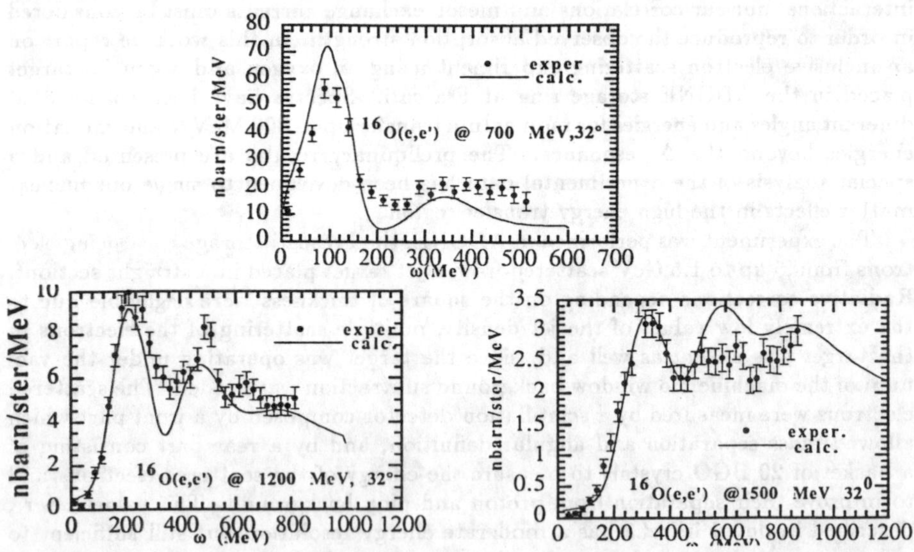

fig1: inclusive spectra in ^{16}O at 32^0 and different initial beam energies: the continuous curve is the result of the calculation discussed in the text.

Some examples of (e,e') data on ^{16}O are reported in fig. 1: the spectra, collected at 32^0 at different initial energies, are reported as a function of the energy transfer. At these momentum transfer the one-nucleon interaction is expected to dominate: the cross section shows in fact two main structures, the quasi-elastic peak and the Δ resonance. However, the cross section on nucleons bound in nuclei is expected to show medium modifications of both resonance and non resonance production mechanisms. To see the nuclear matter effects we compared the experimental data with the calculation where only Fermi-motion and the difference on virtual photon absorption in proton and neutron have been taken into account. Following the prescription of ref [6], the W1 and W2 structure functions in the nucleus were obtained by folding with a realistic momentum distribution given by Ciofi degli Atti[7] and summing proton and neutron contributions. A parametrisation [8] of the experimental electron scattering cross section on free nucleon is assumed as an input quantity : the comparison of this calculation to the experimental data can give therefore an estimation of any effect other than Fermi motion.

This evaluation, reported as a continuous curve in fig. 1, reproduces the general behavior of the experimental data. The strength and the width of the quasi-elastic peak are well accounted for, only weak medium modification being present at low momentum transfer where FSI are expected to give a sizeable contribution.

At higher energy transfer, in the region of the dip, the calculation is not able to reproduce the experimental data at least for beam energies lower than 1.2 GeV. Within a Fermi gas model, the one-body contribution from quasi free nucleon knock out and Δ excitation has a sharp cut-off in energy transfer. In our calculation, indeed, we use a more realistic momentum distribution with a high momentum tail which should move some strength from the quasi elastic peak to this region. It seems, however, that this momentum distribution is not sufficient to fill-in the strength in the dip region : probably two-body effects such as meson exchange currents can account for this difference.

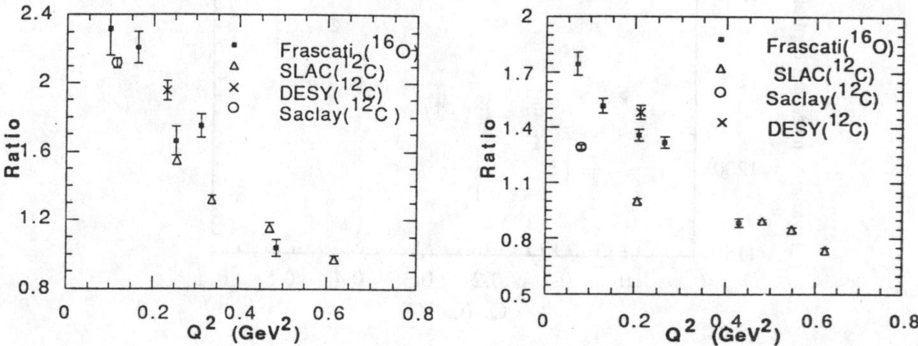

Fig 2: the ratio of the experimental data to the calculation in the dip (a)) and in the Δ (b)) region; our values are plotted together with results from SLAC [9], DESY[12] and Saclay [10]. All data have been measured in forward direction.

118

In fig. 2a the ratio between our experimental cross section integrated in the dip region from W=1.05 GeV to W=1.15 GeV and the calculation, integrated as well, is reported as a function of Q^2 : the strength is reproduced at $Q^2 = 0.5 GeV^2$, a strong enhancement of the experimental cross section being observed at lower momenta.

Comparing the calculation to the experimental cross section in the Δ region, we observe that Fermi motion alone can not account for both strength and width of the resonance. The same ratio defined above was calculated in the Δ region and reported in fig 2b: also in this case a Q^2 dependence is observed. This behaviour with Q^2 both in the dip and Δ region can not therefore be explained only by the broadening of the single nucleon cross section due to Fermi motion, suggesting that it is probably necessary to invoke a specific nuclear background contribution.

For the quasi-elastic peak a fixed shift (25 MeV) with respect to the free proton is observed due to the removal energy, while there is an indication of a Q^2 dependence of the Δ centroid. The cross section, as a function of the invariant mass, has been fitted with a gaussian of 250 MeV FWHM and a continuous background, the centroid being a free parameter. The value of the centroid as a function of Q^2 is reported in fig. 3 and compared to previous measurements obtained in different laboratories [9,10] with different nuclei. The position change with Q^2 seems confirmed by our data: it is , however, difficult to draw conclusions about the observed shift before a quantitative description of the reaction mechanisms is made. As a comparison , the peak position at Q^2=0 obtained from the universal curve of the photoabsorption data in nuclei is also reported[11]: the sharp discontinuity observed at low Q^2 may indicate the onset of some longitudinal background.

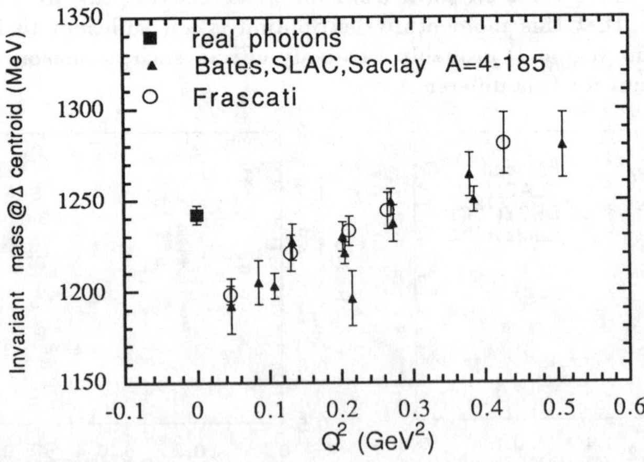

fig. 3: invariant mass of the Δ-region peak vs. Q^2 for our data compared with results of refs. 9 and 10 in different nuclei

As a conclusion,we have measured the inclusive (e,e') cross section in oxygen and argon at several values of the initial beam energy and scattering angles for three momentum transfer up to 800 MeV/c and excitation energy beyond the Δ resonance. An unconventional detection apparatus was used, allowing measurement of the electron spectra with moderate energy resolution, negligible pion contamination and reduced radiative corrections.

The comparison of the experimental data to a calculation where only Fermi motion has been taken into account indicates that pronounced nuclear matter effects are present particularly at low Q^2 where an excess of strength in the dip and in the Δ region is observed; a Q^2 dependence of the Δ peak centroid has also been found in the collected spectra of this experiment.

References

1. For an extensive summary see M.Giannini , Rep. Prog.Phys 52(1989) 1023.

2. M.Taiuti et al., Nucl. Inst. and Meth. A297(1990) 354.

3. M.Anghinolfi et al., Nucl. Inst. and Meth. A324(1993) 191.

4. L.Mo,Y.Tsai, Rev. Mod. Phys. 41(1969) 205.

5. A.Teglia: Thesis ,umpublished

6. W.Atwood et al., Phys. Rev. D7(1973) 773.

7. C.Ciofi degli Atti et al., Phys. Rev. C43(1991) 1155 and private comm.

8. J.Brasse et al., Nucl. Phys. B110(1976) 413.

9. R.M.Sealock et al., Phys. Rev. Let. 62(1989) 1350.

10. P.Burreau et al., Nucl. Phys. A402(1983) 515.

11. J.Ahrens, Phys. Lett. 146B (1984) 303.

12. F.H.Heimlich et al.,Nucl. Phys. A231(1974) 509.

HADRON STRUCTURE
AND HARD EXCLUSIVE PROCESSES

VICTOR CHERNYAK

Institute of Nuclear Physics,
630090 Novosibirsk-90, Russia

1) General properties of hard exclusive processes
2) QCD sum rules and properties of hadron wave functions
3) Comparison with the experimental data
4) Conclusions

1. GENERAL PROPERTIES OF HARD EXCLUSIVE PROCESSES

Because hadrons are composed of quarks and are always on the mass shell, there are two essentially different scales even in the hard amplitudes (i.e. those with large momentum transfers): a) The short distance scale, $\simeq 1/Q$, $Q \to \infty$, characteristic for the hard kernel of the amplitude which describes a transfer of large momenta between constituents and contains propagators far off-shell; b) The large distance scale, $\simeq 1/\mu_o$, characteristic for the nonperturbative confining interaction of initial and final near mass shell constituents which binds them into hadrons. This part of the amplitude is described by the hadron wave functions.

Therefore, the first problem in describing hard exclusive hadronic amplitudes is to separate properly, when possible, the contributions originating from these two scales. Such a separation is of crucial importance in QCD because the hard kernels (which vary from process to process) are well under control within the QCD perturbation theory, and all our ignorance of the nonperturbative QCD dynamics is isolated then in the hadron wave function which is one and the same for a given hadron participating in any process.

The modern approach to calculation of hard exclusive processes in QCD has been developed in papers [1-6] (for reviews see [7-11]). The general scheme, which has emerged for finding the leading term of any hard exclusive amplitude, looks as follows:

1) Each hadron is replaced by a minimal set, $n_{min} = 2$ for mesons and $n_{min} = 3$ for baryons, of suitable quarks.

2) Each quark can be considered as being massless, on the mass shell, having zero momentum components transverse to the momentum of its parent hadron and carrying a fraction x_i of the large hadron 4-momentum, $\Sigma x_i = 1$.

3) The amplitude for such a set of collinear initial and final quarks, i.e. a hard

kernel, is calculated in accordance with the QCD perturbation theory rules. Neglecting perturbative loop corrections, it can be found from a set of connected Born diagrams.

4) The hard kernel is multiplied by the leading twist hadron wave functions, $\phi_j(x_i)$, and the whole expression is integrated over all independent momentum fractions.

5) Because of loop corrections, wave functions begin to run logarithmically with Q^2, analogously to the deep inelastic structure functions. This evolution is very slow and well under control at, say, $Q^2 \geq 1\,GeV^2$. Wave functions of all hadrons tend to the same universal form: $\phi_j(x_i) \rightarrow \phi^{perturb}(x_i)$ in the formal limit $Q^2 \rightarrow \infty$, $\phi^{perturb}(x_i) = 6\,x_1 x_2$ for mesons and $\phi^{perturb}(x_i) = 120\,x_1 x_2 x_3$ for baryons. However, because the evolution is very slow, if the wave function $\phi(x, Q^2 \simeq 1\,GeV^2)$ (which is determined mainly by nonperturbative interactions) differs significantly from $\phi^{perturb}(x)$, then this difference will persist up to enormous values of Q^2.

The above simple recipe can be spoiled by:

a) The power-like infrared singularities (revealing themselves as power divergences of integrals over x_i) which increase the asymptotia by powers of Q. There are no such infrared singularities in the three-particle amplitudes (form factors, two-particle decays) but, as was first emphasized by P.V.Landshoff [12], there are such singularities in the hadron-hadron scattering amplitudes (see below).

b) Selection rules which lead to zero coefficients of the leading terms and, therefore, to a suppression of the asymptotia by powers of Q in comparison with the dimensional counting.

The selection rules for exclusive processes, [1], follow from the facts that: i) The leading twist hadron wave function is the $L_z = 0$ component of the total hadron wave function; ii) The quark helicity is conserved in QCD at high energies.

As a result, the hadron helicity is equal to a sum of the quark helicities in the leading twist components of hadron wave functions, and the sum of hadron helicities is conserved in the hard process. The general formula for the hadron form factors, both elastic and inelastic, has the form (in the Breit frame) [1]:

$$< p_2,\, s_2,\, \lambda_2 \,|\, J_\lambda \,|\, p_1,\, s_1\, \lambda_1 > \, \sim (1/Q)^{|\lambda_1 - \lambda_2| + 2n_{min} - 3}, \tag{1}$$

where p, s, λ are the hadron momentum, spin and helicity, J is the vector or axial vector current with the helicity λ, and $\lambda = \lambda_1 + \lambda_2 = 0, \pm 1$ in the Breit frame. What is important, is that the asymptotic behaviour is independent of hadron spins and only helicities are the relevant quantum numbers.

A useful consequence of the general Eq.(1) for the $e^+ e^-$ annihilation into two hadrons has been pointed out by S.J.Brodsky and G.P.Lepage [13]. Because only the mesons with $\lambda = 0$ and the baryons with $|\lambda| = 1/2$ can be produced in the leading twist at high Q^2, see Eq.(1), the corresponding angular distributions will be of the form: $\sim \sin^2 \theta$ and $\sim (1 + \cos^2 \theta)$ respectively.

2. PROPERTIES OF HADRON WAVE FUNCTIONS

The leading twist pion and $\rho-$ meson wave functions are defined by the matrix elements:

$$< 0|\, \bar{d}(z)\, I\, \gamma_\mu \gamma_5\, u(-z)\, |\pi(p)> = i\, f_\pi\, p_\mu\, \phi_\pi^A(zp,\, z^2 \sim 0) + ...$$

$$< 0|\, \bar{d}(z)\, I\, \gamma_\mu\, u(-z)\, |\rho_L(p)> = f_\rho^V\, (\, M_\rho\, \epsilon_\mu^L \simeq p_\mu)\, \phi_\rho^V(zp,\, z^2 \sim 0) + ...$$

$$< 0|\, \bar{d}(z)\, I\, \sigma_{\mu\nu}\, u(-z)\, |\rho_T(p)> = f_\rho^T\, (\epsilon_\mu^\perp\, p_\nu - \epsilon_\nu^\perp\, p_\mu)\, \phi_\rho^T(zp,\, z^2 \sim 0) + ...$$

$$\phi_i(zp) = \int_0^1 dx_1\, dx_2\, \delta\, (1 - x_1 - x_2)\, e^{i\, z\, (x_1 p)}\, e^{-i\, z\, (x_2 p)}\, \phi_i(x_1,\, x_2),$$

$$\phi_i(zp = 0) = 1\,, \qquad I = exp\,\{i\, g \int_{-z}^{z} d\,\sigma_\mu\, A_\mu(z)\}\,. \tag{2}$$

The dots in Eq.(2) denote higher twist terms which give only power corrections to the hard amplitudes (as well as taking account of the z^2-dependence of wave functions), because $p \sim Q \to \infty$, $z \sim 1/Q \to 0$. In Eq.(2): x_1 and x_2 are momentum fractions carried by two quarks inside the hadron, $p_1 \simeq x_1 p$, $p_2 \simeq x_2 p$, and the wave function $\phi_i(x_1, x_2)$ determines the distribution of the hadron momentum between quarks. The leading twist wave function of any meson with the helicity $\lambda = 0$ or $|\lambda| = 1$ will have the same form.

Expanding Eq.(2) in z one has (for the pion, for example):

$$< 0|\bar{d}(0)\, \hat{z}\gamma_5\, (z_\mu\, \overleftrightarrow{D}_\mu)^n\, u(0)|\pi(p)> = if_\pi\, (zp)^{n+1} \int_{-1}^{1} d\xi\, \xi^n\, \phi_\pi^A(\xi),$$

$$\xi = x_1 - x_2,\quad x_1 + x_2 = 1,\quad z^2 = 0,\quad D_\mu = i\partial_\mu + gA_\mu,\quad \hat{z} = z_\mu\gamma_\mu. \tag{3}$$

Therefore, if you know these matrix elements then you know the wave function moments and can restore, in principle, the wave function itself.
* The method of QCD sum rules, [14], allows one to find out the values of such matrix elements. Let us sketch the corresponding procedure (see [9,15] for details). We considered the correlator:

$$I_{n0}(z, Q) = i \int dx\, e^{iQx}\, < 0|T\, J_n(x)J_0^+(0)|0> = (zq)^{n+1}\, I_{n0}(Q^2),$$

$$J_n(x) = \bar{d}(x)\hat{z}\gamma_5\, (z_\mu\, \overleftrightarrow{D}_\mu)^n\, u(x),\quad \overleftrightarrow{D} = \overrightarrow{D} - \overleftarrow{D},\quad z^2 = 0, \tag{4}$$

and calculated its asymptotic behaviour at large Q^2. Besides the perturbative contributions, there are nonperturbative ones, proportional to various quark and gluon condensates in the QCD vacuum, and there exists a well developed technique for calculation of such contributions. Finally, this is a special kind of the multipole expansion: a small system (of the size $\sim 1/Q$) composed of two currents emits long wavelength ($\sim 1/\mu_o$) quark and gluon fields which are averaged then over their fluctuations in the physical vacuum.

So, on the one hand, we can calculate such correlator from QCD. On the other hand, its spectral density contains the contribution of the meson we are interested in. Finally, the dispersion relation connects these two parts and allows us to find out the value of the meson contribution. A typical sum rule has the form:

$$\frac{1}{\pi} \int_0^\infty ds\, e^{-s/M^2} Im I_{n0}(s) = \frac{3\,M^2}{4\pi^2(n+1)(n+3)} + C_g \frac{<0|(\alpha_s/\pi)G^2_{\mu\nu}|0>}{12M^2} +$$

$$C_q \frac{16\pi}{81} \frac{<0|\sqrt{\alpha_s}\bar{u}u|0>^2}{M^4} + ...,\tag{5}$$

where C_g, C_q, ... are calculable numbers, M is a free parameter used for a fitting procedure, and the values of the quark and gluon vacuum condensates are known from the phenomenology. For the pion [1] [15]:

$$\frac{1}{\pi} Im\, I_{n0}(s) = f_\pi^2 < \xi^n >_\pi \delta(s) + f_A^2 < \xi^n >_A \delta(s - m_A^2) + \theta(s - s_n)\frac{3}{4\pi^2(n+1)(n+3)},$$

$$C_g = 1, \quad C_q = (11 + 4n), \quad < \xi^n > \equiv \int_{-1}^1 d\xi\, \xi^n\, \phi_\pi^A(\xi).\tag{6}$$

Here, the pion and the a_1 meson contributions are explicitly separated in the spectral density and s_n is the effective threshold for a perturbative continuum in a given channel. Because the right hand side of Eq.(5) is known, one can extract the values of the pion and the a_1−meson wave functions moments, $< \xi^n >_\pi$, $< \xi^n >_{a_1}$, using a special fitting procedure in M^2.

The resulting qualitative picture is as follows. The perturbation theory contribution corresponds to the universal perturbative wave function: $\phi^{perturb}(\xi) = 0.75\,(1 - \xi^2)$. The nonperturbative contributions correct for a difference between the true wave function and $\phi^{perturb}$ and depend strongly on the channel considered. Qualitatively, if the perturbative and non-perturbative contributions are of the same sign, then the true wave function will be wider than ϕ^{pert}, and vice versa. Besides, in the sum rules considered the effect of the quark condensate is much more important

---

[1] For the $\rho_{\lambda=0}$: $C_g = 1$, $C_q = (-7 + 4\,n)$; for the $\rho_{|\lambda|=1}$: $C_g = (n-1)/(n+1)$, $C_q = (4 - 4\,n)$.

usually than the gluon one. [2] The nonperturbative contributions are positive and especially large for the pion wave function (see Eq.(6)), so that the distribution of two quarks inside the pion in the relative longitudinal momentum is much wider than in $\phi^{perturb}$ and the pion momentum is rarely divided equally between two quarks: the largest part of the whole momentum is carried by one quark. A similar result was obtained by M. Lavelle from independent sum rules [16].

Duality teaches us that the total (integrated) hadronic contribution to the correlator is the same as the perturbative one. We observe, however, a strong restructuring of the relative contributions in the local spectral density: the pion wave function moments are much larger than the perturbative ones, while those of $a_1^{\lambda=0}(1260)$ are much smaller.

* Before proceeding further, let us point out that a method to improve the standard QCD sum rules has been proposed recently by S.Mikhailov and A.Radyushkin [17]. The idea is to keep as a whole some non-local vacuum condensates without expanding them in powers of small relative distances as it is usually done. From the viewpoint of the operator product expansion, this corresponds to summing up a subset of higher-dimension power corrections.

There are a number of problems, however, inherent to this approach. First, only a certain subset of power corrections is summed up, while all other power corrections are ignored, in spite of the fact that there are not any reasons to neglect them. Moreover, the set of corrections retained is not even gauge invariant. A simple QED analogy is to account for the loop corrections in the electron and photon propagators and to neglect them simultaneously in the vertex.[3] Second, these nonlocal condensates are unknown at present so that the authors are forced to introduce arbitrary models for them, and the results are sensitive to the models used.

V.M. Braun and I.E. Filyanov have written a sum rule for the value of the pion wave function at the middle point $x_1 = x_2 = 1/2$ [18]. The result looks like: $\phi_\pi^A(x_1 = x_2 = 1/2) \sim 1$, in strong contradiction, for instance, with the model form proposed in [15]. A closer inspection shows, however, that the above value of $\phi_\pi^A(x_1 = x_2 = 1/2)$ obtained in [18] results from a wrong application of spectral representations.

* As for the ρ-meson wave functions properties, [19], the nonperturbative contributions into the corresponding sum rules are positive in the $\rho_L = \rho_{\lambda=0}$ case (but more

---

[2] For instance, let us differentiate the Eq.(5) in M^{-2}. On the one hand, the pion contribution will disappear from the spectral density. On the other hand, the quark condensate contribution will become negative. This indicates that the moments of the a_1 meson wave function are smaller than those of $\phi_{pert}(\xi)$, and this is indeed the case. The fit gives: $< \xi^2 >^\pi_{|\bar{M}^2=1.5\,GeV^2} \simeq 0.38$, $< \xi^2 >^{a_1}_{|\bar{M}^2=1.5\,GeV^2} \simeq 0.10$, while $< \xi^2 >^{perturb} = 0.20$.

[3] On the other hand, it is clear that to account properly for all power corrections in the correlator is the same as to write the corresponding Schwinger-Dyson equations, and the problem becomes then intractable.

mild in comparison with the pion) and negative in the $\rho_T = \rho_{|\lambda|=1}$ case, (see above), so that their wave functions are respectively wider and narrower than $\phi^{perturb}(x)$.

* In comparison with their corresponding nonstrange partners, the wave functions of strange mesons are somewhat narrower and asymmetrical: the mean momentum fraction carried by the strange quark is always larger somewhat than those carried by u or d quarks. Clearly, for the charmed and beauty mesons the light quark carries only a small, $\sim \mu_o/M_Q$, fraction of the hadron momentum.

* The octet baryon wave functions differ greatly from their universal perturbative form: $\phi^{perturb}(x_1, x_2, x_3) = 120\, x_1\, x_2\, x_3$, $x_1 + x_2 + x_3 = 1$, which describes a smooth symmetric distribution of the baryon momentum between its three constituents. For the proton [20-23], $\simeq 60\%$ of its momentum is carried out by one u-quark with its spin parallel to the proton spin, while each of two other quarks carries $\simeq 20\%$ (and a similar situation occurs for the Σ and Ξ hyperons [24]). In contrast with the nucleon, the $\Delta^+_{|\lambda|=1/2}$ wave function is nearly completely symmetric and differs insignificantly from $\phi^{perturb}$ [25,26]. Other decuplet baryon wave functions also do not have large asymmetry [27].

* While the leading twist wave functions, $\phi_i(x_j)$, determine the distribution of quarks in longitudinal momenta inside hadrons at $p_z \to \infty$, the higher twist wave functions are connected with the distribution of quarks in transverse momenta and with many-particle components of hadron wave functions. Some of them have also been investigated with the result that their form also differ greatly, in general, from the corresponding perturbative form, [28-31].

3. COMPARISON WITH THE EXPERIMENT

Happily, probabilities of hard exclusive processes appeared to be highly sensitive to the precise form of the hadron wave functions and are changed by orders of magnitude when the wave function form is varied significantly. This allows one to check in detail the wave function properties. Because the theory is unable at present to calculate power corrections reliably, one calculates usually the leading term only and compares it with the experiment. Such a comparison shows simultaneously (within the accuracy of the leading twist calculations) the role of neglected corrections.

A large number of processes have been calculated and compared with experiment when possible. These include:

1) hadron form factors, both elastic and inelastic;
2) heavy quarkonium strong and electromagnetic decays;
3) weak decays of heavy mesons;
4) "$\gamma\,\gamma \to$ two hadrons" large angle production;
5) high $p_\perp$ prompt meson production: $e^+ e^- \to M_\perp + X$, $\pi N \to M_\perp + X$;
6) large angle Compton scattering: "$\gamma N \to \gamma N$";
7) photo (electro) production: "$\gamma N \to M N$";
8) hadron-hadron large angle scattering.

Unfortunately, there is no place to dwell on details so that the discussion will be

126

very short.

The "good examples" (i.e. those for which the leading twist calculations are in reasonable or even good agreement with the experiment) include:

1) The proton and neutron form factors in a spacelike region. The experimental data are presented in [32-34] for the proton and in [35-36] for the neutron. The theoretical calculations using the nucleon wave functions obtained from the QCD sum rules are described in [20, 24, 37, 38] and are in good agreement with the data, both in sign and magnitude.

2) A large set of data on inelastic proton form factors in the spacelike region (up to $Q^2 \leq 21\, GeV^2$) is presented by P. Stoler in [39]. In agreement with the general formula Eq.(1), $N \to N^*(1535)$ and $N \to N^*(1680)$ transition form factors (spin averaged) show the same dipole behaviour as the elastic nucleon form factor. As for the $N \to \Delta(1232)$ form factor, it decay more quickly. But this is just what has been predicted by the theory [24, 26, 40]. The reason is that due to specific properties of the nucleon and $\Delta(1232)$-wave functions predicted by the QCD sum rules (see above), the coefficient of the leading term appeared to be very small: $\simeq (few)\, 10^{-2}\, GeV^4$ for $N \to \Delta(1232)$ in comparison with $\simeq 1\, GeV^4$ for $N \to N$. Therefore, power corrections dominate this form factor up to very large Q^2, and it decays more quickly at available Q^2 values.

Besides, P. Stoler has checked the helicity selection rules described by Eq.(1). It is interesting, that although the data are for $Q^2 \leq 3\, GeV^2$ only, there is agreement with the selection rules: $(\lambda = 1/2 \to \lambda = 1/2)$ form factors are not suppressed and show the dipole behaviour, while $(\lambda = 1/2 \to \lambda = -3/2)$ form factors decay more quickly.

3) The charmonium decay widths: i) $^3P_0 \to \bar\pi\pi$, $\bar K K$, $\bar\rho\rho$; ii) $^3P_1 \to \bar N N$; iii) $^3P_2 \to \bar\pi\pi$, $\bar K K$, $\bar\rho\rho$, $\bar N N$; iv) $J/\Psi \to \bar p p$, $\bar n n$, $\gamma\pi$, $\rho\pi$ [4] have been calculated using various hadron wave functions obtained from the QCD sum rules [9, 28], all in

---

[4] The decay $J/\Psi \to \rho\pi$ is a higher twist process. It has been calculated in [28] in a standard way, i.e. by calculating the hard kernel and using the twist-three wave functions obtained from the QCD sum rules. In this respect, there is no qualitative difference with, for instance, the nucleon form factor calculations. The result agrees with the data and this shows that this decay can naturally be explained within a standard approach. The difficulty appears then with the $\Psi' \to \rho\pi$ decay which is heavily suppressed and can not be explained in a standard way. It has been speculated in [45] that there exists a $J^{PC} = 1^{--}$ gluonium state in a suitable vicinity of J/Ψ which enhances strongly the $(J/\Psi \to \rho\pi)$ width, while the $(\Psi' \to \rho\pi)$ decay is naturally suppressed because it is a higher twist process. We can not exclude, of course, a possible existence of such a gluonium state but would like to emphasize that either the above calculations [28] are in error (which can be easily checked), or this gluonium contribution will spoil then the agreement between the standard calculations and the data for J/Ψ. It is possible to scan at FNAL the $\bar p p$ annihilation in the vicinity of J/Ψ to check the existence of this gluonium state.

reasonable agreement with the data [43, 44]. This is a strong argument that the scale $q^2 \simeq 10 \, GeV^2$ is sufficiently large, even in the timelike region, for power corrections not to be dominant.

4) Large angle cross sections $\gamma\gamma \to \bar{\pi}\pi$, $K\bar{K}$. The experimental data are presented in [46, 47], and the theoretical calculations in [48-51]. Reasonable agreement with the data is achieved only for the pion and kaon wave functions obtained from the QCD sum rules.

5) Large angle Compton scattering. Unlike previous calculations [52], the resent results [53] are in agreement now with the experimental data [54-58].

6) The experimental data [59] on the high $\mathbf{p}_\perp$ ρ^o- meson production in $\pi N \to \rho_{|\lambda|=1} + X$, (being corrected for the $\eta^{'}$ production [60]), agree with the theoretical calculations [61] when the narrow $\rho_{|\lambda|=1}$ wave function obtained from the QCD sum rules is used.

The "bad examples" (i.e. those which disagree with the leading twist predictions or have no explanations) include:

1) The decay widths: $(J/\Psi \to \pi^+ \pi^-)$ and, especially, $(\Psi^{'} \to \pi^+ \pi^-)$ look too large to be described naturally by the pion form factor, [43].

2) The decay widths $(\Psi^{'} \to \rho\pi, \, K^* K)$ are highly suppressed, unlike $(\Psi^{'} \to \bar{p}p)$ and others, [43].

3) In the radiative decays $(J/\Psi \to \gamma f_2^{(\lambda)}, \gamma f_2^{'\,(\lambda)})$ the helicity zero states of the tensor mesons give the asymptotically leading contributions at $M_c \to \infty$ [62]. The experiment shows however [63] that the $|\lambda| = 2$ and, what is even more surprising, the $|\lambda| = 1$ decays are not really suppressed.

4) $Br\,(\eta_c \to \bar{p}p)$ is suppressed by the helicity selection rules but its experimental value is [43]: $(1.2 \pm 0.4) \cdot 10^{-3}$, which looks large in comparison with $Br\,(\chi_{c1} \to \bar{p}p) \simeq Br\,(\chi_{c2} \to \bar{p}p) \simeq 0.8 \cdot 10^{-4}$, [44], which are not helicity forbidden.

5) The measured $\gamma\gamma \to \bar{p}p$ large angle cross section [64-68] exceeds greatly the leading twist calculations [69].

6) There appeared new data [41] on the proton form factor in the timelike region, up to $q^2 \leq 13 \, GeV^2$. In comparison with the corresponding spacelike Q^2- values the form factor is twice as large. Besides, the dipole behaviour joins reasonably well the previous [42] and new data (but, unfortunately, there are no points to check a matching of the new and previous data).

It seems difficult to explain all of this in a simple way. If there is indeed the dipole behaviour, then the leading term dominates and the coefficient must be the same and can not differ two times. If power corrections are dominant at $q^2 \simeq 13 \, GeV^2$, why is there an approximate dipole behaviour? It is natural, of course, that the scale at which the asymptopia sets in is larger in a timelike region in comparison with a spacelike one. But the value $q^2 \simeq 13 \, Gev^2$ appears to be sufficiently large, taking into account that it is $Q^2 \simeq 5 \, GeV^2$ in a spacelike region and that many two-particle charmonium decays are well described by the leading twist terms (see above).

7) Clear understanding, even qualitative, of the hadron-hadron elastic scattering still remains an open problem. On the one hand, many elastic cross sections show the approximate dimensional scaling behaviour at moderate energies. On the other hand, this dimensional behaviour is clearly violated at very high energies and moderate momentum transfers. For instance, as emphasized by P.V. Landshoff, the ISR data for high energy elastic pp scattering have the behaviour: $d\sigma/dt \sim t^{-8}$ at $2\,GeV^2 \leq |t| \leq 10\,GeV^2$, which corresponds precisely to the infrared Landshoff mechanism [12] and exceeds the dimensional behaviour $d\sigma/dt \sim t^{-10}$.

Besides, there are highly non-trivial spin dependencies at moderate energies [70], and this looks surprising at first sight. The reason is that one can try to explain the corresponding spin dependence (absent at the leading term level for really hard processes) as being due to either power corrections to hard kernels, or large distance infrared contributions. But it is then unclear why the cross sections follow approximately the dimensional counting behaviour, which is specific only for the leading terms of hard kernels.

Also, some interesting regularities has been observed in an experiment [71]. It is seen that the processes that allow the quark interchange contributions at the Born level usually have larger cross sections.

All of these features remain unexplained.

4. CONCLUSIONS

At present, there are no understanding of interrelations between the naive constituent quark model and QCD. The difficulties originate, of course, from the complicated QCD dynamics at low energies: the confinement mechanism, the chiral symmetry breaking and a dynamical origin of quark masses, etc. Unfortunately, the notion of the constituent quark has never been clarified in its connection with the QCD: whether it is some coherent soliton-like state or something else.

The difficulties appear from the beginning when, inspired by the naive picture, one models the hadron state as a state with the smallest number of constituents in, say, the hadron rest frame. First, this leads to difficulties with the Lorenz invariance as the particle number is not a Lorenz invariant quantity in a relativistic field theory. States with different particle numbers mix under the general Lorenz transformations, and this mixing is not purely kinematical but is interaction dependent. However, to calculate various amplitudes with hadrons having different momenta we must do the Lorentz transformations.

Second, even in the rest frame, special reasons are needed for the valence state to be a reasonable approximation. For instance, the valence equal-time wave function is a good approximation for heavy nonrelativistic particles with a weak coupling. However, this approximation becomes increasingly poor as the coupling increases and the constituents become relativistic. But even the constituent quarks are relativistic inside the light hadrons. And it looks somewhat naive to expect that by a simple rotation of the quantization axis over $\pi/4$, i.e. using the light-front wave functions instead of the equal time ones, one will ensure dominance of the two-particle component of the meson wave function in such a complicated theory as QCD.

Let us emphasize that there are no problems either with the Lorenz invariance or with the fixed particle number states when one uses the wave functions determined by Eq.(2). It is only a kind of jargon that these wave functions are called two-particle ones sometimes, because the Heisenberg fields entering their definition do not correspond to a fixed particle number. But in some special situations this may become the case. For instance, the asymptotic behaviour of the hadron form factors in QCD was found in [1,2] using the operator expansions and the Lorenz and gauge invariant wave functions defined in Eq.(2). The minimal twist wave functions give the leading contributions at $Q^2 \to \infty$ in this approach. It was shown by S.J. Brodsky and G.P. Lepage, [5], that the same result is obtained within the approach which uses states quantized on the light front and having a definite particle number; the asymptotically leading contributions give those with a minimal particle number. But for this to be the case, one has to choose the special frame and special "physical" gauge because otherwise this simple interpretation will not be valid.

It was described above how the method of the QCD sum rules allowed us to avoid all the above described difficulties and to investigate the real properties of various hadron wave functions. These properties appeared to be highly nontrivial and versatile. It was recognized that:

1) The distribution of light quarks momenta inside the hadrons is wide, in general. This means that quarks are highly relativistic and the models with non-relativistic constituents can't be used seriously;

2) The wave function form depends essentially on the hadron quantum numbers and, in particular, on its helicity because the non-perturbative effects differ strongly in channels with different quantum numbers. This is in contrast with the naive SU(6) constituent quark model where there is only one common wave function for all hadrons entering the same SU(6) multiplet.

It remains unclear, however, what is the physics underlying the QCD sum rule results for the hadron wave functions. Why are the pion and $a_1^{\lambda=0}(1260)$ wave functions respectively much wider and much narrower than $\phi^{perturb}(\xi)$? Why is the proton wave function so asymmetrical while that of $\Delta^+_{|\lambda|=1/2}(1232)$ is nearly totally symmetrical? Why are the wave functions of the $\rho_{\lambda=0}$ and $\rho_{|\lambda|=1}$ mesons so different ?

As for applications, the simple-minded wave functions, like $\phi^{perturb}(x_i)$, fail by orders of magnitude to describe the data, while the above described wave functions obtained from the QCD sum rules gave a large number of successful predictions. [5] However, the typical accuracy of predictions is not high at present, and is within a factor of $\sim 1.5 - 2$ in probabilities.

Today, the main theoretical problem in this field is our inability to calculate

---

[5] Let us emphasize that there is no one free parameter in all these calculations.

reliably power corrections to the leading term contributions, not using the phenomenological models with free adjustable parameters. But on the other hand, the experimental data on various baryon form factors in a spacelike region, both elastic and inelastic, show that the asymptotic scaling sets in at a natural scale $Q^2 \sim (several)\, GeV^2$ and there are no indications that power corrections are dominant (except for special cases like $N \to \Delta(1232)$ transition). [6]

As was described above, the leading twist calculations gave a large number of successful predictions. There is a number of exceptions however, and each one deserves a careful study.

APOLOGIES

I apologize to all those authors whose papers or results were not mentioned due to a lack of space.

ACKNOWLEDGEMENTS

I would like to thank Professors Carl Carlson, Paul Stoler and Mauro Taiuti for organizing this very interesting Workshop and for their hospitality in Elba International Physics Centre.

---

[6]Of course, there is no universal scale at which the asymptopia sets in for different processes. It is natural to expect that this scale is larger, for instance, for the hadron-hadron scattering as compared with those in form factors, etc.

REFERENCES

1. V.L. Chernyak, A.R. Zhitnitsky, JETP Lett. **25** (1977) 510

2. V.L. Chernyak, V.G. Serbo, A.R. Zhitnitsky, JETP Lett. **26** (1977) 594; Yad. Fiz. **31** (1980) 1053,1068

3. G.R. Farrar, D.R. Jackson, Phys. Rev. Lett. **43** (1979) 246

4. A.V. Efremov, A.V. Radyushkin, Phys. Lett. **B 94** (1980) 245; Teor.i Matem.Fiz. **42** (1980) 147

5. G.P. Lepage, S.J. Brodsky, Phys. Lett. **B 87** (1979) 359; Phys. Rev. Lett. **43** (1979) 545, 1625(E); Phys. Rev. **D 22** (1980) 2157

6. A. Duncan, A.H. Mueller, Phys. Rev. **D 21** (1980) 1636; Phys. Let. **B 90** (1980) 159; Phys. Lett. **B 93** (1980) 119

7. V.L.Chernyak, Proc. XV-th LINP Winter School of Physics, Leningrad, 1980, v.**2**, pp. 65-155

8. A.H. Mueller, Phys. Rep. **C 73** (1981) 239

9. V.L. Chernyak, A.R. Zhitnitsky, Phys. Rep. **112** (1984) 173

10. S.J. Brodsky, G.P. Lepage, in "Perturbative QCD", ed. by A.H. Mueller, World Scient., Singapore, 1989

11. V.L. Chernyak, Preprint TPI-MINN-91/47-T, Minneapolis, 1991

12. P.V. Landshoff, Phys.Rev. **D 10** (1974) 1027

13. S.J. Brodsky, G.P. Lepage, Phys. Rev. **D 24** (1981) 2848

14. M.A. Shifman, A.I. Vainshtein, V.I. Zakharov, Nucl.Phys. **B 214** (1979) 385, 448

15. V.L. Chernyak, A.R. Zhitnitsky, Nucl. Phys. **B 201** (1982) 492

16. M.J. Lavelle, Z. Phys. **C 29** (1985) 203

17. S.V. Mikhailov, A.V. Radyushkin, Yad.Fiz. **49** (1989) 794; Yad.Fiz. **52** (1990) 1095

18. V.M. Braun, I.E. Filyanov, Z. Phys., **C 44** (1989) 157

19. V.L. Chernyak, A.R. Zhitnitsky, I.R. Zhitnitsky, Nucl.Phys. **B 204** (1982) 477

20. V.L. Chernyak, I.R. Zhitnitsky, Nucl.Phys. **B 246** (1984) 52

21. M.J. Lavelle, Nucl.Phys. **B 260** (1985) 323

22. I.D. King, C.T. Sachrajda, Nucl.Phys. **B 279** (1987) 785

23. V.L. Chernyak, A.A. Ogloblin, I.R. Zhitnitsky, Yad.Fiz. **48** (1988) 841

24. V.L. Chernyak, A.A. Ogloblin, I.R. Zhitnitsky, Yad.Fiz. **48** (1988) 1398, 1410

25. G.R. Farrar, H. Zhang, A.A. Ogloblin, I.R. Zhitnitsky, Nucl.Phys. **B 311** (1988) 585

26. C.E. Carlson, J.L. Poor, Phys.Rev. **D 38** (1988) 2758

27. A.A. Ogloblin, Yad. Fiz. **50** (1989) 189

28. V.L. Chernyak, A.R. Zhitnitsky, I.R. Zhitnitsky,
Sov.J.Nucl.Phys. **38** (1983) 645, 775; **41** (1985) 127, 284

29. V.G. Geshkenbein, M.V. Terent'ev, Yad.Phys. **40** (1984) 758

30. A.S. Gorsky, Yad. Fiz. **41** (1985) 430

31. V.M. Braun, I.E. Filyanov, Yad.Fiz. **52** (1990) 199

32. R.G. Arnold *et al.*, Phys. Rev. Lett. **57** (1986) 174

33. P.E. Bosted *et al.*, Phys. Rev. Lett. **68** (1992) 3841

34. A.F. Sill *et al.*, Phys. Rev. **D 48** (1993) 29

35. S. Rock *et al.*, Phys. Rev. Lett. **49** (1982) 1139; SLAC-PUB-5239, November 1991

36. A. Lund *et al.*, SLAC-PUB-5861, November 1992

37. C.R. Ji, A.F. Sill, R.M. Lombard-Nelson, Phys. Rev. **D 36** (1987) 1308

38. A. Schafer, Phys. Lett. **B 217** (1989) 545

39. P. Stoler, Phys. Rev. Lett. **66** (1991) 1003; Phys. Rev. **D 44** (1991) 73

40. C.E. Carlson, M. Gari, N.G. Stefanis, Phys. Rev. Lett. **58** (1987) 1308

41. T.A. Armstrong *et al.*, Phys. Rev. Lett. **70** (1993) 1212

42. G. Bardin *et al.*, Phys. Lett. **B 257** (1991) 514

43. Particle Data 1992

44. T. Armstrong *et al.*, Nucl. Phys. **B 373** (1992) 35

45. S.J. Brodsky, G.P. Lepage, San Fu Tuan, Phys. Rev. Lett. **59** (1987) 621

46. J. Boyer *et al.*, Phys. Rev. Lett. **56** (1986) 207

47. H. Aihara *et al.*, Phys. Rev. Lett. **57** (1986) 404

48. S.J. Brodsky, G.P. Lepage, Phys. Rev. **D 24** (1981) 1808

49. V.L. Chernyak, I.R. Zhitnitsky, Nucl. Phys. **B 222** (1983) 382

50. J.F. Gunion, D. Millers, K. Sparks, Phys. Rev. **D 33** (1986) 689

51. M. Benayoun, V.L. Chernyak, Nucl. Phys. **B 329** (1990) 285

52. G.R. Farrar, H. Zhang, Phys. Rev. **D 41** (1990) 3348

53. A.S. Kronfeld, B.Nizic, Phys. Rev. **D 44** (1991) 3445

54. M.A. Shupe *et al.*, Phys. Rev. **D 19** (1979) 1921

55. M. Jung *et al.*, Z. Phys. **C 10** (1981) 197

56. J. Duda *et al.*, Z. Phys. **C 17** (1983) 319

57. Y. Wada *et al.*, Nucl. Phys. **B247** (1984) 313

58. T. Ishii *et al.*, Nucl. Phys. **B 254** (1985) 458

59. M. Benayoun *et al.*, Phys. Lett. **B 183** (1987) 412

60. M. Benayoun, Private communication

61. M. Benayoun, P. Leruste, J.L. Narjoux, R. Petronzio, Nucl. Phys. **B 282** (1987) 653

62. V.N. Baier, A.G. Grozin, Yad. Fiz. **35** (1982) 1021

63. L.-P. Chen *et al.*, SLAC-PUB-5669, 1991

64. M. Althoff *et al.*, Phys. Lett. **B 130** (1983) 449; **B 142** (1984) 135

65. W. Bartel *et al.*, Phys. Lett. **B 174** (1986) 350

66. H. Aihara *et al.*, Phys. Rev. **D 36** (1987) 3506

67. H. Albrecht *et al.*, Z. Phys. **C 42** (1989) 543

68. B. Ong, CLEO coll. Proceedings of 7-th Meeting of the American Physical Society, 1992

69. G.R. Farrar, E. Maina, F. Neri, Nucl. Phys.**B 259** (1985) 702; Nucl. Phys. **B 263** (1986) 746 (E)

70. A.D. Krisch, Preprint UM-HE / 91-22, Michigan, 1991

71. B.R. Baller *et al.*, Phys. Rev. Lett. **60** (1988) 1118

FIXED ANGLE HIGH ENERGY ELASTIC SCATTERING AND PROBLEM OF HELICITY CONSERVATION

M.P.Chavleishvili

Bogolubov Laboratory of Theoretical Physics
Joint Institute for Nuclear Research, Dubna, Russia
E-mail: **chavlei@theor.jinrc.dubna.su**

Perturbative QCD and hard proton-proton scattering. There exists contradiction between perturbative QCD and experiment. This is connected with proton-proton scattering at high energies and large fixed angles. This is just the region where PQCD must works. But, one can say, that "the naiv PQCD" files here in the lowest orders of perturbation theory and it is difficult to wait for improving situation by calculating an anormous number of diagrams of higer order.

The point is that from PQCD was obtained the "helicity conservation" rule [3-5] which gives in particular for assymetry parameter A_{nn} the value 1/3 in the discussed region, and this predictions are in contradiction with experiments.

QCD: Helicity conservation. The use of perturbative QCD for discussed reactiomis based on the following assumptions:
– Factorisation property for proton-proton scattering.
– Simple connection between quark and proton helicities.
– No helicity changes in gluon quark vertex in asymptotical region.
   The result is the helicity conservation rule:

$$\lambda_1 + \lambda_2 = \lambda_3 + \lambda_4. \tag{1}$$

As mensioned above, this relation and its consequances are not in agreement with experiments.

Dynamic amplitudes. The helicity amplitudes which are convenient to describe spin-particle reactions have a clear physical meaning, and physical observables (polarization cross sections, asymmetries, etc.) are simply expressed via them. The corresponding equations are quita simple, but helicity amplitudes contain kinematic singularities and the conservation laws do not fulfill automatically – so kinematics and dynamics are not separated.

In [4,5] we considered an universal formalism for any binary processes with the particles of arbitrary spins and masses based on the principles of symmetry, invariance and conservation laws. Formalism can be used as a framework as the general basis for model building with guaranteed fulfillment of symmetry and conservation laws. This method is interesting for studying general characteristics ofnparticle reaction theory and it is also suits for exploring concrete processes.

We can determine the dynamic amplitudes for alastic processes

$$f_{\lambda_3\lambda_4,\lambda_1,\lambda_2}(s,t) = \left(\frac{\sqrt{-t}}{m+\mu}\right)^{-|\lambda-\mu|} \left(\frac{\sqrt{L^2+st}}{(m+\mu)^2}\right)^{-|\lambda+\mu|} \left(\frac{L}{(m+\mu)^2}\right)^{-2(s_1+s_2)} D_{\lambda_3\lambda_4,\lambda_1,\lambda_2}(s,t). \tag{2}$$

Here $L = [s - (m + \mu)^2][s - (m - \mu)^2]$.

As the connection between the helicity and dynamic amplitudes is one-to-one, every helicity amplitude for elastic scattering is expressed in terms of one dynamic amplitude. Hence it follows that all attractive features of the helicity amplitudes, a clear physical meaning, simple relations with observables, and equal dimensions, are also inherent in the dynamic amplitudes. The formalism of dynamic amplitudes is simple for low spins and remains such also for higher spins: the formalism is simple for any spins.

Kinematic hierarchy. At high energies and fixed angles the observables are espressed by dynamic amplitudes with definite kinematical factors. This kinematical factors in considered region in fact are small parameters in different povers. This gives us the hierarchy of contributions in observables sorted by small parameter. In ferst approximation such "kinematic hierarchy" gives us definite relations betveen observable quantities. For proton-proton scattering we obtain relations between asymetry parameters and making simple and natural assumption even a numerical value for asymptotic values of them. For asymetry parameter A_{nn} it seems that their experimental values just goes up to the obtained asymptotic number.

Proton-proton scattering. The connection between helicity and dynamic amplitudes for nucleon-nucleon scattering (expressing kinematic factors via s and θ can be write as folloving

$$f_1 = \frac{m^2}{p^2} D_1, f_2 = \frac{m^2}{p^2} D_2, f_3 = \cos^2 \frac{\theta_s}{2} D_3, f_4 = \sin^2 \frac{\theta_s}{2} D_4, f_5 = \frac{\sqrt{s}}{2m} \sin \theta_s D_5. \tag{3}$$

Asimmetry parameters [6] are expressed via dinamic amplitudes by relatios ($p = \sqrt{s - 4m^2}/2$):

$$\frac{d\sigma}{dt} A_{nn} = -Re[-\frac{m^4}{p^4} D_1 D_2^* + \frac{1}{2} \sin^2 \theta D_3 D_4^* + \frac{s}{2m} \sin^2 \theta \mid D_5 \mid^2] \tag{4}$$

$$\frac{d\sigma}{dt} A_{ss} = Re[\frac{m^4}{p^4} D_1 D_2^* + \frac{1}{2} \sin^2 \theta D_3 D_4^* + \frac{s}{2m} \sin^2 \theta \mid D_5 \mid^2] \tag{5}$$

$$\frac{d\sigma}{dt} A_{ll} = -\frac{1}{2} \{ \frac{m^4}{p^4} [\mid D_1 \mid^2 + \mid D_2 \mid^2] - \cos^4 \frac{\theta}{2} \mid D_3 \mid^2 - \sin^4 \frac{\theta}{2} \mid D_4 \mid^2 \} \tag{6}$$

In the high-energy large-fixed-angle region the helicity amplitudes are splitted into three classes in the order of smallness determined by the kinematic factors. So we obtain

$$\frac{\sqrt{s}}{2m} \sin \theta D_{1/2,1/2;1/2,-1/2} \gg \cos^2 \frac{\theta}{2} D_{1/2,-1/2;1/2,-1/2}) \sim \sin^2 \frac{\theta}{2} D_{1/2,-1/2;-1/2,1/2} \gg$$

$$\gg \frac{m^2}{p^2} D_{1/2,1/2;1/2,1/2} \sim \frac{m2}{p^2} D_{1/2,1/2;-1/2,1/2}. \tag{7}$$

or, in terms of helicity amplitudes:

$$f_{1/2,1/2;1/2,-1/2} \gg f_{1/2,-1/2;1/2,-1/2} \sim f_{1/2,-1/2;-1/2,1/2} \gg f_{1/2,1/2;1/2,1/2} \sim f_{1/2,1/2;-1/2,1/2}. \tag{8}$$

where "$a \gg b$" means that the contribution of b is suppressed relative to the contribution of a in the observables.

For proton-proton scattering at $\theta_{c.m} = 90°$ we have from $s - u$ crossing symmetry that

$$f_{1/2,1/2;1/2,-1/2}(90°) = 0, f_{1/2,-1/2;1/2,-1/2}(90°) = f_{1/2,-1/2;-1/2,1/2}(90°) \tag{9}$$

Taking into account dominated amplitudes we get for asymmetries

$$A_{nn} = A_{ss} = \frac{2Re f_{1/2,-1/2;1/2,-1/2}f^*_{1/2,-1/2;-1/2,1/2}}{\mid f_{1/2,-1/2;1/2,-1/2}\mid^2 + \mid f_{1/2,-1/2;-1/2,1/2}\mid^2} \to 1. \tag{10}$$

In fact the kinematic hierarchy assumption here is the result of the assumption

$$\frac{1}{2}\sin^2\theta D_3 D_4^* \gg \frac{m^4}{p^4}D_1 D_2^* \tag{11}$$

__Conclusion.__ QCD gives for thise quantities the value 1/3 [1-3]. The massive quark model [7] gives $A_{nn} = 0.97; A_{ss} = -0.01$.

For experiment we have at $90°$, for $A_{nn}(p_\perp^2 (GeV)^2)$ the following values [8]

$$A_{nn}(3.81) = 0.26; A_{nn}(4.79) = 0.52; A_{nn}(5.56) = 0.59. \tag{12}$$

So we see that there is the tendency of rising up the data with $p_\perp^2$ in agreement with our asimptotic predictions.

REFERENCES

1.S.J.Brodsky, G.P.Lapage, *Phys. Rev.*, **D24** (1981) 2848.

2.S.J.Brodsky, C.E.Carlson, H.J.Lipkin, *Phys. Rev.*, **D20** (1979) 2278.

3G.R.Farrar, S.Gottlieb, D.Sivers, G.H.Thomas, *Phys. Rev.*, **D20** (1979) 202.

4.M.P.Chavleishvili, Spin kinematics for binary processes. Dispersion amplitudes. Ludvig-Maximilian University Preprint LMU-03/93, Munich, 1993.

5.M.P.Chavleishvili, Spin kinematics for binary processes. Dynamic amplitudes for elastic scattering. Ludvig-Maximilian University Preprint LMU-02/93, Munich, 1993.

6.C.Bourrely, E.Leader, J.Soffer, *Phys. Reports*, **59** (1980) 96.

7.L.Angeliti, L.Nitti, M.Pellicoro, G.Preparata,G.Valentini, Rivista del Nuovo Cimento, **6** (1983) 3.

8.E.A.Crosbie et al., *Phys. Rev.*, **D23** (1981) 600.

ON PROTON AND DELTA WAVE FUNCTIONS

N. G. STEFANIS and M. BERGMANN

Institut für Theoretische Physik II,
Ruhr-Universität Bochum, D-44780 Bochum, Germany

Abstract

Physical wave functions for the nucleon and the Δ^+ isobar are presented, which unify the best features of previous models. With these wave functions we can calculate elastic form factors and the decays of the charmonium levels $^3S_1$, $^3P_1$, $^3P_2$ into $p\bar{p}$ in agreement with the data. A striking scaling behavior between $R = |G_M^n|/G_M^p$ and the coefficient B_4 of the Appell polynomial decomposition of the nucleon distribution amplitude is found; the implications for elastic nucleon cross sections are discussed.

I. INTRODUCTION

An interesting testing ground for applications of perturbative QCD has emerged in the study of exclusive processes and elastic form factors of few-quark systems. Such systems can be described within a convolution formalism[1] assuming factorization of highly off-shell or large transverse momentum regions of phase space from regions of low momenta necessary to form bound states. Recent progress[2] in Sudakov-suppression techniques provides support for the conjectured infrared protection of the perturbative picture.

For modelling of the nucleon and its low resonances, elastic form factors play a key role because they provide an integrated view of the implications of QCD from low to high Q^2. Thus they offer a powerful link between theoretical concepts and measurements and they can serve to test both the scaling properties as well as the detailed structure of the nucleon wave function. The theoretical tools for such a description are provided by the hard-scattering amplitude which describes the perturbative quark-gluon interaction in a particular process, and the probability amplitude for finding the three-quark valence state in the scattered nucleon or nucleon resonance: $\Phi_N(x_i, Q^2)$. A major theme of this talk will be to examine how QCD deals with the derivation of such distribution amplitudes for the nucleon and the Δ^+ isobar, focusing our attention on recent developments.

II. GENERAL FEATURES

The momentum-scale dependence of $\Phi_N(x_i, Q^2)$ is given by

$$\Phi_N(x_i, Q^2) = \Phi_{as}(x_i) \sum_{n=0}^{\infty} B_n \tilde{\Phi}_n(x_i) \left(\frac{\alpha_s(Q^2)}{\alpha_s(\mu^2)} \right)^{\gamma_n}, \tag{1}$$

in which $\{\Phi_n\}_0^\infty$ are orthonormalized eigenfunctions of the interaction kernel of the evolution equation[1] expressed in a truncated basis of Appell polynomials of maximum degree M, and $\Phi_{as}(x_i) = 120 x_1 x_2 x_3$ is the asymptotic form of the nucleon distribution amplitude. The corresponding eigenvalues γ_n turn out[3] to be the anomalous dimensions of multiplicatively renormalizable $I_{1/2}$ baryonic operators of twist three. Because the γ_n are positive fractional numbers increasing with n, higher terms in this expansion are gradually suppressed. A basis including a total of 54 eigenfunctions ($M = 9$) together with the associated normalization coefficients and anomalous dimensions is given in Ref. 4.

The derivation of the nucleon distribution amplitude from QCD is intimately connected with confinement and employs nonperturbative methods. Using the properties of the Appell polynomials, the inverse of Eq. (1) determines the (nonperturbative) expansion coefficients B_n:

$$B_n(\mu^2) = \frac{N_n}{120} \int_0^1 [dx] \tilde{\Phi}_n(x_i) \Phi_N(x_i, \mu^2), \tag{2}$$

so that the "renormalization-group improved" coefficients $B_n(Q^2)$ are given by

$$B_n(Q^2) = B_n(\mu^2) \exp\left\{ -\int_{\alpha_s(\mu^2)}^{\alpha_s(Q^2)} \frac{d\alpha}{\beta(\alpha)} \gamma_n(\alpha) \right\} \approx B_n(\mu^2) \left\{ \frac{\ln(Q^2/\Lambda_{QCD}^2)}{\ln(\mu^2/\Lambda_{QCD}^2)} \right\}^{-\gamma_n}. \tag{3}$$

In terms of the moments of the nucleon distribution amplitude,

$$\Phi_N^{(i0j)}(\mu^2) = \int_0^1 [dx] x_1^i x_2^0 x_3^j \Phi_N(x_i, \mu^2), \tag{4}$$

Eq. (3) becomes

$$\frac{B_n(\mu^2)}{\sqrt{N_n}} = \frac{\sqrt{N_n}}{120} \sum_{i,j=0}^{\infty} a_{ij}^n \Phi_N^{(i0j)}(\mu^2), \tag{5}$$

where the projection coefficients a_{ij}^n are calculable to any order M. Specifically, those up to order $M = 9$ have been tabulated in Refs. 4, 5.

To determine the moments, a short-distance operator product expansion is performed at some spacelike momentum μ^2 where quark-hadron duality is valid.[6] One considers matrix elements of appropriate three-quark operators which are related to moments of the covariant distribution amplitudes[7] V, A, and T: $\Phi_N(x_i) = V(x_i) - A(x_i)$, $\Phi_N(1, 3, 2) + \Phi_N(2, 3, 1) = 2T(1, 2, 3)$ with V(1,2,3)=V(2,1,3), A(1,2,3)=-A(2,1,3), and T(1,2,3)=T(2,1,3).

III. DISTRIBUTION AMPLITUDES OF THE NUCLEON AND THE Δ^+ ISOBAR

Based on QCD sum-rule calculations, useful theoretical constraints on the moments of baryon distribution amplitudes have been obtained[6,8-11]. Physical wave functions and observables for the nucleon[6,8,9] and the Δ^+ isobar[10,11] are then calculated using the full set of these constraints to determine the first few expansion coefficients

B_n in a truncated basis of Appell polynomials. Depending on the value of Λ_{QCD}, these models predict approximately the right size and Q^2-evolution of G_M^p, while they give $R = |G_M^n|/G_M^p \leq 0.5$. An alternative nucleon distribution amplitude was proposed[12] to give $|G_M^n| \ll G_M^p$, in accordance to phenomenological data analyses[13] and the latest high-Q^2 SLAC data[14] at the expense that some of the amplitude moments cannot match the sum-rule requirements[6] in the allowed saturation range[15].

However, several crucial questions have to be resolved: For instance, does the *optimum* solution to the sum rules automatically yield *best* agreement with the data? Do solutions exist with characteristics *distinctive* from those of the COZ and the GS amplitudes? If so, what are the fundamental ordering parameters to classify these solutions? In recent works[16,17] we have shown that it is indeed possible to amalgamate the best features of COZ-type[8] and GS-type[12] nucleon distribution amplitudes into a hybrid-like amplitude, we termed the "heterotic" solution (see Fig. 1).

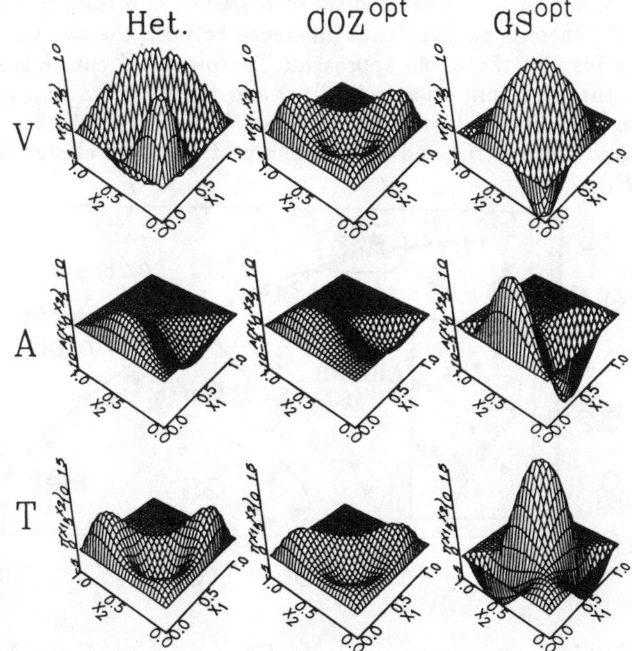

FIG. 1. Covariant distribution amplitudes V, A, and T of the optimized versions of the COZ and GS models vs. the novel heterotic model.

In order to develop a credible nucleon distribution amplitude, we employ a χ^2 criterion which parametrizes the deviations from the sum-rule intervals according to the moment order. This "hierarchical" treatment of the sum rules takes into account the higher stability of the lower-level moments[15] and does not overestimate the significance of the still unverified constraints[8] for the third-order moments. [For more details, see Ref. 19.]

The underlying assumption is that contributions of higher-order terms are either negligible or of minor importance relative to those of second-order. Then the model space is also truncated at states with bilinear correlations of fractional momenta and the pattern of solutions found in this order should dominate the (orthonormalized) Appell polynomial series at every order of truncation. In this way the parameter space of the Appell decomposition coefficients can be systematically scanned seeking for local minima of χ^2. Using for the first and second order moments either the COZ or the KS sum-rule constraints in conjunction with those of COZ for the third-order moments, a simple *scaling relation* between the ratio R and the expansion coefficient B_4 emerges as one progresses through the generated solutions.[20]

We have plotted in the (B_4, R) plane interpolating solutions to the COZ sum rules (+ labels) and such to a combined set of KS/COZ sum rules (o labels). As it turns out (Fig. 2), there is no significant difference between the two treatments and this insensitivity justifies the whole approach. The presented curves are fits to the local minima of the COZ sum rules (solid line) and the KS/COZ sum rules (dotted line). They constitute an *orbit* with respect to χ^2, beginning in the heterotic region (small R and large positive B_4) and terminating past the COZ cluster (large R and large negative B_4).

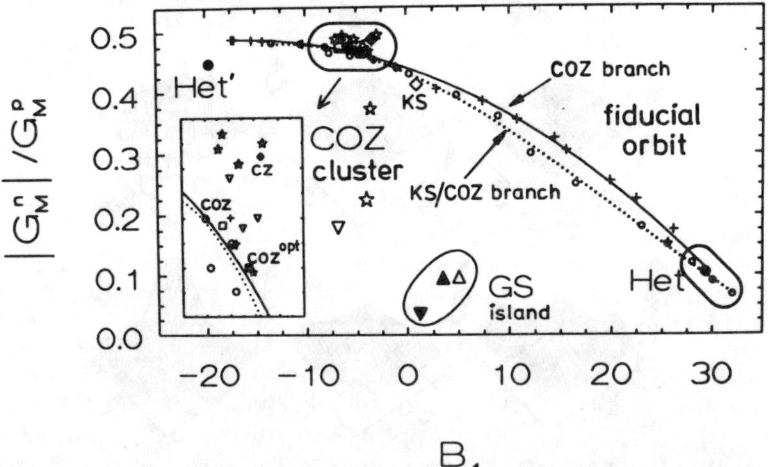

FIG. 2. Scaling relation between the ratio R of the magnetic nucleon form factors and the coefficient B_4 of the Appell polynomial decomposition of the nucleon distribution amplitude. The various models of nucleon distribution amplitudes discussed in the text are compiled in Table I. The inset at the lower left expands the vertical scale between 0.455 and 0.495 (corresponding to the B_4 interval $[-10, 0]$) to show the close agreement between the fiducial orbit and a variety of nucleon distribution amplitudes *not* included in the fit.

The lower part of the orbit is associated with the heterotic solution which corresponds to the smallest possible ratio still compatible with the sum-rule constraints. The upper region of the orbit controls COZ-type amplitudes and contains a cluster of solutions densely populating the orbit in the R-interval $0.455 \div 0.495$ (see the inset in Fig. 2). This cluster contains the amplitudes COZ^{opt}, KS/COZ^{opt} which are associated with the absolute minima of χ^2 and play the role of strange attractors for all other solutions with similar features.[19] GS-type amplitudes correspond to local minima of χ^2 at considerably lower levels of accuracy and thus they constitute in the (B_4, R) plane an isolated region (an "island") that is separated from the characteristic orbit by a large χ^2 barrier. The profiles of the distribution amplitudes across the orbit change in an orderly sequence of gradations with some mixture of COZ and GS characteristics until the COZ amplitude is transmuted into the heterotic solution.[19]

TABLE I. Theoretical parameters defining the nucleon distribution amplitudes discussed in the text. The "hybridity" angle ϑ is discussed in [19].

Model	B_1	B_2	B_3	B_4	B_5	ϑ[deg]	R	χ^2	Symbol
Het	3.4437	1.5710	4.5937	29.3125	-0.1250	-1.89	.104	33.48	●
Het'	4.3025	1.5920	1.9675	-19.6580	3.3531	24.44	.448	30.63	●
COZ^{opt}	3.5268	1.4000	2.8736	-4.5227	0.8002	9.13	.465	4.49	■
COZ^{up}	3.2185	1.4562	2.8300	-17.3400	0.4700	5.83	.4881	21.29	+
COZ	3.6750	1.4840	2.8980	-6.6150	1.0260	10.16	.474	24.64	□
CZ	4.3050	1.9250	2.2470	-3.4650	0.0180	13.40	.487	250.07	◆
KS^{low}	3.5818	1.4702	4.8831	31.9906	0.4313	-0.93	.0675	36.27	o
KS/COZ^{opt}	3.4242	1.3644	3.0844	-3.2656	1.2750	9.47	.453	5.66	o
KS^{up}	3.5935	1.4184	2.7864	-13.3802	2.0594	13.82	.482	40.38	o
KS	3.2550	1.2950	3.9690	0.9450	1.0260	2.47	.412	116.35	◇
GS^{opt}	3.9501	1.5273	-4.8174	3.4435	8.7534	80.87	.095	54.95	▲
GS^{min}	3.9258	1.4598	-4.6816	1.1898	8.0123	80.19	.035	54.11	▼
GS	4.1045	2.0605	-4.7173	5.0202	9.3014	78.87	.097	270.82	△

Let us now turn to models with functional representations which make use of higher Appell polynomials in connection with additional *ad-hoc* cutoff-parameters.[21,22] The inset in Fig. 2 shows how such models[21] (stars) and[22] (light upside-down triangles) group around the optimum amplitudes COZ^{opt} and KS/COZ^{opt}, thus establishing the scaling relation between R and B_4 in a much more general context. This result suggests that the inclusion of higher-order Appell polynomials in the nucleon distribution amplitude is a marginal effect, as conjectured above. Those model amplitudes[21,22] which appear as isolated points scattered towards the GS island are unacceptable on physical grounds, either because they exhibit unrealistic large oscillations in the longitudinal momentum fractions[21] or because they yield a wrong evolution behavior for the nucleon form factors.[22]

One place to test these results is in the data for the elastic cross sections σ_{p} and σ_{n}. For small scattering angles, where the terms $\propto tan^2(\theta/2)$ can be neglected, there are two main possibilities for the ratio $\sigma_{\mathrm{n}}/\sigma_{\mathrm{p}}$. If the Dirac form factor F_1^{n} is zero or small compared to the Pauli form factor F_2^{n},[13] then σ_{n} should be due only to the higher-order term F_2^{n}. At large Q^2 the ratio would become (M_{N} is the nucleon mass)

$\frac{\sigma_n}{\sigma_p} \Rightarrow \left(\frac{C_2^n}{C_1^p}\right)^2 \frac{1}{4M_N^2 Q^2}$ and would decrease with increasing Q^2 due to the extra power of $1/Q^2$ of the Pauli form factor. Alternatively, if F_1^n is comparable to F_2^n, then σ_n would eventually be due to F_1^n at large Q^2. Then the ratio σ_n/σ_p would be given by some constant determined by the nucleon wave functions $\frac{\sigma_n}{\sigma_p} \Rightarrow \left(\frac{C_1^n}{C_1^p}\right)^2$. In these expressions, the wave-function characteristics are parametrized by the (dimensionful) coefficients C_i, which are functions of the expansion coefficients B_n and the "proton decay constant" $|f_N| = (5.0 \pm 0.3) \times 10^{-3} GeV^2$.

The principal result from the above discussion is that in the intermediate Q^2 domain, σ_n/σ_p should be within the range 0.238 and 0.01. Comparing with available data[23], we see that the measured σ_n/σ_p enters the estimated range already at $Q^2 \approx 8 GeV^2/c^2$ (see Fig. 3).

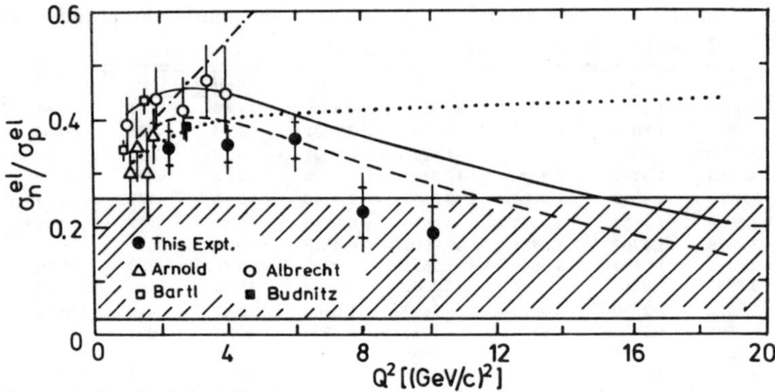

FIG. 3. Bounds on the elastic cross sections σ_n/σ_p from QCD sum rules (shaded area). The data are from 23.

In view of these results, it is worth remarking that the present accuracy of QCD sum rules seems to be sufficient to limit σ_n/σ_p within the observed region. Fig. 3 shows that the available data in the range $Q^2 \approx (8 \div 10)GeV^2/c^2$ are well below the calculated upper bound and still decreasing. This indicates that distribution amplitudes which give $|G_M^n|/G_M^p \approx 0.5$ may be in contradiction to experiment because they yield a Dirac form factor F_1^n which starts to overestimate the data already at $Q^2 \approx 8 GeV^2/c^2$. On the contrary, models which give a small value of $|G_M^n|/G_M^p$ can explain the data only under the assumption that in this Q^2 region the Pauli contribution is still dominant. Fig. 4 serves not only to amplify the preceeding discussion but also to advertise the consistency of the heterotic model with the form factor data. A similar good agreement with the data is found also for the axial form factors.[16,18] We now turn our attention to the Δ^+ isobar. It was pointed out[24] that model amplitudes for the nucleon are characterized by an anticorrelation pattern between G_M^n and G_M^*. COZ-like models yield $|G_M^n|/G_M^p \leq 0.5$ and $|G_M^*|/G_M^p$ small, while GS-like models lead to the reverse situation. This pattern was derived, under the assumption that the Δ amplitude can be crudely modelled by the symmetric part of the nucleon distribution amplitude. Fig. 5 shows that the transition form factor calculated with the nucleon heterotic amplitude and more realistic Δ amplitudes, derived from QCD sum rules,[10,11] is positive with a magnitude between those of previous models. In order to obtain an optimum distribution amplitude for the Δ, we try to comply with the constraints of the CP[10] and FZOZ[11] analyses simultaneously.[17] This concept leads to a hybrid-like amplitude, denoted again "heterotic". This solution fulfills all FZOZ constraints and provides the best possible compliance with the CP constraints. In addition, it gives the best agreement with the data (see Fig. 5 and Ref. 25). In particular, when including the effect of perturbative (i.e., *logarithmic*) Q^2 evolution of the expansion coefficients B_n, the combined use of the heterotic amplitudes for the nucleon and the Δ^+ yields a form factor behavior which conforms with the observed decrease of available data within their quoted errors.[17]

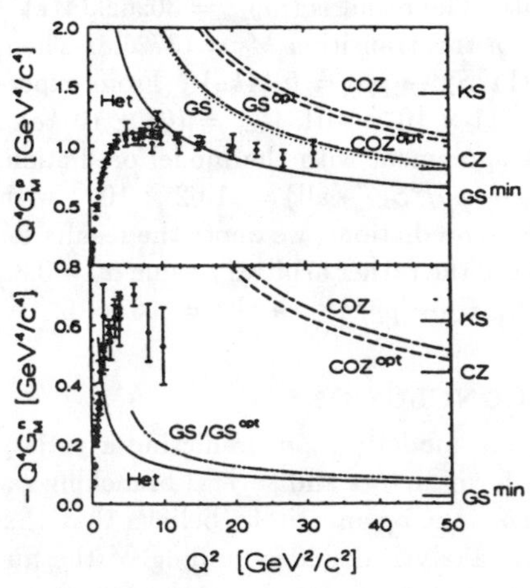

FIG. 4. Magnetic form factor of the proton and the neutron, calculated with different models.

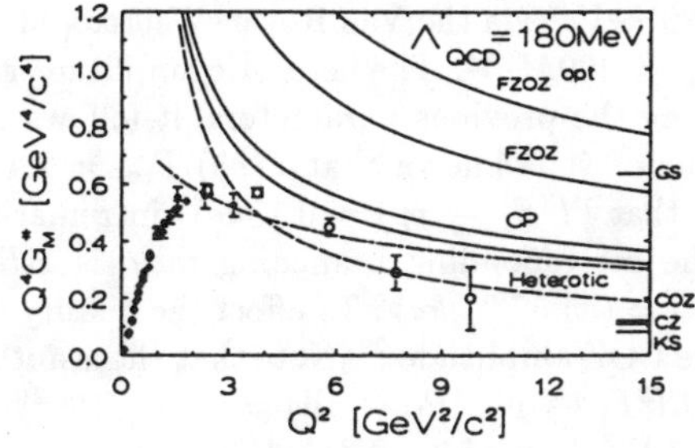

FIG. 5. Comparison with available data of the transition form factor $\gamma p\Delta^+$ calculated with the heterotic nucleon distribution amplitude, modelling the Δ^+ isobar by different distribution amplitudes labeled by their acronyms (solid lines). The data (open circles) are taken from Ref. 25. The form factor at $Q^2 = 15 GeV^2/c^2$ using different nucleon distribution amplitudes in conjunction with the heterotic Δ^+ amplitude is given on the right axis.

There is yet another type of solution for the Δ amplitude—compatible with the sum rules[10,11]—but in sizeable disagreement with the data. This solution ($FZOZ^{opt}$) is obtained by demanding that G_M^* calculated with COZ^{opt} is positive. Thus, as in the nucleon case, optimum agreement with the (existing) sum rules *does not* automatically entail best agreement with the data.

Exclusive decays of charmonium levels to $p\bar{p}$ are very sensitive to the nucleon distribution amplitude. The branching ratio for the decay of the χ_{c1} state ($J^{\mathrm{CP}} = 1^{++}$) into $p\bar{p}$ is proportional to the decay amplitude M_1, which involves Φ_N and f_N. Inputing the heterotic amplitude, M_1 is computed using an elaborated integration routine which accounts for contributions near singularities.[5] Thereby we find $M_1^{\mathrm{het}} = 99849.6$ and as a result $BR(^3P_1 \to p\bar{p}/^3P_1 \to \mathrm{all}) = 0.77 \times 10^{-2}\%$, which is in excellent agreement with the recent high-precision experimental value[26] ($0.78 \pm 0.10 \pm 0.11 \times 10^{-2}$ of the E760 Colaboration at FNAL.

Analogously for the χ_{c2} state ($J^{\mathrm{PC}} = 2^{++}$), we find $M_2^{\mathrm{het}} = 515491.2$. Setting[27] $\alpha_s(m_c) = 0.210 \pm 0.028$, we then obtain $BR(^3P_2 \to p\bar{p}/^3P_2 \to \mathrm{all}) = 0.89 \times 10^{-2}\%$ in excellent agreement with the FNAL value[26] $0.91 \pm 0.08 \pm 0.14 \times 10^{-2}\%$.

Similar considerations apply also to the charmonium decay of the level $^3S_1$ with $J^{\mathrm{PC}} = 1^{--}$. The partial width of J/ψ (or χ_{c0}) into $p\bar{p}$ is $\Gamma(^3S_1 \to p\bar{p}) = (\pi\alpha_s)^6 \frac{1280}{243\pi} \frac{|f_\psi|^2}{M} \left| \frac{f_N}{M^2} \right|^4 M_0^2$, where f_ψ determines the value of the $^3S_1$-state wave function at the origin. Its value can be extracted from the leptonic width $\Gamma(^3S_1 \to e^+e^-) = (5.36 \pm 0.29)keV$[28] via the Van Royen-Weisskopf formula. The result is ($m_{J/\psi} = 3096.93 MeV$) $|f_\psi| = 409 MeV$. The heterotic amplitude gives for this transition $M_0 = 13726.8$. Then, using the previous parameters, it follows that $\Gamma(^3S_1 \to p\bar{p}) = 0.14 keV$. From experiment[28] it is known that $\Gamma(p\bar{p})/\Gamma_{\mathrm{tot}} = 2.16 \pm 0.11 \times 10^{-3}$ with $\Gamma_{\mathrm{tot}} = (68 \pm 10)keV$, so that $\Gamma(^3S_1 \to p\bar{p}) = 0.15 keV$ in remarkable agreement with the model prediction. The corresponding branching ratio is $BR(^3S_1 \to p\bar{p}/^3S_1 \to \mathrm{all}) = 1.62 \times 10^{-3}$ with $\Gamma_{\mathrm{tot}} = (85.5^{+6.1}_{-5.8})keV$. To effect the quality of these predictions, we quote the results for the COZ amplitude:[29] [Note that these authors use the rather arbitrary value $\alpha_s = 0.3$.] $BR(^3P_1 \to p\bar{p}/^3P_1 \to \mathrm{all}) = 0.50 \times 10^{-2}\%$, $BR(^3P_2 \to p\bar{p}/^3P_2 \to \mathrm{all}) = 1.6 \times 10^{-2}\%$, and $\Gamma(^3S_1 \to p\bar{p}) = 0.34 keV$.

IV. SUMMARY AND CONCLUSIONS

Given the apparent success of the heterotic model[16–18] in predicting a variety of observables such as magnetic and transition form factors and several branching ratios of exclusive decays of charmonium into $p\bar{p}$, it is optimistic to believe that this approach—albeit approximative for a complete analytical understanding of the nucleon distribution amplitude—is sufficient of reproducing the observed phenomena. Since higher than order $M = 3$ expansion coefficients are unspecified by the present knowledge of QCD sum rules, the model does not depend on unconstrained (higher-order) parameters. While higher-order effects on the nucleon distribution amplitude itself are found to be large,[21,22] the agreement with the data is actually not improved.[4] This is also true for the optimized version of the COZ amplitude[30], we have derived, which represents the global minimum of χ^2. We emphasize that the (normalized) coefficients B_n calculated via the central values of the 10 independent sum rules of Ref. 8 *do not* correspond to a solution with $\chi^2 = 0$. Although such a solution *must* exist, its determination is not a trivial task. Furthermore, at relatively large distances probed in present experiments, still uncalculable contributions of higher twists are presumably more significant than higher-order terms of the Appell polynomial series.

This work was supported in part by the Deutsche Forschungsgemeinschaft and the COSY-Jülich project.

REFERENCES

[1] G. P. Lepage and S. J. Brodsky, Phys. Rev. **D22** (1980) 2157.

[2] H.-N. Li and G. Sterman, Nucl. Phys. **B381** (1992) 129.

[3] M. Peskin, Phys. Lett. **B88** (1979) 128.

[4] M. Bergmann and N. G. Stefanis, Bochum Report RUB-TPII-45/93 (1993); and these Proceedings.

[5] M. Bergmann, Ph.D thesis, Bochum University, 1993.

[6] V. L. Chernyak and I. R. Zhitnitsky, Nucl. Phys. **B246** (1984) 52.

[7] A. B. Henriques, B. H. Kellet, and R. G. Moorhouse, Ann. Phys. (N.Y.) **93** (1975) 125.

[8] V. L. Chernyak, A. A. Ogloblin, and I. R. Zhitnitsky, Z. Phys. **C42** (1989) 569.

[9] I. D. King and C. T. Sachrajda, Nucl. Phys. **B279** (1987) 785.

[10] C. E. Carlson and J. Poor, Phys. Rev. **D38** (1988) 2758.

[11] G. R. Farrar *et al.*, Nucl. Phys. **B311** (1988/89) 585.

[12] M. Gari and N. G. Stefanis, Phys. Lett. **B175** (1986) 462; Phys. Rev. **D35** (1987) 1074.

[13] M. Gari and W. Krümpelmann, Z. Phys. **A322** (1985) 689; Phys. Lett. **B173** (1986) 10; and references therein.

[14] R. G. Arnold *et al.*, Phys. Rev. Lett. **57** (1986) 174.

[15] N. G. Stefanis, Phys. Rev. **D40** (1989) 2305; *ibid.* **D44** (1991) 1616 (E).

[16] N. G. Stefanis and M. Bergmann, Phys. Rev. **D47** (1993) R3685.

[17] N. G. Stefanis and M. Bergmann, Phys. Lett. **B304** (1993) 24.

[18] N. G. Stefanis, in *Proceedings of the 4th Hellenic School on Elementary Particle Physics*, Corfu, Greece, 2-20 September, 1992.

[19] M. Bergmann and N. G. Stefanis, Bochum Report RUB-TPII-36/93 (1993).

[20] M. Bergmann and N. G. Stefanis, Bochum Report RUB-TPII-37/93 (1993).

[21] A. Schäfer, Phys. Lett. **B217** (1989) 545.

[22] J. Hansper, R. Eckardt, and M. F. Gari, Z. Phys. **A341** (1992) 339; R. Eckardt, J. Hansper, and M. F. Gari, Z. Phys. **A343** (1992) 443.

[23] S. Rock *et al.*, Phys. Rev. **D46** (1992) 24.

[24] C. E. Carlson, M. Gari, and N. G. Stefanis, Phys. Rev. Lett. **58** (1987) 1308.

[25] L. Stuart, these Proceedings.

[26] T. A. Armstrong *et al.*, Nucl. Phys. **B373** (1992) 35.

[27] R. Barbieri, R. Gatto, and E. Remiddi, Phys. Lett. **B106** (1981) 497.

[28] Particle Data Group, K. Hikasa *et al.*, Phys. Rev. **D45** (1992) 1.

[29] V. L. Chernyak, A. A. Ogloblin, and I. R. Zhitnitsky, Z. Phys. **C42** (1989) 583.

[30] M. Bergmann and N. G. Stefanis, Bochum Report RUB-TPII-49/93 (1993).

EVOLUTION EFFECTS ON THE NUCLEON DISTRIBUTION AMPLITUDE

M. Bergmann and N. G. Stefanis

Institut für Theoretische Physik II

Ruhr-Universität Bochum, D-44780 Bochum, Germany

Abstract

We study the Brodsky-Lepage evolution equation for the nucleon and construct an eigenfunction basis by including contributions of up to polynomial order 9. By exployting the permutation symmetry P_{13} of these eigenfunctions, a basis of symmetrized Appell polynomials can be constructed in which the diagonalization of the evolution kernel is considerably simplified. The anomalous dimensions are calculated and found to follow a power-law behavior. As an application, we consider the Brodsky-Huang-Lepage ansatz. An algorithm is developed to properly incorporate such higher order contributions in a systematic way.

I. GENERAL FRAMEWORK

The momentum scale dependence of the nucleon distribution amplitude is given by the Brodsky-Lepage evolution equation[1]

$$x_1 x_2 x_3 \left[\frac{\partial}{\partial \xi} \tilde{\Phi}_i(x_i, \xi) + \frac{3}{2} \frac{C_F}{\beta} \tilde{\Phi}(x_i, \xi) \right] = \frac{C_B}{\beta} \int_0^1 [dy] V[x_i, y_i] \tilde{\Phi}(y_i, \xi), \qquad (1)$$

with $\xi = \ln \ln \frac{Q^2}{\Lambda_{QCD}^2}$, $C_B = \frac{N_c+1}{2N_c} = \frac{2}{3}$ and $C_F = \frac{N_c^2-1}{2N_c} = \frac{4}{3}$ the bosonic and fermionic Casimir-operators of $SU(N)_{color}$, respectively; $\beta = 11 - \frac{2}{3} N_F = 9$ being the Gell-Mann and Low function. The asymptotic solution of this equation is $\Phi_{AS} = 120 \, x_1 x_2 x_3$. To leading order in α_s, the interaction kernel between quark pairs $\{i, j\}$ is given in Ref. 1. Factorization of Eq. (1) leads to

$$-\eta \, x_1 x_2 x_3 \, \tilde{\Phi}(x_i, \xi) = \int_0^1 [dy] V[x_i, y_i] \tilde{\Phi}(y_i, \xi) \quad \text{and} \quad x_1 x_2 x_3 \left[\frac{\partial}{\partial \xi} \tilde{\Phi}(x_i, \xi) + \gamma \tilde{\Phi}(x_i, \xi) \right] = 0 \qquad (2)$$

with $\gamma = \frac{3}{2} \frac{C_F}{\beta} + \eta \frac{C_B}{\beta}$ and the integration measure is defined by $\int_0^1 [dx] \equiv \int_0^1 dx_1 \int_0^{1-x_1} dx_2 \int_0^1 dx_3 \, \delta(1 - x_1 - x_2 - x_3)$. [In the following we use the operator $\int_0^1 [dy] \, V[x_i, y_i] \mapsto \hat{V}$, which commutes with the permutation operator $\hat{P}_{13}$ $(x_1 \Leftrightarrow x_3)$.] The evolution behavior of the eigenfunctions comes from the sensitivity on the transverse momentum integration[1] arising from the gluon exchange kernels. It can be expressed in the form $\tilde{\Phi}(x_i, Q^2) = \ln^{-\gamma_F} \left(\frac{Q^2}{\Lambda_{QCD}^2} \right) \tilde{\Phi}(x_i)$. The representation of the evolution kernel is conveniently described in terms of Appell polynomials, $\tilde{\mathcal{F}}_{mn}(x_1, x_3)$, which constitute an orthogonal basis with weight[1] $\Phi_{AS}(x_i)/120$. It was shown in Ref. 2 that in this basis,

$\hat{V}$ becomes block-diagonal with respect to different polynomial orders $m + n$. Because of $[\hat{P}_{13}, \hat{V}] = 0$, it is useful[3,4] to define a symmetrized basis of Appell polynomials

$$\tilde{\mathcal{F}}_{mn}(x_1, x_3) = (1/2)\,(\mathcal{F}_{mn}(x_1, x_3) \pm \mathcal{F}_{nm}(x_1, x_3)) \quad \text{for } (m \geq n \ / \ m < n). \tag{3}$$

The particular importance of this basis lies in the fact that $\hat{V}$ is block-diagonal within a particular order for different symmetry classes $(S_n = \pm 1)$ with respect to $\hat{P}_{13}$. As a result, it is possible to analytically diagonalize $\hat{V}$ up to order 7 (cf. Tab. I) [up to order $(\mathcal{O}(n) =) M = 9$ this is done in Refs. 3,4]. The eigenvalues of any order $\mathcal{O}(n)$ and for a specific symmetry class S_n form a "multiplet"-like set (see Fig. 1), which follows an approximate power-law: $\gamma_n = 0.37\,\mathcal{O}(n)^{0.565}$ that differs from that of scalar (S=0) and vector (S=1) mesons[3,4]. This observation is in contrast to the claims of Ref. 5. The eigenfunctions of the evolution equation have properties of a commutative group

$$\tilde{\Phi}_k(x_i)\,\tilde{\Phi}_n(x_i) = \sum_{l=0}^{\infty} F_{kn}^l\,\tilde{\Phi}_l(x_i) \quad \text{with} \quad |\mathcal{O}(k) - \mathcal{O}(n)| \leq \mathcal{O}(l) \leq \mathcal{O}(k) + \mathcal{O}(n) \tag{4}$$

and obey the above triangle relation. The structure coefficients[3,4] F_{kn}^l of the group are calculated by $F_{kn}^l = N_l \int_0^1 [dx]\,x_1 x_3 (1 - x_1 - x_3)\,\tilde{\Phi}_k(x_i)\tilde{\Phi}_n(x_i)\tilde{\Phi}_l(x_i)$, with the particularly important case $F_{kk}^0 = \frac{N_0}{N_k}$.

II. APPLICATIONS

The evolution effect of the nucleon distribution amplitude is important for the calculation of various form factors at intermediate values of the momentum transfer $Q^2 \approx 10 \div 30\,GeV^2/c^2$ (see Ref. 1, 3–5). In this range, $\alpha_S(Q^2)$ tends to diverge whereas the Q^2 evolution of the distribution amplitude, given by

$$\Phi_N(x_i, Q^2) = \Phi_{AS}(x_i)\left(\sum_{n=0}^{n_{max}} B_n(Q^2)\tilde{\Phi}_n(x_i)\right), \tag{5}$$

significantly reduces the value of the form factors by more than $\approx 30\%$. In the above representation of Φ_N, contributions of eigenfunctions up to polynomial order 3 have been studied[5,6]. By incorporating on the rhs of eq. (5) (c.f. Ref. 7) the factor $f(x_i, \lambda_j) = e^{-\lambda_1^2\left(\sum_{i=1}^3 \frac{1}{x_i} - \lambda_2^2\right)}$, eigenfunctions of higher orders can be taken into account having recourse to a Brodsky-Huang-Lepage type of ansatz. This extended ansatz for $\lambda_1 = 0.03$ and $\lambda_2 = 3$ has been studied in Ref. 5. In order to determine its evolution behavior, one has to project[7] Φ_N on the eigenfunctions $\tilde{\Phi}_n$. For this purpose, f can be expanded in the basis of eigenfunctions of the nucleon evolution equation

$$f(x_i, \lambda_j) = \sum_k c_k(\lambda_j)\,\tilde{\Phi}_k(x_i), \quad \text{with} \quad c_k(\lambda_j) = N_k \int_0^1 [dx]\,x_1 x_2 x_3\,f(x_i, \lambda_j)\tilde{\Phi}_k(x_i). \tag{6}$$

This expansion and the properties of products of eigenfunctions given by the structure coefficients F_{nk}^l lead to

148

$$\Phi_N(x_i) = \Phi_{AS}(x_i) \left(\sum_l \hat{B}_l \, \tilde{\Phi}_l(x_i) \right) \quad \text{with} \quad \hat{B}_l = \left(\sum_{n,k} B_n c_k F^l_{nk} \right). \tag{7}$$

In the last equation the evolution of the distribution amplitude is fully determined by the scale-dependence of the expansion coefficients $\hat{B}_l(Q^2) = \hat{B}_l(\mu^2) \left(\frac{\alpha_S(Q^2)}{\alpha_S(\mu^2)} \right)^{\gamma_l}$, which are a complicated mixture of the original expansion coefficients B_n and the projection coefficients c_k of the extended ansatz. In contrast to the assumption[5]

$$\Phi_N(x_i, Q^2) = \Phi_{AS}(x_i) \left(\sum_{n=0}^{n_{max}} B_n(\mu^2) \left(\frac{\alpha_S(Q^2)}{\alpha_S(\mu^2)} \right)^{\gamma_n} \tilde{\Phi}_n(x_i) \right) f(x_i, \lambda_j), \tag{8}$$

the evolution of the distribution amplitude is *not* found to be determined by the scale-dependence $B_n(Q^2) = B_n(\mu^2) \left(\frac{\alpha_S(Q^2)}{\alpha_S(\mu^2)} \right)^{\gamma_n}$. The deviation of the approximation[5] from the correct evolution behavior can be expressed by the Q-dependent ratio

$$\mathcal{M}_l(Q^2) = \left(\hat{B}_l(\mu^2) \left(\frac{\alpha_S(Q^2)}{\alpha_S(\mu^2)} \right)^{\gamma_l} \right) \Big/ \left(\sum_{n,k} B_n(\mu^2) \left(\frac{\alpha_S(Q^2)}{\alpha_S(\mu^2)} \right)^{\gamma_n} c_k F^l_{nk} \right). \tag{9}$$

TABLE I. Orthogonal eigenfunctions $\tilde{\Phi}_n(x_1, x_2, x_3) = \sum_{lk} a^n_{kl} x_1^k x_3^l$ of the nucleon evolution equation (represented by the coefficient matrix a^n_{kl} with $a^n_{kl} = S_n a^n_{lk}$; $a^n_{22} = 0$ for all n). The normalization is given by $\int_0^1 [dx]\, x_1 x_2 x_3 \, \tilde{\Phi}_k(x_i)\tilde{\Phi}_n(x_i) = (N_n)^{-1} \delta_{kn}$.

n	M	S_n	γ_n	η_n	N_n
0	0	1	$\frac{2}{27}$	-1	120
1	1	-1	$\frac{26}{81}$	$\frac{2}{3}$	1260
2	1	1	$\frac{10}{27}$	1	420
3	2	1	$\frac{38}{81}$	$\frac{5}{3}$	756
4	2	-1	$\frac{46}{81}$	$\frac{7}{3}$	34020
5	2	1	$\frac{16}{27}$	$\frac{5}{2}$	1944
6	3	1	$\frac{115-\sqrt{97}}{162}$	$\frac{-(-79+\sqrt{97})}{24}$	$\frac{4620\left(485+11\sqrt{97}\right)}{97}$
7	3	1	$\frac{115+\sqrt{97}}{162}$	$\frac{79+\sqrt{97}}{24}$	$\frac{4620\left(485-11\sqrt{97}\right)}{97}$
8	3	-1	$\frac{559-\sqrt{4801}}{810}$	$\frac{-(-379+\sqrt{4801})}{120}$	$\frac{27720\left(33607-247\sqrt{4801}\right)}{4801}$
9	3	-1	$\frac{559+\sqrt{4801}}{810}$	$\frac{379+\sqrt{4801}}{120}$	$\frac{27720\left(33607+247\sqrt{4801}\right)}{4801}$
10	4	-1	$\frac{346-\sqrt{1081}}{405}$	$\frac{-(-256+\sqrt{1081})}{60}$	$\frac{196560\left(7567-13\sqrt{1081}\right)}{1081}$
11	4	-1	$\frac{346+\sqrt{1081}}{405}$	$\frac{256+\sqrt{1081}}{60}$	$\frac{196560\left(7567+13\sqrt{1081}\right)}{1081}$

n	a^n_{00}	a^n_{10}	a^n_{20}	a^n_{11}	a^n_{30}	a^n_{21}	a^n_{40}	a^n_{31}
0	1	0	0	0	0	0	0	0
1	0	1	0	0	0	0	0	0
2	-2	3	0	0	0	0	0	0
3	2	-7	8	4	0	0	0	0
4	0	1	$-\frac{4}{3}$	0	0	0	0	0
5	2	-7	$\frac{14}{3}$	14	0	0	0	0
6	1	-6	$\frac{41+\sqrt{97}}{4}$	$\frac{3\left(31-\sqrt{97}\right)}{4}$	$\frac{-5\left(17+\sqrt{97}\right)}{16}$	$\frac{-5\left(31-\sqrt{97}\right)}{8}$	0	0
7	1	-6	$\frac{41-\sqrt{97}}{4}$	$\frac{3\left(31+\sqrt{97}\right)}{4}$	$\frac{-5\left(17-\sqrt{97}\right)}{16}$	$\frac{-5\left(31+\sqrt{97}\right)}{8}$	0	0
8	0	1	-3	0	$\frac{601+\sqrt{4801}}{264}$	$\frac{59-\sqrt{4801}}{44}$	0	0
9	0	1	-3	0	$\frac{601-\sqrt{4801}}{264}$	$\frac{59+\sqrt{4801}}{44}$	0	0
10	0	1	-5	0	$\frac{379+\sqrt{1081}}{48}$	$\frac{61-\sqrt{1081}}{8}$	$\frac{-\left(159+\sqrt{1081}\right)}{40}$	$\frac{-\left(61-\sqrt{1081}\right)}{8}$
11	0	1	-5	0	$\frac{379-\sqrt{1081}}{48}$	$\frac{61+\sqrt{1081}}{8}$	$\frac{-\left(159-\sqrt{1081}\right)}{40}$	$\frac{-\left(61+\sqrt{1081}\right)}{8}$

This deviation at intermediate Q^2 is found to be of order 0.9 for $\mathcal{O}(l) \leq 3$ and increases exponentially for higher orders. Since in the ansatz of Ref. 5 $\gamma_0 \geq \gamma_n \geq \gamma_9$, one can approximate

$$\mathcal{M}_l(Q^2) \mapsto \tilde{\mathcal{M}}_l(Q^2) = \left(\hat{B}_l(\mu^2) \left(\frac{\alpha_S(Q^2)}{\alpha_S(\mu^2)} \right)^{\gamma_n} \right) \bigg/ \left(\left(\frac{\alpha_S(Q^2)}{\alpha_S(\mu^2)} \right)^{\bar{\gamma}} \sum_{n,k} B_n(\mu^2) c_k F_{nk}^l \right). \quad (10)$$

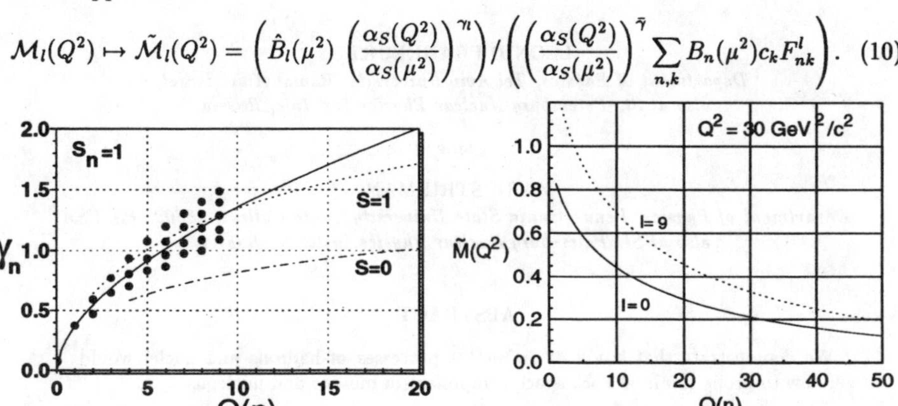

FIG. 1. The eigenvalues of the evolution equation (for $S_n = 1$) vs. the corresponding order $\mathcal{O}(n) = M$ (solid line).

FIG. 2. Deviation measure $\tilde{\mathcal{M}}_l(Q^2 = 30$ $GeV^2/c^2)$ for the two extreme cases $\bar{\gamma} = \gamma_0$ and $\bar{\gamma} = \gamma_9$.

Using the power-law behavior for γ_l, Fig. 2 shows the exponetially increasing deviation of the approximation of Ref. 5 compared to the Q^2-scaling behavior based on the renormalization group equation. This comparison shows that it is important to project the nucleon distribution amplitude Φ_N, as in the case of mesons[7], on the eigenfunctions of the evolution equation. For this purpose, powerful tools have been developed and the basis of eigenfunctions $\tilde{\Phi}_n$ has been extended up to polynomial order 9. Using analytical and numerical algorithms developed in Refs. 3, 4 the calculation of higher order eigenfunctions up to any desired precision is shown to be possible.

This work was supported in part by the Deutsche Forschungsgemeinschaft and the COSY-Jülich project.

REFERENCES

[1] G. P. Lepage und S. J. Brodsky, Phys. Rev. **D22** (1980) 2157.

[2] K. Tesima, Phys. Lett. **110B** (1982) 319.

[3] M. Bergmann, Ph.D. thesis, Bochum University (1993).

[4] M. Bergmann und N. G. Stefanis, Bochum Report RUB-TPII-45/93 (1993).

[5] R. Eckardt, J. Hansper, und M. F. Gari, Z. Phys. **A343** (1992) 443; *ibid.* **A341** (1992) 339.

[6] A. Schäfer, Phys. Lett. **B217** (1989) 545.

[7] T. Huang und Q.-X. Shen, Z. Phys. **D50** (1991) 139.

[8] N. G. Stefanis and M. Bergmann, these Proceedings.

PROBING MINIMAL FOCK SPACE COMPONENTS IN MESONS
IN DIFFRACTIVE HADRON-NUCLEON (NUCLEUS) PROCESSES *

LEONID FRANKFURT

Department of Physics, Tel Aviv University, Ramat Aviv, Israel
also at St. Petersburg Nuclear Physics Institute, Russia

and

MARK STRIKMAN

Department of Physics, Pennsylvania State University, State College, 16802, PA, USA
also at St. Petersburg Nuclear Physics Institute, Russia

ABSTRACT

We demonstrate that study of diffractive processes of hadrons and nuclei would allow to probe minimal Fock space component in mesons and nucleons

1. Introduction

Physics of hard processes is well understood now in terms of pQCD. It has been established experimentally that at space-time intervals $\mid r^2 - t^2 \mid \ll (1-2)GeV^{-2}$ and $r \ll 1Fm$ hard processes are well described through the interactions between quarks and gluons. So it seems promising to use well understood physics of hard processes to investigate softer ones where space -time interval is small but longitudinal (in the direction of projectile in the target rest frame) distances r are large: $r = 1/2m_N x$.

The aim of this talk is to demonstrate that combination of ideas and methods of physics of coherent processes, and physics of hard processes -pQCD produces rather powerful method to calculate and to predict new phenomena,to investigate distribution of color in hadrons. In particular, they provide new ways of measuring minimal Fock components in the wave functions of hadrons, including their x-dependence, which is not accessible to the studies using elastic form factors. We will review our recent works [1-5] which contain references to the previous relevant works. The talk is organized as following. First we consider formalism of scattering eigenstates and introduce idea of color fluctuations in hadron interactions. In section 3 we summarize information about magnitude of color fluctuations in hadrons and calculate the probability for pion to interact with small cross section through $q\bar{q}$ component of the pion Fock wave function, $\Psi_\pi(x)$, and that this calculation agrees reasonably with phenomenological determination of this quantity from analysis of diffractive processes. New evidence for presence of color fluctuations in hadrons from coherent diffraction of nuclei is described in section 4. In section 5 we consider new process of coherent diffractive dissociation of pion into two high p_t jets. We demonstrate that this cross

---

*presented by M.Strikman

section can be expressed through $\Psi_\pi(x)$, allowing for the first time to measure directly the x -dependence of this quantity. In section 6 we outline two possibilities for studying minimal components of pion and nucleon wave function at the collider experiments. Section 7 provides summary of the analysis of the color transparency phenomena in quasielastic reactions at intermediate energies.

2. Scattering-eigenstates formalism

Studies of strong interaction at high energies have established several features of high-energy QCD which have to be incorporated in the realistic picture of hadron collisions:

i) **At high energies various quark-gluon configurations in the projectile hadron can be considered as frozen.** It has long been known that the time scale, given by the uncertainty principle, for quantum fluctuations from a hadronic state h into a state X is given by $\frac{2E_h}{m_X^2 - m_h^2}$. Such fluctuations are inhibited at large enough energies, so that one may treat the hadron as frozen in its initial configuration if coherence length

$$\frac{2E_h}{m_X^2 - m_h^2} \geq R_A. \tag{1}$$

Here R_A is a typical size of a target (nucleus).

ii) **Configurations of very different spatial size and different interaction cross section exist in hadrons.** In fact, inelastic diffraction which has been observed experimentally would be impossible otherwise. To illustrate this point let us consider a two component model of the projectile $\mid h \rangle = a \mid \alpha \rangle + b \mid b \rangle$. After passing through an absorber the wave package is modified to $\mid h \rangle = \epsilon_1 a \mid \alpha \rangle + \epsilon_2 b \mid b \rangle$. If two components are absorbed with equal strength, $\epsilon_1 = \epsilon_2$ and the final state is just $\mid h \rangle$ and no inelastic states are produced. On the other hand if $\epsilon_1 \neq \epsilon_2$ the final state does not coincide with $\mid h \rangle$ and inelastic diffraction takes place.

The constructive way to incorporate this physics is provided by the Good and Walker scattering-eigenstates formalism which accounts for the fact that the projectile is a composite object and its absorption by the target depends on its internal coordinates, contains information about the internal structure of hadrons. This can be seen explicitly by introducing cross-section eigenstates.

We expand the hadronic state $\mid \Psi \rangle$, $\mid \Psi \rangle = \sum_k c_k \mid \psi_k \rangle$ in eigenstates $\mid \psi_k \rangle$ of the scattering amplitude T, which we assume to be purely imaginary: $\mathrm{Im}T \mid \psi_k \rangle = t_k \mid \psi_k \rangle$ The c_k are normalized by $\sum_k \mid c_k \mid^2 = 1$. The small t elastic and diffractive scattering cross sections are given by

$$\frac{d\sigma_{el}}{dt} = \frac{1}{16\pi} \left(\sum_k \mid c_k \mid^2 t_k \right)^2 = \frac{1}{16\pi} \langle \mathrm{Im}T \rangle^2; \quad \left(\frac{d\sigma_{diff}}{dt} \right)^{pp} = \frac{1}{16\pi} \left(\langle \mathrm{Im}T^2 \rangle - \langle \mathrm{Im}T \rangle^2 \right). \tag{2}$$

From the optical theorem $(\mathrm{Im}T \mid_{t=0} = \sigma)$ Miettinen and Pumplin relation:

$$\left(\frac{d\sigma_{diff}}{dt} \right)^{pp}_{t=0} = \frac{1}{16\pi} \left(\langle \sigma^2 \rangle - \langle \sigma \rangle^2 \right) \tag{3}$$

152

As a result, at high energies, the projectile can be treated as a coherent superposition of scattering eigenstates, each with an eigenvalue σ. The probability that a given configuration interacts with a nucleon with a total cross section σ is $P(\sigma)$. It turns out that for many high-energy applications it is sufficient to know $P(\sigma)$ and not more complicated $t_k(t)$. Above formulae follow within the inelastic eikonal approximation from the possibility to apply closure over diffractively produced hadron states. Beyond the region $-t \ll R_N^{-2}$ this approximation looks suspicious since at larger t coherence between contributions of states with different masses is lost-scattering eigen states depend on t.

iii) **Small objects have small cross sections.** The color field of a small color neutral (singlet) object is suppressed because fields of individual closely separated quarks and gluons cancel each other. Thus the interactions between the small color singlet wave packet and nucleons are smaller than those between ordinary hadrons and nucleons. This decoupling theorem is a consequence of gauge invariance and color neutrality. Color screening provides one of the dynamical mechanisms for the fluctuations of the cross section. Another mechanism is the nucleon configurations with and without pion cloud which have different interactions with the target. Since color dynamics determines all the interactions, we use the term *color fluctuations.*

3. Summary of current information on color fluctuations in hadrons

What is known about the $P(\sigma)$? The convenient procedure is to consider moments of this quantity:$\langle \sigma^n \rangle = \int d\sigma \sigma^n P(\sigma)$. The zeroth moment is unity, by conservation of probability, and the first is the total cross section.

The second moment has been determined using current diffractive dissociation data from the nucleon using Eq.(3). Complementary determination is possible using information on inelastic corrections to the total hadron-deuteron cross section. These two determinations give consistent values of $\langle \sigma^2 \rangle$. The value of $\langle \sigma^3 \rangle$ has been determined from the analysis of the data on coherent diffraction dissociation off the deuteron: $p + D \to X + D$, and it corresponds to $\langle (\sigma - \langle \sigma \rangle)^3 \rangle \approx 0$ which is natural for the distribution nearly symmetric around $\langle \sigma \rangle$.[2]. Further information about $P(\sigma)$ for $\sigma \to 0$ can be obtained from the fact that in QCD

$$P(\sigma) \sim \sigma^{N_q-2}, \quad \text{for } \sigma \ll \sigma, \tag{4}$$

where N_q is the number of valence quarks. Thus the nucleon distribution $P(\sigma)$ is $\sim \sigma$ for small σ. Based on this information we approximated $P(\sigma)$ in the form

$$P(\sigma) = N \frac{\sigma/\sigma_0}{\sigma/\sigma_0 + a} e^{-((\sigma-\sigma_0)/(\Omega\sigma_0))^n}, with \ n = 2, 4, 6. \tag{5}$$

Two tests of this picture are possible at the moment. One is theoretical - to calculate $P(\sigma)$ at high energies for small σ for the pion projectiles. Indeed the physics at small σ is dominated by small-size valence $q\bar{q}$ Fock state configurations of the pion, measured by the wave function $\Psi(x, b)$, where x is the light-cone momentum fraction, and b is the transverse distance between quark and antiquark. Thus

$$P_\pi(\sigma = 0) = \int_0^1 dx |\Psi(x, b = 0)|^2 \left(\frac{db^2}{4d\sigma}\right)_{b=0}. \tag{6}$$

The difference from similar formulae of Bertsch et al. [6] is that above b is the transverse distance between quarks. The asymptotic QCD wave function of the pion of Ref. [7] is:

$$\Psi(x, b = 0) = \sqrt{48}\pi x(1 - x)f_\pi, \tag{7}$$

where $f_\pi = 93$ MeV is the pion-decay constant.This component of the wave function is normalized to reproduce the width of pion decay into muon and neutrino [8]. Using this wave function, we have

$$P_\pi(\sigma = 0) = \frac{2\pi^2}{5} f_\pi^2 \left(\frac{db^2}{d\sigma}\right)_{b=0}. \tag{8}$$

With the more realistic Chernyak-Zhitnitski wave function [9] based on QCD sum rules, we obtain a factor of 25/21 higher value of $P_\pi(\sigma = 0)$.

Since color effects dominate for small transverse separation, b, of the valence $q\bar{q}$ pair in the pion, we can in fact apply perturbative QCD directly to calculate $\sigma(b)$ at $b^2 \ll \langle b^2 \rangle$. For small-size configurations, the dominant process in πN scattering is exchange of two gluons, a process analogous to the box diagram for γ-hadron scattering, where instead of the $\gamma \to q\bar{q}$ coupling, the $\pi \to q\bar{q}$ wave function enters. Taking the leading gluon exchange diagrams into account in the limit in which the center of mass energy $\sqrt{s}$ is large compared with b^{-1}, we find for pion projectile [3],

$$P(\sigma = 0) = \frac{3f_\pi^2}{5\alpha_s(4k_t^2)\bar{x}G_N(\bar{x}, 4k_t^2)}, \tag{9}$$

where $\alpha_s(4k_t^2)$ is the QCD running coupling constant, and $G_N(\bar{x}, 4k_t^2)$ is the gluon distribution in the nucleon; $\bar{x} = 4k_t^2/s_{\pi N}$; $k_t^2 \approx 1/b^2$ where f_π is the constant for $\pi \to \mu\nu$ decay. For realistic value of $\alpha_S(1 GeV^2) \approx 0.4$ we find $P(0) \sim 0.017 - 0.023 mb^{-1}$ which is quite close to the value of about 0.015-0.02 which we obtained from fitting $P(\sigma)$. We find also [3] the $q\bar{q}N$ cross section if $q\bar{q}$ pair is in a small size configuration:

$$\sigma(b^2) = \frac{2\pi^2}{3} \left(b^2\alpha_s(Q^2)\bar{x}G_N(\bar{x}, Q^2)\right)_{\bar{x}=1/sb^2, Q^2=1/b^2}, \tag{10}$$

For simplicity we give this result for the case $x_q \sim x_{\bar{q}} \sim 1/2$ in the pion. Since $G_N(x, Q^2)$ can be measured independently, Eqs. (9,10) are exact, parameter-free result in the limit $1/s \ll b^2 \ll \langle b^2 \rangle$.

154

4. Inelastic diffraction off nuclei as evidence for color fluctuations near average value

Another nontrivial experimental test of the idea of color fluctuations is provided by the inelastic coherent diffractive hadron production off nuclei. It is quite straightforward to calculate σ_{diff} through $P(\sigma)$ as [5]:

$$\sigma_{diff}(A) = \int d^2 B$$
$$\left[\int d\sigma P(\sigma) \sum_n [< h \mid F(\sigma, B) \mid n >^2] - [\int d\sigma P(\sigma) < h \mid F(\sigma, B) \mid h >]^2 \right], \qquad (11)$$

where $F(\sigma, B) = 1 - e^{-\frac{1}{2}\sigma T(B)}$ and $T(B) = \int_{-\infty}^{\infty} \rho_A(B, Z)dz$. Here the direction of the beam is $\hat{Z}$ and the distance between the projectile and the nuclear center is $\vec{R} = \vec{B} + Z\hat{Z}$. The advantage of Eq. (11) as compared to the related equation of Ref. [10] within the two gluon exchange model is that we do not need to assume the validity of pQCD at average interquark distances in hadrons where α_S is large (with an extra prescription for dealing with gauge noninvariant effects due to introduction of nonzero gluon mass and use constituent quark model wave functions instead of parton wave functions).

It is instructive to consider the extreme black disk (bd) limit of this formula (which would be a reasonable model if color fluctuations were small). In this limit, the function $F(\sigma, B)$ is unity for positions inside the nucleus and zero otherwise, so that σ_{diff} vanishes! In particular, the black disk model leads to the expressions $\sigma_{tot}^{bd} = 2\pi R_A^2$, $\sigma_{el}^{bd} = \pi R_A^2$ and $\sigma_{diff}^{bd} = 0$. Another way to show that color fluctuations cause $\sigma_{diff}(A)$ is to ignore the composite nature of the nucleon and take $P(\sigma)$ to be a delta function, e.g. $P(\sigma) = \delta(\sigma - \bar{\sigma})$ gives a zero result for $\sigma_{diff}(A)$. This is in marked difference from the calculation using Eq. (11) and realistic $P(\sigma)$ which leads to $\sigma_{diff}^{pA}(A) \propto A^{0.80}$ for $A \sim 16$ and $A^{0.4}$ for $A \sim 200$. For the pion projectile we find even faster A-dependence. The results of the calculation agree well with the A-dependence of semiinclusive data of [11]on $n + A \rightarrow p\pi^- + A$ for $p_n \sim 300 GeV$ and of [12] on $\pi^+ + A \rightarrow \pi^+ + \pi^+ + \pi^- + A$ for $p_{\pi+} = 200$ GeV. In both cases fluctuations of σ near average value dominate. Our numerical results (using $P(\sigma)$) for $\sigma_{diff}(A)$ are reasonably close to the calculation of the preQCD model of [13]. The reason is that this calculation used similar values of ω_σ.

In the limit that A becomes infinite configurations of small size, can be expected to dominate since the nucleus acts as a black disk for all other configurations. Indeed Ref. [6,10] suggested that the effects of such small-sized configurations would dominate the pion-nucleus inelastic diffractive cross section. This early result is inherent in our Eq. (11). The integration over small values of σ in Eq.(11) gives a result $\propto 1/T(B)$ for values of B within the nucleus. The integration over $d^2 B$ leads to $\sigma_{diff} \propto R_A \propto A^{1/3}$. Numerical evaluations of Eq.(11) show that the behavior is close to $A^{0.33}$ for fantastic values of A greater than about 10000.

An additional result indicating that small-sized configurations play a small role is obtained by simply cutting off the integrals over σ at a maximum value of $\sigma_{max} = 5mb$. This contribution varies from 2% to 5% as A increases from about 12 to 200. Thus pQCD contribution into inelastic diffraction suggested in Ref [6,10] don't exceed 2% to 5%. The present result does not mean that small-sized configurations are impossible to find (see discussion in the next section).

To summarize, presence of significant color fluctuations in hadrons is well established now. On the other hand, it is known from the Reggeon Calculus that diffractive processes are intimately related to the processes with high multiplicities - the AGK cutting rules - they correspond to different cuts of the same diagrams. Consequently, observation of significant inelastic coherent diffraction as well as of inelastic shadowing correction to $\sigma_{tot}(hA)$ implies that color fluctuations should play significant role in hadron collisions as well.

We focus on using a pion beam because experimental investigations seem imminent. Note also that pion diffraction to jets could be studied in the multi-TeV energy range using proton-proton colliders, if pion exchange is selected by observing the highest energy Δ's or neutrons (n) produced at low p_t (B.Walker). Indeed, for the reaction: $pp \rightarrow \Delta(n) + p + 2$ jets, the diffractively produced jets would appear in the central detector, so studying this reaction at FNAL may be feasible.

5. Hard π-Nucleon Diffraction

At high pion momentum P_π (500 GeV), the π breaks into a $q\bar{q}$ pair well before hitting the target. In that case, the momentum transfer to the target is transverse $(t \approx -q_t^2)$, so the $q\bar{q}$-nucleon interaction is essentially independent of the longitudinal momentum of $q\bar{q}$ pair.

The frozen approximation, in which the transverse $q\bar{q}$ separation b is a constant, should be valid at such a high energy. The observation of two final state high p_t jets carrying a total momentum very close to $\vec{P}_\pi$ allows us to consider only $q\bar{q}$ components of the pion, $\psi_\pi(x, \vec{b})$. The invariant amplitude $\mathcal{M}(N)$ for a nucleon target is then given at large invariant energies s by

$$\mathcal{M}(N) = \int d^2 b \psi_\pi(x, \vec{b}) f(b^2)/2 e^{i\vec{\kappa}_t \cdot \vec{b}}. \tag{12}$$

Here $f(b^2)$ is the forward $q\bar{q}$ scattering amplitude normalized according to the optical theorem $Im\, f(b^2) = s\, \sigma(b^2)$, and $t = 0$. We neglect the small real part of $f(b^2)$. The factor $e^{-i\vec{\kappa}_t \cdot \vec{b}}$ accounts for the final hadronic wave function of two jets with a high relative transverse momentum.

At sufficiently small b the two gluon exchange term is the dominant term at large s. The QCD calculation of $\sigma_{PQCD}(b^2)$ has been discussed above.

We make a numerical estimate using $t = 0$ and $\kappa_t^2 \gg 1 GeV^2$

$$\frac{d^3 \sigma_N}{dx dM_j^2 \cdot d^2 P_{N_t}} = 0.7 \text{ GeV}^{-6} \left(\frac{\text{GeV}}{\kappa_t}\right)^8 \phi^2(x), \tag{13}$$

For $\kappa_t = 2$ GeV and at the maxima of $\phi(x)$ we find $\frac{d^3 \sigma_N}{dx dM_T^2 d^2 P_{N_t}} = 0.4(0.8) \times 10^{-3}$ GeV^{-6} where the larger (smaller) value is obtained using ϕ_{cz} (ϕ_{as}). These values show that the diffractive production of jets is a small but measurable cross section.

156

We now compute the nuclear amplitude $\mathcal{M}(A)$. Eq. (12) for a nuclear target applies at an energy large enough so that time dilation effects prevent the expansion of the $q\bar{q}$ pair as it moves through the target, i.e. $2P_\pi/(M_J^2 - m_\pi^2) \gg R_A$, where M_J is the invariant mass of the $q\bar{q}$ system. $M_J^2 = \kappa_t^2/x(1-x) \approx 4\kappa_t^2$, so $2P_\pi/(M_J^2 - m_\pi^2) \approx 12 Fm$ for $P_\pi = 500 GeV/c$ and $\kappa_t = 2 GeV/c$. $\mathcal{M}(A)$ is obtained by including nuclear multiple scattering effects. The result is

$$\mathcal{M}(A) = i \, s \int d^2 b \, \psi_\pi(x, b) e^{-i\vec{\kappa}_t \cdot \vec{b}} \int d^2 B \, e^{i\,\vec{q}_t \cdot \vec{B}} \left(1 - e^{-\frac{\sigma(b^2)}{2} T_A(B)} \right), \qquad (14)$$

where B is the nuclear impact parameter, $t = -\vec{q}_t^{\,2}$ and $T_A(B) = \int_{-\infty}^{\infty} \rho_A(\sqrt{B^2 + Z^2})dZ$. A Wood-Saxon form is used for the nuclear density $\rho_A(R)$.

The factor $f_A(b^2)$, with

$$f_A(b^2) \equiv \int d^2 B \left(1 - e^{-\frac{\sigma(b^2)}{2} T_A(B)} \right), \qquad (15)$$

accounts for the $q\bar{q}$ multiple scattering series of interactions on different nucleons. For small b^2, $f(b^2) \approx \frac{\sigma(b^2)}{2} \int d^2 B \, T(B) = \frac{\sigma(b^2)}{2} A$, and $\mathcal{M}(A) \propto A$. The "image" function $f_A(b^2)$ is very sensitive to the form of $\sigma(b^2)$ for large values of b. For Eq. (11) the asymptotic value of f_A at large b is $\approx \pi R_A^2$, while for ordinary eikonal theory $f_A \approx \pi R_A^2$ for all b. Thus eikonal theory would lead to $\mathcal{M}(A) \propto A^{2/3}$.

We keep some terms of infinite order in $1/\kappa_t^2$, while neglecting other higher twist terms of the same κ_t dependence. This is justified because the series we keep involves terms of order $\frac{A^{1/3}}{\kappa_t^2}$, and $A^{1/3}$ is a large parameter.

We may obtain the nuclear cross section by replacing N by A everywhere in Eq. (12). For small t and for sufficiently large κ_t, we find that

$$\frac{d^3\sigma_A}{dx \, dM_J^2 \, d^2 P_{A_t}} = \frac{d^3\sigma_N}{dx \, dM_J^2 \, d^2 P_{N_t}} \, F_A^2(t) A^2 \,, \qquad (16)$$

where $F_A(t)$ is the nuclear form factor.

This A^2 dependence is a strong signal of CT. The cross section for the coherent nuclear process is larger than for the nucleon target by a factor of A^2 (or more), but falls rapidly as t decreases from 0. No other process (except the Coulomb excitation process) has this behavior, so using this property it should be possible to observe the coherent process.

Equations (12) and (14) allow simple evaluations of cross sections. Uncertainties in input parameters allow only estimates of the absolute values. But the ratio $\mathcal{M}(A)/\mathcal{M}(N)$ is less model dependent than either amplitude.

It would be useful to study $d^3\sigma_N$ and $d^3\sigma_A$ separately. Those are proportional to $\phi^2(x)$ and the x dependence may allow a determination of $\phi(x)$. Indeed, the asymptotic wave function has a maximum at x=1/2, while the CZ wave function vanishes.

At sufficiently large energies the $q\bar{q}$ interaction with nuclei is corrected by the effects of gluon shadowing. To see this, simply observe that in leading twist approximation Eq. (10) is applies for a nuclear target (A). So far we have used $G_A = AG_N$. But at small enough $x_N = 4\kappa_t^2/s$, the interactions between emitted gluons spread out in space even when the quarks and anti-quarks are closely separated. Then gluon shadowing occurs. This is a relatively small correction for the present kinematics. But at TeV energies x_N can be smaller than 10^{-3} and the gluon shadowing should become much more important. The gluon shadowing effect may be even larger than for sea quarks observed in deep inelastic scattering because the QCD coupling constant is larger for the octet representation than for the triplet one.

6. Collider experiment options

There are several possibilities to prove minimal Fock state components in hadrons at colliders:

One is to measure photon scattering off pion at HERA in the processes: $\gamma + p \rightarrow (jet(p_t) + jet(p_t + k_t)) + n(k_t)$. The idea is that high p_t jets with quantum numbers of meson guarantee that meson is in small size configuration. Final neutron with small momentum k_t and large longitudinal momentum would signal that scattering occurs off pion field of a proton (In principle one could also try to measure this reaction in coherent diffractive dissociation of a meson into 2 high p_t jets off Coulomb field of nucleus. One should observe Coulomb peak to distinguish this process. This would be useful also for estimating background to the process of coherent pion dissociation to two jets which we discussed above)

The three quark component of nucleon can be measured in double parton scattering hard processes at the $p\bar{p}$ collider. One has to select kinematics where two interacting partons of hadron 1, x_{11} and x_{21} are large - $x_{11} \sim 0.5 - 0.6$ and $x_{21} \sim 0.3$ which are typical for Chernyak - Zhitnitski $|3q >$ wave function and x_{12} and x_{22} are small. In this case the major background which is due to bremsstrahlung process $2g \rightarrow 4g$ is suppressed by the factor $F_{2N}(x_{11} + x_{21})/F_{2N}(x_{11}) \sim 0.02 - 0.03$ as compared to the case of scattering at moderate x where these two processes are comparable. Study of rapidity gaps in this process will help to measure the transverse size of $|3q >$ configuration.

In this talk we have concentrated on the case of mesonic projectiles. This logic can be extended to calculate $P(\sigma)$ for small σ for the case of proton projectiles and also measure 3 quark component of the proton wave function through diffractive proton dissociation to 3 high p_t jets.

7. Estimating color transparency at intermediate energies

Diffractive processes we discussed above are closely related to the physics of color transparency (CT), see [14,1]. Here we want to comment briefly on the studies of this phenomenon at intermediate energies in $A(e, e\prime p)$ and $A(p, p\prime p)$ which were reported at this meeting.

For sufficiently hard processes pQCD unambigously predicts onset of CT but with calculable corrections. However at moderate Q^2 new interesting phenomenon becomes important-space-time evolution of bubble of approximately perturbative phase within the nonperturbative QCD medium. Really the small object created in a high Q^2 reaction is a coherent superposition of physical states. This is a wave packet and not a stationary state, so it undergoes time evolution. Its size increases, since (by definition) the PLC starts as small. Consider a small object in its rest frame. It expands with a characteristic time defined as τ_0. Now suppose the object moves with high energy E in the lab. Time dilation increases this time by a factor of E/m, so the relevant expansion time τ is given by $\tau = E/m\tau_0$. For sufficiently large energies, τ is long enough so that the object can leave the nucleus while small enough to avoid final state interactions. Then CT occurs. So the expansion length l_h can be written as $l_h = \frac{2p}{\Delta M^2}$. Different estimates based on the quark model, Regge model, connection with resonance spectra in the pp diffractive processes lead to the values for the parameter ΔM^2 within the range $0.7 - 1.2 GeV^2$. So for presently available energies, $l_h \approx (0.3 - 0.5) fm \cdot p_N (GeV/c)$ is small enough so that the PLC expands significantly as it moves through the nucleus. Thus the final state interaction is suppressed but does not disappear. In pQCD one can derive $\sigma_{eff}(l) \propto l/l_h$ for $l \ll l_h$ [15] and one can expect this pattern to be reasonable interpolation for $l \le l_h$ as indicated by the observation of precocious Bjorken scaling.

One can also consider expansion effects in hadronic basis [16,1,17]. In this model small cross section of interaction for the outgoing PLC is ensured by imposing sum rules which connect matrix elements of soft and hard processes [14,17]. In practice this involves making ad hoc assumptions about values of overlaping integrals between the states produced in large Q^2 nearthreshold $e + p \rightarrow e\prime X$ processes and in high energy $p + N \rightarrow p\prime X$ processes. Usual assumption is that this integral is close to one. However distributions of hadrons produced in these processes are quite different. In particular, one has to ignore in the practical calculations substantial M_X dependence of the t-dependence of the diffractive processes in $p + N \rightarrow p\prime X$ processes. Typical for deep inelastic processes (as well as other hard processes) correlations between spectra of hadrons produced in current and target fragmentation regions are absent in soft diffraction processes. This mismatch between soft and hard physics is relevant for the smallness of overlapping integral and rapid convergence of integral over M_X- triple reggeon formulae do not work in this limit. Thus at x=1 masses of hadron states X produced in intermediate states in rescattering processes are $M_X^2 \ll Q^2$. This condition guarantee coherence between states X since effective momentum needed to produce mass M_X:

$$k_3 = \frac{M_X^2 - m^2}{2q_3} \tag{17}$$

decreases with Q^2 . Under these conditions one can use completeness over hadronic states to deduce pQCD results in terms of hadronic basis. Distinctive feature of color transparency regime is the decrease with Q^2 of energy deposited by PLC in nuclear medium as a result of weakening of interaction. Remember that charge transparency has been observed by D.Perkins (1956) just as decrease of energy transferrred by $PLCe^+e^-$ pair to emulsion near the point where PLC is produced in the decay $\pi^0 \to \gamma + e^+e^-$. Result of calculations in hadronic basis is $\sigma_{eff}(l) \propto (l/l_h)^n$ with n=1-2 depending on the number of hadronic states used and assumptions on the amplitudes of their soft interactions. In spite of above mentioned shortcomings this is a useful model for describing color tranparency phenomena at intermediate Q^2 since it allows alternative treatment of the time evolution of the wave packet which was in PLC in the initial moment.

Important advantage of the quasielastic reactions is that in this case we have a basic Distorted Wave Impulse Approximation model (DWIA) which is known to produce a quantitative description of the data at the intermediate energies where CT effects are absent. Since accuracy of DWIA improves with increase of E_{inc} one can interpret small deviations from DWIA as precursors of new high-energy physics. Most of the current calculations of the color transparency include only two effects - decrease of the interaction size with the momentum transfer and the expansion of PLC after/before the interaction. However if one wants to match low-energy data with the data obtained in the color transparency searches it is necessary to account in addition for several phenomena [18]: i) substantial dependence of the elementary pN total cross section on E_p below 2 GeV, ii) Influence of soft final state interaction on transparency in different kinematic ranges of the intranuclear proton momentum. iii) Effect of suppression of small size configurations in bound nucleons - color screening effect in nuclei [19]. Note that in the diffusion model Fermi motion effects in the elementary amplitude are small since in this approximation configurations of average mass $M^2 \propto l_h/l$ are effectively involved in the interaction. We find that in the kinematics of the $NE - 18$ SLAC $(e, e'p)$ experiment transparency,T, with Q^2 due to variation of the total pN cross section with energy (dashed curve in Fig.1). At the same time, our calculation of CT in the quantum diffusion model [20] including three above mentioned effects leads to practically the same T at $Q^2 = 1$ GeV^2 and $Q^2 = 7$ GeV^2 - see solid curves for [21] DWIA predicts a substantial decrease of the $\Delta M^2 = 0.7$ and $1.1 GeV^2$ in Fig.1. For $Q^2 = 1 GeV^2$ we give range of values for T reflecting medium effects (like Pauli blocking) which reduces $\sigma_{pN}^{free} \approx 36mb$ to $\sigma_{pN}^{eff} \approx 30mb$. In the case of $(p, 2p)$ reaction we find that selection of different momentum intervals for the struck nucleon substantially changes the value of T since low-momentum nucleons , in average, are closer to the surface. This trend is present in DWIA and somewhat enhanced in the CT regime (see dashed and solid curves in Fig.2). It is consistent with behavior of T in the BNL experiment [22] at $p_N = 6$ and 10 GeV/c. At the same time the data at 12 GeV/c require an additional phenomenon, presumably due to contribution of soft (large transverse size) amplitude [23,24]. Strong dependence of T on the momentum of struck nucleon indicates that the future analyses of the data should be performed up to the standards of intermediate energy nuclear physics, where this phenomenon was implicitly taken into account in all DWIA studies.

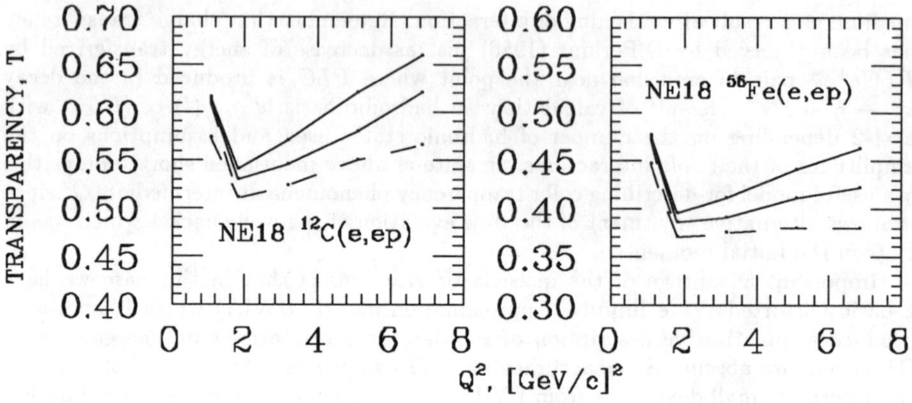

Fig.1 Transparency in $A(e,e\prime p)$ reactions for NE-18 kinematics.

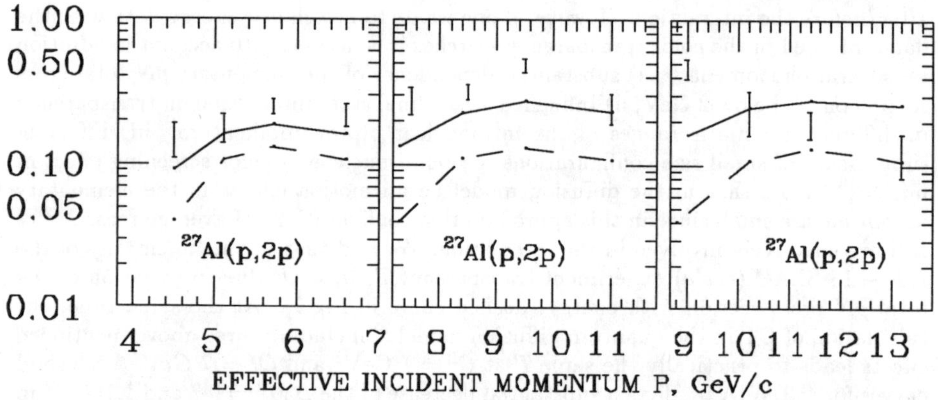

Fig.2 Transparency in $A(p,2p)$ reactions for BNL kinematics.

1. L. Frankfurt, G.A. Miller, and M. Strikman, Comments on Nuclear and Particle Physics **21**,1(1992).
2. G. Baym, B. Blättel, L. L. Frankfurt, H. Heiselberg, and M. Strikman, Phys. Rev D47, 2761 (1993).
3. B. Blättel, G. Baym, L.L. Frankfurt, and M. Strikman, Phys. Rev. Lett.71, 896 (1993).
4. L. Frankfurt, G.A. Miller and M.Strikman, Phys. Lett. 304B,1 (1993).
5. L. Frankfurt, G.A. Miller, and M. Strikman, Phys.Rev.Lett., in press.
6. G. Bertsch, S.J. Brodsky, A.S. Goldhaber and J.G. Gunion, Phys. Rev. Lett. **47**, 297 (1981).
7. G.P. Lepage and S.J. Brodsky, Phys. Rev. **D22**, 2157 (1980).
8. Note that the present result for $\Psi(x, r_\perp = 0)$ differs from that given in Ref. [6] for $\Psi(x, b = 2r_\perp = 0)$ by a factor of 2.
9. V.L. Chernyak and A.R. Zhitnitski, Phys. Rep. **112**, 173 (1984).
10. A.B. Zamolodchikov, B.Z. Kopeliovich, L.I. Lapidus JETP Lett. 33,595 (1981).
11. M.Zielinski et al., Z.Phys.C, **16** ,197 (1983).
12. W.Mollet et al., Phys.Rev.Lett. **26**,1646 (1977).
13. V.M. Braun and Yu. M. Shabel'skiĭ, Sov. J. Nucl. Phys. 37,599 (1983).
14. L.L. Frankfurt and M. Strikman, Prog. Part. Nucl. Phys. **27**, 135 (1991).
15. L. Frankfurt and M. Strikman, Phys. Rep. **160**,235 (1988).
16. B.K. Jennings and G.A. Miller, Phys. Lett. **B236**,209(1990); Phys. Rev. **D44** ,692 (1991); Phys. Rev. Lett. **70**,3619 (1992); Phys. Lett. **B274**,442 (1992); Phys.Rev¿lett. **69** 3619 (1992).
17. L.Frankfurt, W.R.Greenberg, G.A.Miller and M.Strikman, Phys.Rev.,**C46** 2547,(1992).
18. L. Frankfurt, M. Strikman, and M.Zhalov, in Proceedings of VI workshop on perspectives in Nuclear Physics at Intermediate Energies, 1993, to be published
19. L. Frankfurt and M. Strikman, Nucl.Phys., **B250**,147(1985).
20. G.R. Farrar, H. Liu, L.L. Frankfurt & M.I. Strikman, Phys. Rev. Lett. **61** (1988) 686.
21. A.Lung, talk at the workshop.
22. A.S. Carroll et al., Phys. Rev. Lett. **61**, 1698(1988).
23. S.J.Brodsky and G.F.de Teramond, Phys.Rev.Lett.,**60**,1924 (1988).
24. J.P.Ralston and B.Pire, Phys.Rev.Lett. **61**,1823 (1988).

Strangeness and Charm Components
in
Heavy Meson Production

J.M. LAGET

Service de Physique Nucléaire, Centre d'Etude de Saclay
F91191, Gif-sur-Yvette CEDEX, France

Abstract

Exclusive electroproduction of heavy vector mesons (ϕ, J/ψ, ...) at large momentum transfer appears to be a promising way to undertake the study of the structure of the hard part of the Pomeron and the short range structure of nuclei. These topics address two important questions : how does color neutralizes in bound hadronic systems and what is the nature of quark correlations in hadronic matter?. The production of ϕ meson appears to be particularly appealing.

1   Heavy Flavors

The production and the propagation of heavy flavors (strangeness and charm) in hadronic systems provide us with new tools to study their structure. On the one hand the specific flavor of the corresponding quark is a powerful tag to follow how the transferred energy and momentum are distributed in the system. On the other hand, the corresponding reactions do not involve the valence quarks of the target, and probe its sea quark and gluon distributions.

The basic diagram, which drives the production of heavy flavors, is depicted in Fig. 1. To the extent that the nucleon ground state does not contain a significant amount of heavy flavor admixture, the incoming photon must fluctuate into an heavy quark pair ($s\bar{s}$ or $c\bar{c}$) which exchanges a gluon with one of the valence quarks of the target. This pair may recombine into a heavy vector meson (hidden flavor production) : it should emit a second gluon to conserve the color. Or, each heavy quark may recombine with light quarks of the target and leads to a strange or charmed hadron in

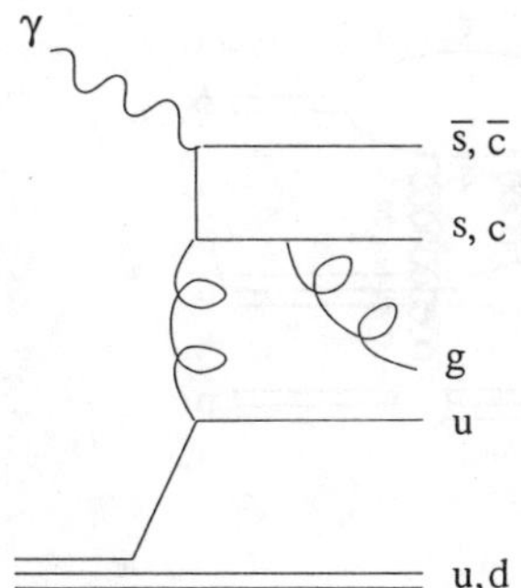

Figure 1: *The basic diagram which drives the production of heavy flavors.*

the final state (open flavor production). It should be noted that, in the hard scattering regime, this diagram can be viewed as a pertubative way to generate a strange or charmed intrinsic component in the proton ground state : one of the proton valence quarks radiates a gluon which eventually converts into an heavy quark pair, which interacts with the incoming photon. It is only in the non pertubative regime that the concept of intrinsic heavy flavor component may be relevant.

With respect to charm, strangeness production [1] presents two advantages. On the one hand, the threshold for strangeness (0.9 GeV for $p(\gamma, k)\Lambda$, or 1.54 GeV for $p(\gamma, \phi)p$, for instance) is much lower than the threshold for heavier flavors (8.2 GeV for $p(\gamma, J/\Psi)p$, for instance). Beams of lower energy allow to probe hard mechanisms sooner. On the other hand, for a comparable excess of energy above threshold, the cross-section of strangeness production is larger (400 nb for the $p(e, e'\phi)p$ reaction in the range $3 < E_\gamma < 10 GeV$) than the cross-section of charm production (5 nb for the $p(e, e'J/\Psi)p$ at 20 GeV). The experiments are more easy and higher values of P_T can be reached.

Among the possible channels, ϕ meson *exclusive* production deserves a particular attention. In order to conserve the color of the strange quark pair, two gluons must be exchanged (Fig. 2). On the nucleon, this two gluon exchange mechanism provides us with a model of the hard part of the Pomeron. On few body targets, it opens up an almost virgin field : the study of quark correlations in hadronic matter. More particularly, this is a *direct* way to reveal possible hidden color component [2] : each exchanged gluon can couple to a quark belonging to a different colored cluster and connect it to a white asymptotic state.

These topics are expected to shed some light on two important issues : the

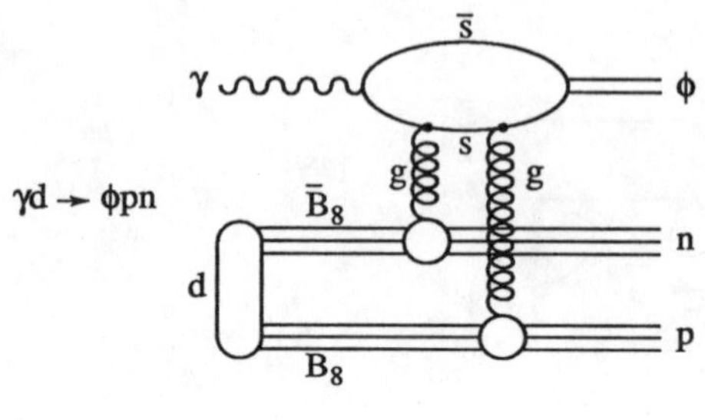

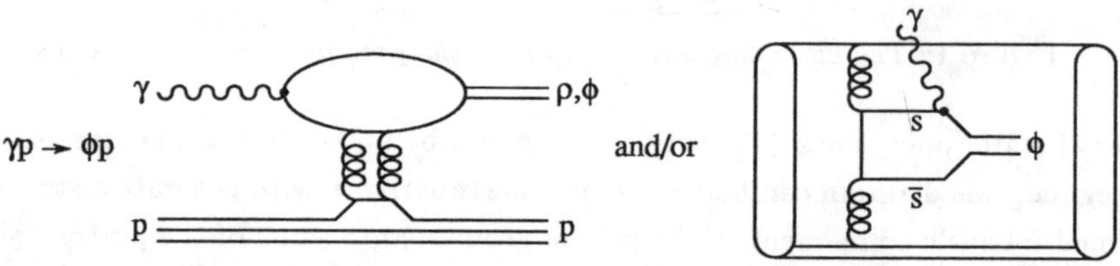

Figure 2: *Possible two gluon exchange mechanisms which could be responsible for ϕ photo and electroproduction on deuterium (top) and free nucleon (bottom). The notations B_8 and $\overline{B_8}$ stand for colored quark clusters. The contribution of a possible intrinsic strange component is depicted on the right.*

short range structure of hadronic matter and the neutralization of the color in bound hadronic systems.

2 ϕ exclusive photoproduction on the nucleon

At low momentum transfer $(-t \leq 0.4 \ (\text{GeV/c})^2)$, the cross section of the $\gamma p \to \phi p$ reaction is purely diffractive and takes the form [3,4] :

$$\frac{d\sigma}{dt} = \left(\frac{d\sigma}{dt}\right)_{t=0} exp(Bt) \tag{1}$$

The variation with the incoming photon energy of the forward angle $(t = 0)$ cross section exhibits a characteristic threshold behavior at low energy and becomes almost constant above $E_\gamma = 6$ GeV. A very good fit is obtained with the following expression :

$$\left(\frac{d\sigma}{dt}\right)_{t=0} = C \left(\frac{p_\phi}{k}\right)^2_{CM} \tag{2}$$

with $C = 2.93 \pm 0.08 \mu b/GeV^2$.

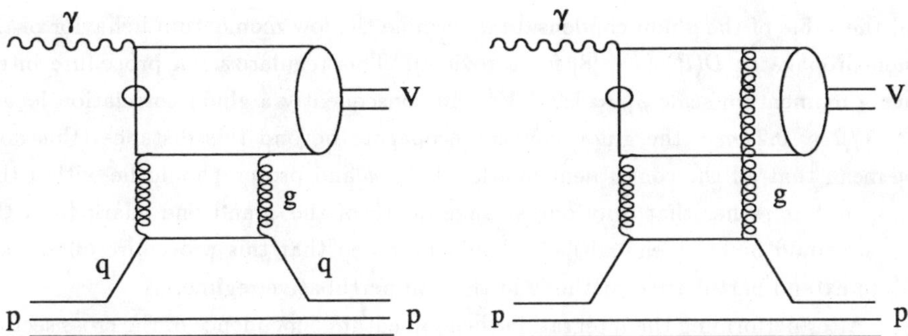

Figure 3: *The two gluon exchange mechanisms which survive at high energy in ϕ photo and electroproduction on free nucleon.*

When the photon energy is large enough ($E_\gamma \geq 5$ GeV) to avoid trivial threshold behavior, the cross section exhibits an universal momentum dependence and a change of slope around $-t = 0.4(GeV/c)^2$: below $-t = 0.4(GeV/c)^2$, the slope parameter take the value $B = 6 \pm 0.5(GeV/c)^{-2}$, above it is smaller ($\simeq 4(GeV/c)^2$).

Above $-t = 1(GeV/c)^2$, no experimental data exist and the exchange of two gluons ("pertubative Pomeron") is a good candidate (Fig. 2). In the limit of large energy transfer ($\nu \to \infty$), the amplitude reduces to the simple form [5,6,7] :

$$A = K F_1(t) \int dl^2 \frac{-4l^2 + t}{(m_\phi^2 + Q^2 - 4l^2)(m_\phi^2 + Q^2 - t)} \left[D(l^2 + \frac{1}{4}t) \right]^2 \qquad (3)$$

It is made of three part. The first is a constant factor K which depends on the various coupling constants and the static properties of the ϕ. The second is the proton form factor $F_1(t)$ which insures that the outgoing proton remains in its ground state. The third is the integral, over the transverse momentum, which describes the exchange of two gluons. The two terms which appear in the denominator correspond to the propagators of the far off-shell quarks marked by a circle in each graph of Fig. 3 : in Eq. 3 their amplitudes have been combined in a compact expression.

When performing the loop integral, it happens that the exchanged gluons become on-shell. This induces infrared singularities, but this is also unphysical since the gluons are confined and cannot propagate far away. Their propagators $D(l^2)$ must be regularized at low momentum. The analysis of the asymptotic high energy behavior of pp scattering, of the small x behavior of the nucleon structure function and the value of the gluon condensate determine the low momentum behavior of the gluon propagator $D(l^2)$ (see [8] for a review). This regularization procedure introduce a momentum scale $\mu_0^2 \approx 1.2 GeV^2$ and consequently a gluon correlation length $a \approx 1/\mu \approx 0.2 fm$: the gluon can not propagate beyond this distance (this does not mean that all the constituent quarks of the ϕ and proton should lie within this distance, but rather that only one strange quark of the ϕ and one quark from the proton should be close enough). It should be noted that this procedure offers us a way to extend pertubative methods in the non pertubative regime.

After performing the integration the momentum dependence of the cross section takes the form :

$$\frac{d\sigma}{dt} \propto \left| \frac{2\mu_0^2 + t}{(m_\phi^2 + Q^2 + 2\mu_0^2)(m_\phi^2 + Q^2 - t)} F_1(t) \right|^2 \tag{4}$$

where the proton form factor has the following expression :

$$F_1(t) = \frac{4m_p^2 - 2.8t}{(4m_p^2 - t)(1 - t/0.7)^2} \tag{5}$$

When integrated against t, these expressions lead to a fair agreement not only with both the magnitude and the Q^2 dependence of the cross section of the $p(e, e')\rho p$ reaction [9,10], but also with the density matrix of the final decay $\rho \to \pi^+ \pi^-$.

The t dependence has never been checked and this remains to be done. It turns out that this model reproduces fairly well the t dependence, as well as the magnitude, of the cross section of the $\gamma p \to \phi p$ in the range $0.4 \leq -t \leq 1 (GeV/c)^2$. It follows from Eq. 4 that the cross section exhibits a node at $t = -2\mu_0^2 = -2.4(GeV/c)^2$, in Fig. 4. It results from the cancellation between the direct two gluon exchange graph (Fig. 3), where the two gluons couple to the same strange quark in the ϕ, and the crossed graph where each gluon couples to a different strange quark in the ϕ.

The physical reason is the following. Around $-t = 1 (GeV/c)^2$, the size of the interaction volume is about $0.2 fm$ and is comparable to the gluon correlation length : the distance over which a gluon can propagate (within which it is confined). The more likely mechanism is therefore the mechanism where one strange quark, of the ϕ, and one valence quark, of the proton, are close enough to be able to exchange two gluons. This does not mean that all the five quarks are in the small interaction volume, but only two. When the momentum transfer increases, components of small spatial extension are selected in the ϕ wave function, *since in this exclusive process we require the proton to stay intact*. The strange quarks of ϕ can now be both at

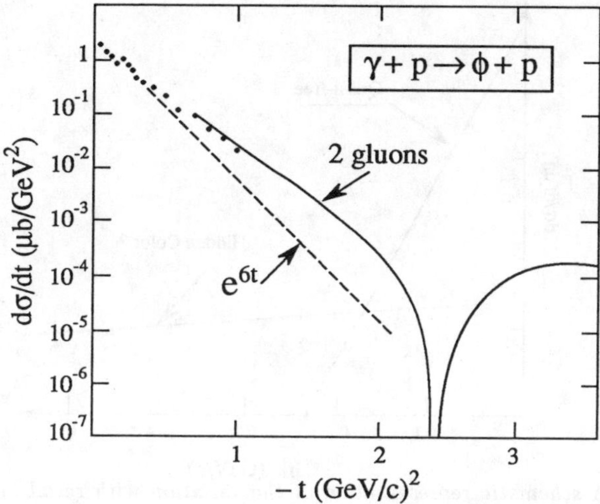

Figure 4: *The cross section of the $\gamma p \rightarrow \phi p$ reaction at $E_\gamma = 10 GeV$ is plotted against t. The dashed line corresponds to the parameterization of the existing experimental data below $-t = 0.4(GeV/c)^2$, in the framework of VDM. The full line corresponds to the two gluon exchange mechanism.*

the same time in the interaction volume, and each of the exchanged gluons can now couple to a different strange quark. At large t, this diagram overwhelms the diagram where the two gluons couple to the same strange quark. At intermediate t, both diagram interfere, providing us with an interesting signature of two gluon exchange mechanisms. Although non leading diagrams, such as those which imply a possible strange intrinsic component in the proton ground state, may shift or partially fill in this dip, its experimental observation would be particularly rewarding since its position is almost fixed by the size of the gluon condensate.

3   The nuclear case

When the nucleon is embedded in the nuclear medium, two mechanisms govern the photo- and the electroproduction of ϕ mesons. The first is trivial : this is the quasi-free production mechanism. The second is much more interesting : this is the exchange of two gluons, each of them being coupled to a different quark. It provides us with a way to look for hidden color components (Fig. 2), or more generally to the correla-

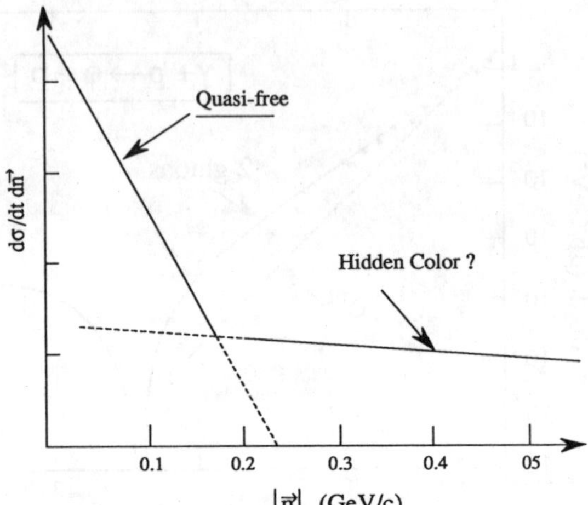

Figure 5: *A schematic representation of the variation with recoil momentum of the cross section of the $D(\gamma, \phi p)n$ reaction.*

tion between two quarks in hadronic matter. A coincidence experiment, for instance $D(\gamma, \phi p)n$ or $^3He(\gamma, \phi 2p)n$, must be performed to disentangle them.

The cross section of the quasi-free process in the $D(\gamma, \phi p)n$ reaction takes the form

$$\frac{d\sigma}{dtd\vec{n}} \propto \left. \frac{d\sigma}{dt}\right|_{\gamma p \to \phi p} \rho(|\vec{n}|) \tag{6}$$

where $\rho(|\vec{n}|)$ is nothing but the nucleon momentum distribution in Deuterium. It decreases very quickly when the undetected neutron momentum increases : selecting high values of the neutron momentum $(n > 200 MeV/c)$ is the way to suppress the contribution of such a trivial mechanism.

The cross section corresponding to the exchange of each of the two gluons with a different quark cluster is expected to exhibit a more flat momentum distribution, since the recoil momentum is shared between the two nucleons. This is schematically depicted in Fig. 5. However it is very difficult to make a quantitative estimate of its magnitude.

A first crude way would be to write the corresponding cross section as :

$$\frac{d\sigma}{dtd\vec{n}} \propto \left. \frac{d\sigma}{dt}\right|_{\gamma p \to \phi p} \left[\varphi_{cc}(\frac{\vec{n}}{2})\right]^2 \frac{F_1^4(\frac{t}{4})}{F_1^2(t)} \tag{7}$$

where the fourth power of the nucleon form factor comes from the fact that two nucleons have to recombine, each at the momentum transfer $t/4$. It is assumed that the recoil momentum is equally shared between the two colored clusters whose the wave function is $\varphi_{cc}(\frac{\vec{n}}{2})$: this component of the deuterium wave function is unknown, and experiments must be performed to determine it, or at least to put some constraint on it.

A more educated guess would be to introduce the relative wave function of two quarks $\varphi_{qq}(l^2)$ in the expression 3 :

$$A = K \frac{F_1^2(\frac{t}{4})}{F_1(t)} \int dl^2 \varphi_{qq}(l^2) \frac{-4l^2 + t}{(m_\phi^2 + Q^2 - 4l^2)(m_\phi^2 + Q^2 - t)} \left[D(l^2 + \frac{1}{4}t) \right]^2 \qquad (8)$$

But again this wave function is not known and the purpose of experiments is to determine it.

In order to increase the probability of the hidden color component, or to enhance the effect of correlations between quarks, one would like to increase the density of the target nucleus. The reactions $^3He(\gamma, \phi 2p)n$ and $^4He(\gamma, \phi 2p)nn$ are good candidates. A further advantage comes from the fact that the target pp pair is almost in a pure $^1S_0$ state, contrary to the deuterium which has a sizable D wave component : the high momentum tail of the quasi-free process is reduced accordingly, leaving more room for the two gluon exchange quark rearrangement processes.

4   Conclusion

The production of heavy quarks on hadronic target is driven by gluon exchange mechanisms and induces color rearrangement in hadronic matter. More specifically, the production of heavy vector mesons at high t selects the exchange of two gluons. While this process is fundamental on its own, to understand how the Pomeron arises from gluon exchange, it provides us with a *direct* way to reveal and study quark correlations in hadronic matter.

While the study of this almost virgin field requires a continuous beam of electrons in the energy range around $15\ GeV$ [11], it should be started at Cebaf [12].

References

[1] J.M. Laget, Lectures given at the 5th Summer School on Nuclear Physics, Seoul, June 29—July 4, 1992, *Journal of the Korean Physical Society* **26** (1993) S244.

[2] Y. Yamauchi and M. Wakamatsu, *Nucl. Phys.* (1986) **A457** 621.
A.M. Kusainov, V.G. Neudatchin and I.T. Obukhovsky, *Phys. Rev.* **C44** (1991) 2343.

[3] H.J. Behrends *et al.*, *Nucl. Phys.* **144** (1978) 22;
D.C. Fries *et al.*, *Nucl. Phys.* **143** (1978) 408.

[4] D.P. Barber *et al.*, *Z. Phys.* **C12** (1982) 1.

[5] A. Donnachie and P.V. Landshoff, *Phys. Lett.* **B185** (1987) 403.

[6] J.R. Cudell, *Nucl. Phys.* **B336** (1990) 1.

[7] P.V. Landshoff and O. Nachtman *Z. Phys.* **C35** (1987) 405.

[8] P.V. Landshoff *Nucl. Phys.* **B** (Proc. Suppl.) **18C** (1990) 211.

[9] D.G. Cassel *et al.* *Phys. Rev.* **D24** (1981) 2787.

[10] J.J. Aubert *et al.* *Phys. Lett.* **B161** (1985) 203; **B133** (1983) 370.

[11] G. Audit *et al.*, *Proposal to the European Electron Facility* (March 1993).

[12] G. Audit *et al.*, *CEBAF Proposal #93-031* (April 1993).

Two pion decay of electroproduced Baryon resonances

M.Ripani

INFN Sezione di Genova

Via Dodecanneso 33, Genova, I-16146, Italy

V. Burkert

Continuous Electron Beam Accelerator Facility

12000 Jefferson Avenue, 23606 Newport News, Virginia, U.S.A

and

The N* group in the CLAS collaboration

Abstract

The two pion electroproduction process on the nucleon ($eN \rightarrow e'N\pi\pi$) will be used in the mass region between threshold and 2.2 GeV both to investigate poorly known baryon resonances and to search for "missing" resonances that are predicted by quark models but have not yet been found experimentally. The electron beam allows to vary both the energy and momentum transfer exploring the Q^2 dependence of the couplings. Many of these resonances should have a large branching ratio in the $\Delta\pi$ and in the $N\rho$ channel, both giving a $N\pi\pi$ final state. Using the CEBAF Large Acceptance Spectrometer CLAS, the hadronic products will be detected in coincidence with the scattered electron. This will make it possible to measure the differential cross section for production of $\Delta\pi$ and $N\rho$ and the decay angular distribution of the Δ and the ρ, which are both expected to be especially sensitive to resonance contributions.

1   Introduction

Understanding the structure of baryons in terms of the fundamental interactions of the constituent quarks and gluons is one of the challenges in strong interaction physics. This interaction is governed by QCD. However, solutions of this theory in the non-perturbative domain of the interaction are extremely difficult to achieve. The lattice gauge approach offers the best hope for exact calculations, but results seem to be far in the future. Thus, models will continue to play an important role. Microscopic models that utilize ingredients from QCD relate the internal structure of baryons to the strong interaction of the confined constituent in an "effective" description. Probing the baryon structure with electromagnetic probes at distance scales of 0.5 fm or less, will give us detailed insight into this interaction. The excitation of baryon resonances is genuinly a non-perturbative phenomenon. Measurements of the transition amplitudes from the ground state nucleon to excited states are sensitive to the internal spatial structure of the excited state, as well as to the spin structure of the transition. Resonance transitions to states in the lower mass region M < 1.7 GeV may be studied effectively using the single-pion or eta decay channel. Nearly all of our knowledge about the electromagnetic coupling of light quark baryons comes from photo- or electroproduction of these channels. In the mass region around and above 1.7 GeV the single-pion and eta channels become less effective in the study of resonance excitations as many states tend to decouple from these channels. Measurement of states such as $S_{31}(1650)$, $D_{13}(1700)$ and $D_{33}(1700)$ has been very difficult in the single-pion channel due to the small branching ratio. However, measurement of the transition form factors of these states is crucial for testing symmetry properties of the quark model, such as the assumption that resonance excitation proceed through single quark transition[1]. From πN scattering experiments we know that many of the higher mass states decay dominantly in multipion channels, such as $\Delta\pi$ or ρN, leading to $N\pi\pi$ final states (see Table I).

Table I. N^* resonances with branching ratio in multipion channels more than 30% (data from *Particle Data Group*, 1988

Res.	Mass (MeV)	Γ (MeV)	B($N\pi\pi$) (%)	B($\Delta\pi$) (%)	B($N\rho$) (%)	B(p.s.) (%)
P11 1/2+	1400-1480	120- 350	30 - 50	10- 20	10 - 15	5 - 20
D13 3/2-	1510-1530	100- 140	40 - 50	20- 30	15 - 25	< 5
S31 1/2-	1600-1650	120- 160	65 - 75	60- 70	10 - 20	-
D15 5/2-	1660-1690	120- 180	60 - 65	55- 60	< 10	< 5
F15 5/2+	1670-1690	110- 140	35 - 45	10- 15	10 - 20	15 - 20
D13 3/2-	1670-1730	70 - 120	80 - 90	15- 70	< 20	< 70
D33 3/2-	1630-1740	190- 300	80 - 90	50- 90	< 35	< 40

The SU(6) symmetric quark model[2] predicts many more states than have been found in well studied channels such as $\pi N \to \pi N$. In the QCD improved version of this model this fact is explained by QCD mixing effects leading to the decoupling of many of these states from the single pion channel. Consequently, they are not expected to be seen in elastic πN scattering experiments. However, other models such as the Quark Cluster Model have been developed, which predict a fewer number of states than the symmetric model in accordance with the experimental observation. Clearly, existence or non-existence of these states will be a crucial test of these symmetry properties underlying the light quark baryon spectroscopy. If these states did not exist, our current picture of light quark baryon structure could change dramatically. Search for at least some of these states is called for and is crucial in discriminating between alternative approaches for modeling the baryon structure. How can we search for these states ? The QCD improved quark model predicts many of these states to couple strongly to the two-pion channel through decays such as $N^* \to \Delta\pi$ or $N^* \to \rho N$ (see Table II).

Table II. Resonances predicted by quark models which are <u>not</u> coupled to the πN channel, but which should be coupled to the $\pi\pi N$ channel. (from ref. 3)

Predicted resonance	Mass (MeV)	Multiplet	Predicted dominant decay mode(s)
F15	1955	$(70,2^+)$	ωN, $\Delta\pi$, ρN
F15	2025	$(70,2^+)$	ωN, $\Delta\pi$
P13	1955	$(70,2^+)$	$\Delta\pi$ (P-wave)
P13	1980	$(70,2^+)$	$\Delta\pi$ (F-wave)
P31	1875	$(56,2^+)$	ρN
F35	1975	$(70,2^+)$	ρN, $\Delta\pi$

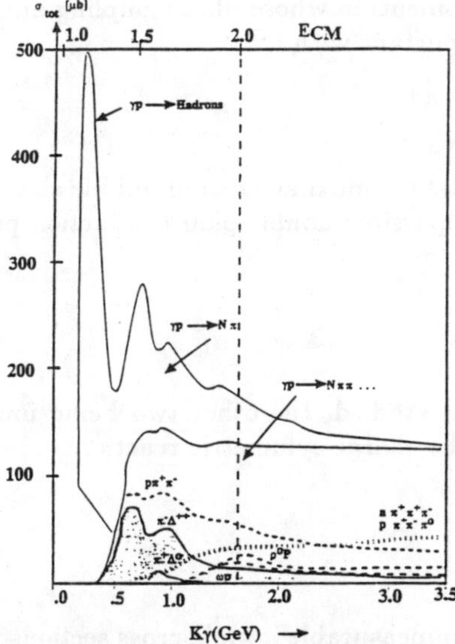

Fig. 1. Total photohadron cross section on proton

In the total photoabsorption cross section on the proton (see Fig. 1) above 400 MeV of photon LAB energy two pion production becomes possible, as shown clearly by several data from bubble chamber experiments[4] . Therefore we suggest that measuring the process

$$eN \to e'N'\pi\pi$$

on protons and neutrons, it will be possible to extract informations on poorly known and "missing" baryon resonances. Main contributing channels to this process are:

a) $\Delta(1236)$ production : $eN \to e'\Delta\pi \to e'(N\pi)\pi$

b) ρ meson production : $eN \to e'N\rho \to e'N(\pi\pi)$

(where the brackets indicate that a peak in the invariant mass of the pair should be expected) There is also a phase space contribution

c) $eN \to e'N\pi\pi$.

All these reactions can proceed through continuum processes, described at low energies by Born terms, or through the excitation of baryon resonances; for example:

$$eN \to e'N^* \to e'\Delta\pi \to e'(N\pi)\pi$$

Exploiting the Q^2 degree of freedom in electroproduction is an important aspect in this study since many of the "missing" resonances are states of highly excited 3-quark orbital angular momentum whose photocoupling amplitudes are predicted to become increasingly important with Q^2.

2 Existing Data

The two-pion-production data come mainly from bubble chamber experiments with real photons[4]; of the three possible double pion production processes on proton

$$\gamma p \to p\pi^+\pi^-$$
$$\gamma p \to p\pi^0\pi^0$$
$$\gamma p \to n\pi^+\pi^0$$

only the first one has been studied, the other two being unmeasurable in bubble chamber experiments; of the charge symmetric reactions

$$\gamma n \to n\pi^+\pi^-$$
$$\gamma n \to n\pi^0\pi^0$$
$$\gamma n \to p\pi^-\pi^0$$

only the second one was "unmeasurable", while cross sections for the other two ones have been deduced from photoproduction experiments on deuterium[5].

In the analysis of the bubble chamber data, the different contributing processes have been separated using combined fits in the invariant masses of the hadronic pairs; the main results are that the $p\pi^+\pi^-$ channel is dominated by strong $\Delta^{++}\pi^-$ production for photon LAB energies between 400 MeV and about 1 GeV (corresponding to W between 1.4 and 1.7 GeV) with a cross section that steeply rises above threshold and then decreases smoothly with increasing photon energy, while above 1 GeV ρ^0 production becomes the main process with a cross section nearly independent of energy within a few GeV. The CMS angular distribution of the π^- produced together with the Δ^{++} turns out to be nearly isotropic near the threshold, while becoming more forward-peaked as the energy increases. The angular distribution of the ρ^0 shows at the contrary a diffraction-like behavior being strongly forward-peaked.

Only a few data exist for two pion electroproduction[6], with W around 1.4 GeV and no attempt has been made to search for resonance contributions.

3 Cross Section Calculations

We have carried out model calculations to determine the sensitivity of the two pion production channel to the presence of resonances. The production of a Δ together with a charged pion[7,8] is dominated by the *"contact"* interaction and the *"pion-in-flight"* process, while these diagrams are absent in the case of production of a neutral pion. In the ρ meson production, the diffraction scattering[9] dominates the forward CMS angles of production [4,10]; in both cases, the resonant part of the hadronic current involves the photocouplings $A_{1/2}$, $A_{3/2}$, $S_{1/2}$, and an effective decay coupling[11,12,13].

Concerning the Q^2 dependence of the Born amplitudes, we have followed the Vector Meson Dominance indications of ref. 6. Concerning the Q^2 dependence of the photocouplings of the resonances, a behavior deduced from ref. 13 has been used for the known states, while for the "missing" ones quark model predictions by Koniuk and Isgur have been used, with a prescription by F. Foster and G. Hughes[14] for the Q^2 evolution.

The decay angular distribution of the produced state (Δ or ρ) can give further information on the contributing processes[4]. Therefore, we have studied the density matrix of the Δ^{++} and its sensitivity to the presence of resonances. It turns out that this quantity is sensitive to the presence of the resonance.

In a second step, a "missing" resonance was introduced in the calculation to study the sensitivity for this case. As a candidate, the $F_{15}(1955)$ has been chosen, having both a strong photocoupling and a strong decay coupling in the $\Delta \pi$ and the ρN channels, as predicted by quark models.

Two observations are worth noting:

1) very interesting channels are those containing a neutral pion or a charged ρ meson, like $\Delta^+ \pi^0$ and $\rho^+ n$ on proton, because in these cases the Born terms contribution is considerably reduced or almost absent.
2) Increasing Q^2 helps separating the resonant contribution from the Born continuum.

4 Experimental Method

This experiment will use the *CEBAF Large Acceptance Spectrometer* to measure $\Delta \pi$ and ρN electroproduction off protons and neutrons as a function of the mass of the hadronic system W, of Q^2 and of the production angles. Differential cross sections will be measured detecting the hadronic products in coincidence with the scattered electron, and using invariant mass analysis to separate the $\Delta\pi$ and ρN contributions. A subset of data will contain the information on the decay correlation of Δ and ρ, allowing to get additional sensitivity to resonance formation.

Fig. 2 shows a single $e^- p \pi^+ \pi^-$ event plot on the CLAS detector.

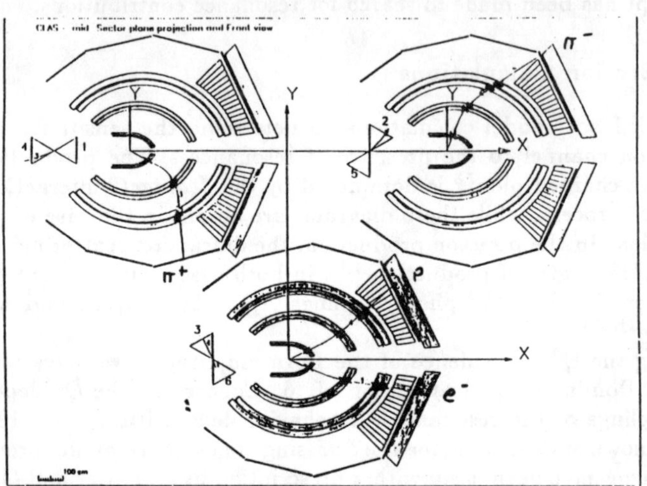

Fig. 2. Single $ep \rightarrow ep\pi^+\pi^-$ event as seen in the CLAS detector; $E = 4\,GeV$, $W = 1.7\,GeV$, $Q^2 = 1\,GeV^2$.

The full kinematics will be measured at one fixed settings, therefore eliminating uncertainties related to time dependent factor, like beam energy, charge integration, beam position, etc.

5 Simulation of the Experiment

To simulate the experiment and to study the response of the CLAS detector (acceptance, resolution, mass separation), an event generator, already developed by the Genova group of INFN for CEBAF experiments simulations[15] and containing the contribution of almost all relevant channels has been extended to all two pion channels.

It's important to remark that measuring simultaneously well known reactions like elastic ep scattering and using the large geometrical coverage of the detector will allow to check the detector acceptance and uniformity, the track and mass reconstructions and efficiencies and to determine the absolute values of the cross sections.

Fig. 3 (a) shows the missing mass reconstruction on the CLAS when measuring $p\pi^+\pi^-$ in coincidence with the electron. Owing to the good momentum and angular resolution of CLAS, one, two and three pion production processes can be well separated.

Fig. 3 (b) shows the the $p\pi^+$ invariant mass reconstruction.

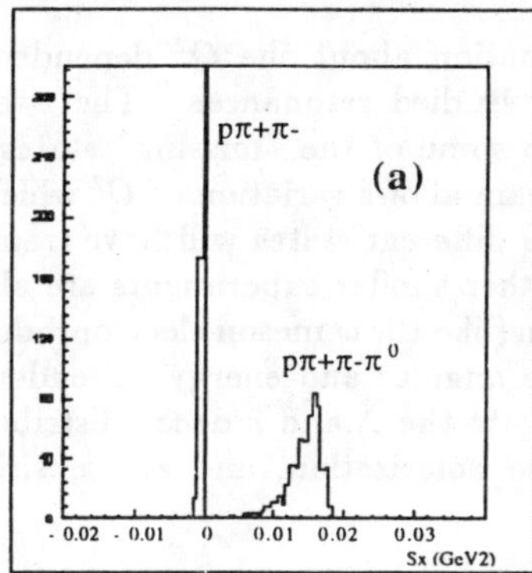

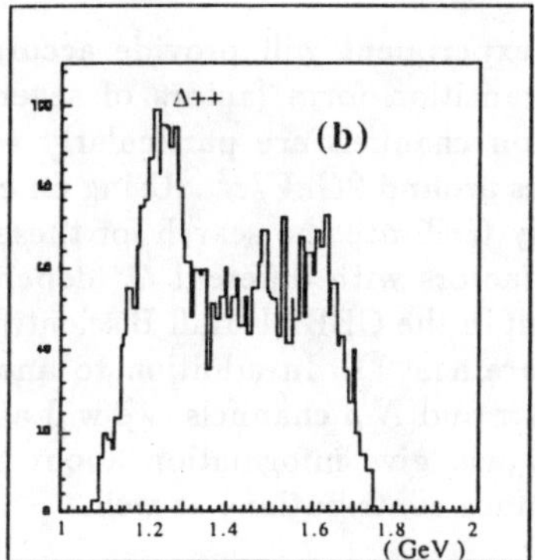

Fig. 3 (a) missing mass reconstruction when $p\pi^+\pi^-$ are detected; (b) invariant mass reconstruction of the $p\pi^+$ pair in the same case; $W = 2\,GeV$, $Q^2 = 1\,GeV^2$, full magnetic field

The detection of neutrals will allow measurement of channels which are of particular interest as they are little affected by Born background; in this case efficiencies are lower, due to the fact that neutrons and photons are detected only in the shower calorimeter, with smaller geometrical acceptance.

In order to illustrate the sensitivity of the measurement to resonance contributions, we have chosen one of the predicted "missing" states, the $F_{15}(1955)$. In Fig. 4 are displayed the expected counts with their statistical error evaluated under the assumptions reported in the figure itself, for the angular distributions of the ρ meson channel and in comparison with the old DESY experiment: the improvement in statistics is quite impressive.

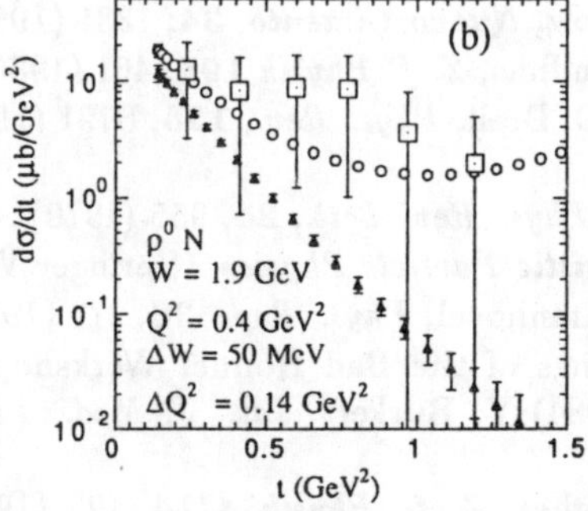

Fig. 4 Expected counts in the measurement of the angular distribution of production of the ρ meson; black triangles: only non- resonant terms; open points: including the "missing" state $F_{15}(1955)$; pointed squares: experimental data from ref. 6; assumed running conditions: beam energy 4 GeV, luminosity of $10^{34}\,cm^{-2}\,sec^{-1}$, 21 days run, resolution of about 70 MeV on W and about 0.16 GeV^2 on Q^2.

6 Summary

This experiment will provide accurate information about the Q^2 dependence of
the transition form factors of several poorly studied resonances. The two-pion
reaction channels are particularly sensitive to some of the "missing" states with
masses around $2GeV/c^2$. Using an electron beam allows variation of Q^2 which will
greatly facilitate the search for these states, as different states will have transition
form factors with different Q^2 dependence. Other similar experiments are already
present in the CEBAF Hall B scientific program (like the ω meson electroproduction
measurement[16]). In addition to analyzing the angular and energy dependence of
the $\Delta \pi$ and $N \rho$ channels, we will also investigate the Δ and ρ decay distributions
which will give information about the particle polarization, and are sensitive to
resonance contributions as well.

References

1. A. J. G. Hey and J. Weyers, *Phys. Lett.* **48B**, 69 (1974); W.N. Cottingham
   and I.H. Dunbar, *Z. Phys.* **C2**, 41 (1979)

2. N. Isgur and G. Karl, *Phys. Rev.*, **D23**, 817 (1981); R. Koniuk and N. Isgur,
   Phys. Rev., **D21**, 1868 (1980); M.M. Giannini, *Rep. Prog. Phys.*, **54**, 453
   (1990)

3. D. M. Manley, *CEBAF internal report* (not published)

4. Cambridge Bubble Chamber Group, *Phys. Rev.*, **155**, 1477 (1967); ABBHHM
   Collaboration, *Phys. Rev.*, **175**, 1669 (1968); D. Lüke and P. Söding - *Springer
   Tracts in Mod. Phys. - 59* (1971)

5. A. Piazza et al., *Lettere al nuovo cimento*, /bf Vol. III N. 12, 403 (1970)

6. P. Joos et al., *Phys. Lett.*, /bf 52B, 481 (1974) V. Eckart et al.,*Nucl. Phys.*,
   B55, 45 (1973) K. Wacker et al. /it Nucl. Phys., **B144**, 269 (1978)

7. P. Stichel and M. Scholz, *Nuovo Cimento*, **34**, 1381 (1964)

8. M. Locher and W. Sandhas, *Z. f. Physik* **195**, 461 (1966)

9. S. M. Berman and S. D. Drell, *Phys. Rev.*, **133**, B791 (1964); P. Söding, *Phys.
   Lett.* **19**, 702(1966)

10. H.H. Bingham et al., *Phys. Rev. Lett.*, **24**, 955 (1970)

11. H.M. Pilkuhn, *Relativistic Particle Physics*, (Springer Verlag 1979), p. 164

12. H. J. Weber and H. Arenhövel, *Phys. Rep.*, **36**, 277 (1978)

13. V. Burkert, Proceedings of the Bad Honnef Workshop, *Springer Tracts in
    Mod. Phys.*, **234** (1984) V. Burkert, *Int. J. Mod. Phys.*, **E, Vol. I**, 421
    (1992)

14. F. Foster and G. Hughes, *Z. f. Physik*, **C14**, 123 (1982); V. Burkert and
    Zhujun Li, private communication

15. P. Corvisiero, L. Mazzaschi, M. Ripani, private communication

16. *CEBAF proposal;* PAC number 91-024

SIGNATURES FOR HYBRIDS

T.Barnes

Physics Division and Center for Computationally Intensive Physics
Oak Ridge National Laboratory, Oak Ridge, TN 37831-6373, USA
and
Department of Physics and Astronomy, University of Tennessee
Knoxville, TN 37996, USA

Abstract

In this review talk I summarize theoretical expectations for properties of hybrid mesons and baryons, and discuss the prospects for identifying these states experimentally.

1. Introduction: Why hybrids rather than glueballs?

Since QCD is a theory which contains both quarks *and* gluons as dynamical degrees of freedom, we would expect to see evidence of both these building blocks in the spectrum of physical color-singlet hadrons. There is much indirect evidence of gluonic basis states in mixing effects, for example in the Breit-Fermi one-gluon-exchange Hamiltonian used in potential models and in the η and η' masses. It is remarkable however that of the hundreds of hadronic states now known, most can be accurately described as states made only of quarks and antiquarks in the nonrelativistic quark model, and the remaining problematic resonances show no clear evidence for states with dominant gluonic valence components. Reviews of candidate gluonic states and other unusual hadronic states from the viewpoints of theorists [1] and experimentalists [2] can be found in the proceedings of recent meetings on hadron spectroscopy.

A priori one might expect that a search for gluons in the spectrum should concentrate on dominantly pure gluon states, since the properties of these might be expected to differ maximally from quark and antiquark states. This naive expectation does not survive closer investigation. The lightest color-singlet glueball basis states one can form from transverse gluons are $|gg\rangle$, and the quantum numbers allowed for these states are $I = 0, J^{PC} = 0^{\pm+}, 2^{\pm+}, 3^{++}, 4^{\pm+}$, and so forth. Since $q\bar{q}$ states can also be made with these quantum numbers, there is a danger of confusion and one would need to identify all such $q\bar{q}$ states in the mass range anticipated for glueballs. Probably the most reliable glueball mass estimates are derived from lattice gauge theory. These QCD simulations anticipate that the scalar should be the lightest glueball, with a mass of about 1.4 GeV; other glueball states are expected to lie near or above 2 GeV [3]. A 1.4 GeV scalar glueball should be evident in the $I = 0$ $\pi\pi$ S-wave phase shift, which is well-established experimentally to about 2 GeV [4]. This phase shift shows a single narrow state, the $K\bar{K}$-molecule candidate $f_0(975)$ [5], and in addition only a slowly rising phase underneath this state. If there are more scalar resonances coupled to this channel they are evidently very broad states (a broad $^3P_0$ $q\bar{q}$ state $f_0(1250)$ is also expected here and has recently been identified in $\gamma\gamma$ [6]). Thus glueball spectroscopy may involve a search for quite broad resonances with conventional I=0 $q\bar{q}$ quantum numbers, which could be a very unproductive or at best ambiguous exercise. Although one can make J^{PC}-exotic states from $|ggg\rangle$ basis states, which would be much more characteristic experimentally, these should appear at rather higher masses. A final problem is that expectations for the preferred decay modes of glueballs are rather obscure theoretically, since glueballs naively have flavor-singlet couplings, but these can be masked by phase space and wavefunction effects. Thus we are led to ask whether gluons might appear elsewhere in the spectrum, in states which might be more distinct experimentally.

This leads us to the subject of hybrids, which are resonances in which both quarks and gluons are present in the dominant basis state. Since a single gluon is a color octet, it may be combined with $q\bar{q}$ and qqq color-octet quark states to make overall color-singlet hybrid basis states. These new basis states will lead to additional resonances beyond those expected by the $q\bar{q}$ and qqq quark model assignments; although physical resonances are linear combinations of conventional and hybrid basis states, even if the mixing is large we still will find more levels than the quark model alone expects. The detailed mass and quantum number predictions for these additional states depend somewhat on the model used to study them. Nonetheless, as we shall see, there is general agreement that these states have characteristic features that make them most attractive experimentally than the broad isosinglet glueballs, so that the hybrid sector is where we may first see clear evidence of resonances with large or dominant gluonic components.

2. Hybrid mesons; masses and quantum numbers

Both meson and baryon hybrids are anticipated theoretically, since we can form color-singlet basis states from $q\bar{q}g$ and $qqqg$. Although these hybrid basis states mix with ordinary quark-model basis states such as $|q\bar{q}\rangle$ and $|qqq\rangle$ to form physical hadrons, we will in any case have more physical states than the quark model would anticipate, and if the mixing is not large the dominantly hybrid states may have unusual properties.

We will first consider hybrid mesons because they have a very attractive feature: Some hybrid mesons have exotic-J^{PC} quantum numbers forbidden to ordinary $q\bar{q}$ mesons, and for these states the possibility of confusion with $q\bar{q}$ states due to mixing does not arise. Identification of dominantly hybrid states or other unusual states which have quantum numbers accessible to conventional quark model hadrons will remain an ambiguous exercise until the quark states in the relevant mass region are well established. Establishing these $q\bar{q}$ states in the $\approx 1.5 - 2.5$ GeV mass region of interest for hybrids is a problem which will undoubtedly require considerable future effort. In contrast, identification of a state with exotic J^{PC} quantum numbers would at least indicate that we have found a state beyond the conventional quark model. One could then determine decay modes and search for other members of a multiplet, to see whether the new state agrees with expectations for a hybrid or glueball or perhaps a molecular multiquark state. Hadronic molecules such as the $K\bar{K}$ candidates $f_0(975)$ and $a_0(980)$ and other possibilities such as vector-vector states will complicate the identification of hybrids somewhat by contributing more non-$q\bar{q}$ resonances to the meson spectrum, and these molecules can also have exotic quantum numbers. Fortunately molecules should have rather characteristic features [7] such as S-wave meson-meson quantum numbers and masses not far below the associated two-meson threshold, so it should be easy to distinguish them from hybrids.

We begin our discussion of hybrid mesons with a summary of predictions for the spectrum. The masses of light hybrid mesons ($q = u, d$ and s) have been estimated using the MIT bag model [8, 9], QCD sum rules [10], the flux tube model [11] and heavy-quark lattice gauge theory [12]. In the bag model the lightest gluon mode is TE, with $J^P = 1^+$; combining this with the $J^P = 0^-$ and 1^- of $q\bar{q}$ in their lowest bag modes gives

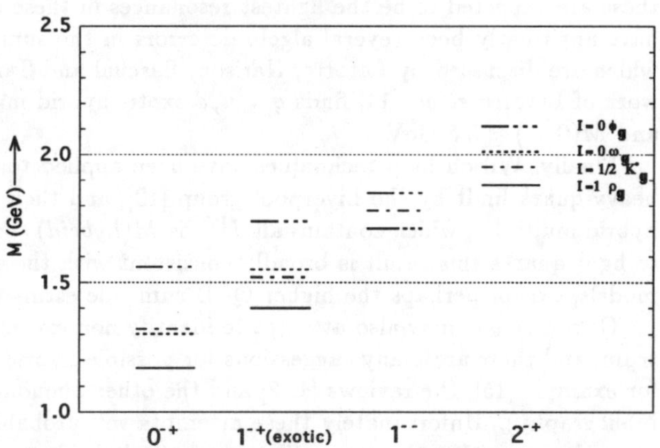

Figure 1: Light hybrid mesons in the bag model [8].

$$J^{PC_n}(q\bar{q}g) = \begin{cases} 0^{-+}, 1^{-+}, 2^{-+}, & S_{q\bar{q}} = 1; \\ 1^{--} & S_{q\bar{q}} = 0 \ . \end{cases} \tag{1}$$

Consideration of other multiplets shows that all J^{PC} can be made from $q\bar{q}g$ basis states. The $J^{PC} = 1^{-+}$ is of special interest because this is the lightest J^{PC}-exotic predicted by the bag model. The exact mass depends on the details of the bag parameters chosen, but typically the mass of this exotic is estimated to be about

182

$M(1^{-+}) \approx 1.5$ GeV. The spectrum of physical hybrids ($q\bar{q}g$ states mixed perturbatively with $q\bar{q}$ and $q\bar{q}gg$ components) found by Barnes, Close and deViron [8] in this multiplet is shown in Fig.1. Similar results for the bag model hybrid spectrum were been reported by Barnes and Close, Chanowitz and Sharpe and Flensberg, Peterson and Sköld [9].

The predictions of the flux tube model for hybrids are especially interesting because this model gives good results for both the spectrum and decays of conventional quark states. The lightest hybrid flux-tube multiplet is quite rich, and contains the quantum numbers $J^{PC} = 1^{\pm\pm}, 2^{\pm\mp}, 1^{\pm\mp}$ and $0^{\pm\mp}$, all approximately degenerate. Note the presence of the 1^{-+} exotic, as in the bag model, and the additional exotics 0^{+-} and 2^{+-}. The mass of this multiplet (with $q = u, d$) is somewhat higher than bag model expectations, and has varied between ≈ 1.7 GeV and 2.0 GeV in the flux tube literature [11].

QCD sum rules can be used to estimate the masses of the exotic hybrids, since these are expected to be the lightest resonances in these J^{PC}-exotic channels. There have apparently been several algebraic errors in the sum rule literature in the past, which are discussed by Latorre, Narison, Pascual and Tarrach [13]. The most recent work of Latorre *et al.* [14] finds $q = u, d$ exotic hybrid masses of $M(1^{-+}) \approx 2.1$ GeV and $M(0^{--}) \approx 3.8$ GeV.

Finally, Wilson loop techniques have been applied to the study of hybrids in the heavy-quark limit by the Liverpool group [12], and their result for the lightest "E_u" hybrid multiplet, which contains all J^{PC}, is $M(hybrid) \approx m_{Q\bar{Q}} + 1$ GeV. If applicable to light quarks this result is broadly consistent with the estimates found using other models, except perhaps the higher QCD sum rule estimate.

Of course one may also attempt to identify non-exotic hybrids in the meson spectrum, and there are many suggestions for possible hybrid states in the literature (see for example [15], the reviews [1, 2] and the other phenomenological references in the bibliography). Unfortunately these attempts will probably remain controversial due to confusion with $q\bar{q}$ states until true exotic hybrids are identified, following which the identification of non-exotic partners in hybrid multiplets should be more straightforward.

There has recently been considerable interest in searches for heavy-quark hybrids. The advantage of these systems is that they should be relatively pure in Hilbert space, because the mixing between $|q\bar{q}g\rangle$ and other basis states is driven by $\vec{j}^{a} \cdot \vec{A}^{a}$, which is reduced due to the lower velocities of heavy quarks. Since the spectrum of heavy quarkonium is less complicated than in light hadronic systems, experimental identification of heavy hybrids may be correspondingly straightforward.

There are several model calculations of heavy-quark hybrid masses in the literature. The mass estimated for the lowest $c\bar{c}$-hybrid multiplet has varied over the range $\approx 4.2 - 4.5$ GeV in flux tube references [11]. Perantonis and Michael [12] find 4.04

GeV for the lightest $c\bar{c}$-hybrids in quenched heavy-quark lattice gauge theory, and estimate 4.19 GeV as a full QCD result. They also note that the effective $Q\bar{Q}$ potential for hybrids is rather shallow, so radially and orbitally excited hybrids should lie not far above the hybrid ground state. Finally, Narison [10] quotes a QCD sum rule result of 4.1 GeV for the 1^{-+} exotic $c\bar{c}$-hybrid, consistent with lattice gauge theory and the lower flux-tube results. Theoretical estimates of the mass of the lightest $c\bar{c}$-hybrid multiplet are thus typically about 4.2 ± 0.2 GeV. Detection of charm hybrids would be an obvious goal of a high luminosity e^+e^- facility such as a Tau-Charm Factory [21]. Typical mass estimates for $b\bar{b}$-hybrids are 10.5 GeV (Narison, sum rules [10]) and up to 11.1 GeV (Merlin and Paton, flux tube model [11]).

In summary, although there is considerable variation in detail, all the approaches applied to the hybrid spectrum predict that the lightest exotic hybrid mesons have masses of $\approx 1.5 - 2.0$ GeV, and in heavy-quark systems (from the flux-tube model and lattice gauge theory) lie near the lightest $Q\bar{Q}$ mass plus ≈ 1 GeV. The exotic quantum numbers $J^{PC} = 1^{-+}$ are often suggested for experimental searches in light-quark systems, because all techniques find a hybrid with these quantum numbers, and the flux tube model predicts that the $I = 1\ 1^{-+}$ in particular should be relatively narrow.

3. Signatures for hybrid mesons: decays and couplings

In addition to general searches for extra or J^{PC}-exotic resonances, one can use theoretical expectations for hybrid decay modes to motivate experimental searches in particular strong final states or in electromagnetic processes.

Theoretical models predict rather characteristic two-body decay modes for hybrids. Both constituent gluon [16] and flux tube [11] models find that the lightest hybrids decay preferentially to pairs of one $L_{q\bar{q}}{=}0$ and one $L_{q\bar{q}}{=}1$ meson, for example πf_1 and πb_1. These unusual modes have received little experimental attention and may explain why hybrids were not discovered previously. Several experiments are planned which will look at these final states, including E818 (to study $\pi^- f_1$ [17]) and E852 at BNL (to study πf_1 and $\pi\eta$ [18]). There is already some evidence for an $I = 1$ 1^{-+} exotic at 1.775 GeV from a SLAC photoproduction experiment [19], which will be studied by E687 at Fermilab. Finally, E781 at Fermilab plans a sensitive search for hybrids using the Primakov effect [20].

Much of the experimental work on possible evidence for hybrids has concentrated on the $\pi\eta$ system, since this is easy to analyze (all odd-L $\pi\eta$ waves are exotic) and there have long been indications of an important P-wave contribution in the angular distribution near and above 1.3 GeV [22]. Of course the important question is whether this P-wave amplitude is resonant (this is complicated by the interference of the P-wave with the resonant $a_2(1320)$). There are recent experimental studies of this reaction by GAMS [23] and E179 at KEK [24] which suggest that this amplitude is indeed resonant. An IHEP group however finds a nonresonant P-wave amplitude, however (also reported in [24]), and several other experiments have reported various results for $\eta\pi$ and related channels such as $\eta'\pi$; see for example the HADRON93 proceedings for recent summaries of this work. In view of the disagreements between experiments and complications in the analyses the nature of this P-wave amplitude should probably be considered an open question.

Heavy-quark hybrids, in particular charmonium hybrids, may be accessible at high-luminosity e^+e^- machines. One may search for the "extra" 1^{--} hybrid states directly through a high-statistics scan of R; the hybrids are expected to appear relatively weakly since they must couple to the photon through their $c\bar{c}$ components, but as they may be below their preferred S+P decay threshold, they may appear as relatively narrow peaks. It may be possible to produce charm hybrids with other quantum numbers than 1^{--} through cascade decays, in which an initial high-mass 1^{--} $c\bar{c}$ pair is produced which then cascades to a charm hybrid plus a light hadronic system such as η or $\pi\pi$. The charm hybrid and light hadronic system should be produced in a relative P-wave, which may assist in identification of hybrid states with specified quantum numbers. The prospects for detecting heavy-quark hybrids using these techniques have recently been discussed by Barnes and Close [21].

4. Hybrid baryons

One may also form color-singlet quark-and-gluon basis states from $qqqg$, and these additional basis states are expected to lead to additional hybrid baryon resonances not predicted by the naive qqq quark model. Since all J^P can be made from conventional qqq quark model baryon states, there are no hybrid baryon exotics, and one must instead hope to identify baryon hybrids as additional states in a well established qqq background. Of course one may hope to identify a dominantly $qqqg$ hybrid baryon by anomalous couplings relative to those of conventional qqq baryons; possibilities include both strong and electromagnetic couplings. There are indications that some familiar predictions for light baryon properties in the qqq quark model are relatively insensitive to admixtures of $qqqg$ basis states [29], and this mixing may be rather large. In this case it may prove difficult to identify the extra hybrid baryon resonances. Identification of hybrid baryons will be an especially difficult problem if the mixing angles are large, since the distinction between ordinary and hybrid baryons will be lost, and the only evidence for hybrids will be an overpopulation of levels relative to the qqq model.

To date the masses of hybrid baryons have been calculated in detail only in the bag model. In the bag model the lowest unmixed hybrid baryon basis states span a **70** under SU(6), which after mixing contribute a **70** of baryon resonances in addition to the states expected by the qqq quark model.

The spectrum of light, physical (mixed with qqq and $qqqgg$) hybrid levels with $q = u, d$ found by Barnes and Close [26] is shown in Fig.2. A very similar pattern of multiplet splittings was found by Golowich, Haqq and Karl [27], albeit with an overall mass scale about 200 MeV lower due to their choice of more conventional bag model parameters. This work was extended to $q = u, d, s$ by Carlson and Hansson [28].

Since the nonstrange baryon spectrum is quite well established to about 2 GeV, one might expect to confirm or refute the bag model description of the hybrid baryon spectrum easily. Remarkably this has not been possible, because there are qqq experimental candidates already known near each of the light hybrid levels predicted. This may imply hidden hybrid levels near the dominantly qqq levels, or perhaps some resonances normally assigned to qqq are actually hybrids. The Roper is often cited as a possible misidentified hybrid, because the lightest hybrid baryon in the bag model has Roper quantum numbers, and Golowich, Haqq and Karl [27] predicted a mass for this hybrid of about 1400-1450 MeV, consistent with the Roper mass. Note that the expectation of a hybrid baryon near the Roper mass is closely linked to the bag prediction of a 1^{-+} exotic hybrid meson near 1.5 GeV, since these predictions use similar bag model parameters.

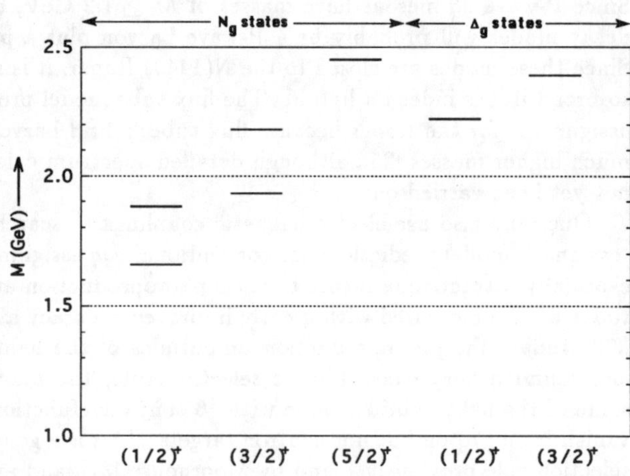

Figure 2: Bag model spectrum of light nonstrange hybrid baryons [26].

Since we cannot distinguish between dominantly qqq and $qqqg$ baryons by their quantum numbers, it may be necessary to use other properties such as strong decay modes or electromagnetic couplings to distinguish these states. The two-body decay modes of hybrid baryons in the flux tube model should satisfy a selection rule similar to that for hybrid mesons. In the baryon case the hybrid flux tube basis state has a spatially-odd wavefunction for reflection of the flux tube through the qqq plane, at least in the heavy-quark limit. When a $q\bar{q}$ pair is formed through flux tube breaking, with qqq and $q\bar{q}$ final states, the initial odd spatial symmetry will lead to a small matrix element unless the final meson or baryon has an internal orbital excitation. Since P-wave $q\bar{q}$ mesons have masses of $M \geq 1.2$ GeV, the preferred hybrid baryon decay modes will probably be a P-wave baryon plus a pion, for example $N(1520)\pi$. Since these modes are closed to the N(1440) Roper, it is surprising that the Roper is so broad if it is indeed a hybrid. The flux tube model probably supports a radial-qqq assignment for the Roper because flux tube hybrid baryons are expected to occur at much higher masses [25], although detailed spectrum calculations in this model have not yet been carried out.

One may also use electromagnetic couplings to search for hybrid baryons and to test quark model predictions for conventional qqq assignments. This approach will be especially attractive in future because photoproduction and electroproduction amplitudes will be measured with greatly improved accuracy at CEBAF. Barnes and Close [30] studied the photoproduction amplitudes of the light hybrid multiplet in Fig.2, and found a very characteristic selection rule; the photoproduction amplitudes of some of the light hybrids, those with ^{4}8 spin wavefunctions in their $qqqg$ component, vanish from proton but not neutron targets. This is a generalization of an excited-qqq selection rule previously found by Moorhouse [31], and applies to the lightest hybrid baryon, which is the candidate Roper state. Since the Roper does not satisfy this selection rule, the bag model hybrid baryon does not appear to be a good description of the Roper. Alternatively it has been suggested that the Roper may be a different combination of basis states than these simple bag model calculations find [32]. To distinguish between hybrid and excited-qqq assignments for states such as the Roper it may be more useful to have accurate quark model predictions for photoproduction and electroproduction amplitudes, since the qqq quark model wavefunctions are reasonably well established. At moderate Q^2, early results by Close and Li [33] suggested that the radial-qqq assignment disagreed with existing Roper data [34], but subsequent quark model calculations by Warns, Pfeil and Rollnik [35] and Capstick [36] found that more accurate wavefunctions alter these conclusions, leading to quite small radial-qqq photoproduction and electroproduction amplitudes. More recently, Capstick and Keister [37] have found that relativistic effects are quite important in radial-qqq electroproduction amplitudes, so previous nonrelativistic calculations may be inaccurate. Evidently careful theoretical studies of electroproduction amplitudes are required for comparison with the accurate experimental data expected from CEBAF. In the higher-Q^2 regime, where we may expect to see an increasingly important contribution from perturbative QCD processes, it may also be possible to distinguish between dominantly qqq and $qqqg$ baryons. Carleson and Mukhopadhyay [38] note that the leading power in the transverse electroproduction form factor should distinguish between these cases, and expect $G_+(Q^2) \propto 1/Q^3$ for a qqq state but $G_+(Q^2) \propto 1/Q^5$ for electroproduction of a dominantly-$qqqg$ state. To test these predictions it will be important to determine resonance electroproduction functions

187

over a wide range of Q^2.

5. Summary and Conclusions

In this review we have discussed theoretical expectations for the properties of hybrid mesons and baryons. Hybrids are theoretical resonances in which gluonic excitations are the dominant basis states. Although mixing between pure quark basis states and excited-glue basis states is anticipated and may be large, the additional gluonic basis states will lead to more resonances than the quark model anticipates. This mixing with quark states can be avoided in the meson sector, since one can form J^{PC}-exotic combinations such as 1^{-+} from $q\bar{q}g$ basis states. Experimental searches for hybrids with these or other exotic quantum numbers offer the best prospects for identifying hybrids, since these cannot be confused with conventional $q\bar{q}$ states. We also discussed predictions for the spectrum of light-quark and heavy-quark hybrid mesons and their expected two-body decay modes. The favored mode has one L=0 and one L=1 $q\bar{q}$ meson; this unusual final state may explain why hybrids have not yet been reported. Finally we discussed light hybrid baryons and the prospects for detecting these experimentally, through studies of the light baryon spectrum and decays and their photoproduction and electroproduction amplitudes.

6. Acknowledgements

I would like to thank C.E.Carlson and the organizers of the Elba conference on Exclusive Reactions at High Momentum Transfers for their kind invitation to attend this meeting and for the opportunity to discuss the status of hybrids and related topics with my fellow participants. I would also like to thank D.V.Bugg, S.Capstick, F.E.Close, N.Isgur, Z.P.Li, J.Paton and E.S.Swanson for additional discussions of material presented here. This research was sponsored in part by the United States Department of Energy under contract DE-AC05-840R21400, managed by Martin Marietta Energy Systems, Inc, and by the United Kingdom Science Research Council through a Visiting Scientist grant at Rutherford Appleton Laboratory.

References

[1] See for example N.Isgur, CEBAF report CEBAF-TH-92-31, in Proceedings of the XXVI International Conference on High Energy Physics (Dallas, August 1992); S.Godfrey, in Proceedings of the BNL Workshop on Glueballs, Hybrids and Exotic Hadrons (AIP, 1989), ed. S.-U. Chung; F.E.Close, Rep. Prog. Phys. 51, 833 (1988).

[2] See for example A.Dzierba, Indiana University report IUHEE-93-2, in Proceedings of the BNL meeting on Future Directions in Particle and Nuclear Physics at Multi-GeV Hadron Facilities (Brookhaven, N.Y. 4-6 March 1993); D.Herzog, in Proceedings of the Second Biennial Conference on Low Energy Antiproton Physics (Courmayeur, 14-19 Sept. 1992), pp.499-518.

[3] See for example C.Michael, University of Liverpool report LTH-286, and Proceedings of the Workshop on QCD: 20 Years Later (Aachen, 9-13 June 1992).

[4] G.Grayer *et al.*, Nucl. Phys. B75, 189 (1974).

[5] J.Weinstein and N.Isgur, Phys. Rev. Lett. 48, 659 (1982); Phys. Rev. D27, 588 (1983); Phys. Rev. D41, 2236 (1990); J.Weinstein, Phys. Rev. D47, 911 (1993).

[6] J.K.Bienlein (for Crystal Ball Collaboration), in Proceedings of the Ninth International Workshop on Photon-Photon Collisions (LaJolla, 22-26 March 1992), eds. D.O.Caldwell and H.P.Paar (World Scientific, 1992), pp.247-257, and references cited therein.

[7] Signatures for molecules are discussed by T.Barnes, Oak Ridge National Laboratory report ORNL-CCIP-93-4, Proceedings of the BNL meeting on Future Directions in Particle and Nuclear Physics at Multi-GeV Hadron Facilities (Brookhaven, N.Y. 4-6 March 1993).

[8] T.Barnes, F.E.Close and F.deViron, Nucl. Phys. B224, 241 (1983).

[9] T.Barnes and F.E.Close, Phys. Lett. 116B, 365 (1982); M.Chanowitz and S.R.Sharpe, Nucl. Phys. B222, 211 (1983); M.Flensburg, C.Peterson and L.Sköld, Z. Phys. C22, 293 (1984).

[10] I.I.Balitsky, D.I.Dyakanov and A.V.Yung, Phys. Lett. 112B, 71 (1982); J.Govaerts, F.deViron, D.Gusbin and J.Weyers, Phys. Lett. 128B, 262 (1983); J.I.Latorre, S.Narison, P.Pascual and R.Tarrach, Phys. Lett. 147B, 169 (1984); J.I.Latorre, P.Pascual and S.Narison, Z. Phys. C34, 347 (1987); S.Narison, "QCD Spectral Sum Rules", Lecture Notes in Physics Vol.26, p.375 (World Scientific, 1989). Some of the earlier references were subsequently reported to have algebraic errors.

[11] N.Isgur, R.Kokoski and J.Paton, Phys. Rev. Lett. 54, 869 (1985); J.Merlin and J.Paton, J. Phys. G11, 439 (1985); Phys. Rev. D35, 1668 (1987).

[12] S.Perantonis and C.Michael, Nucl. Phys. B347, 854 (1990), and references cited therein.

[13] J.I.Latorre, S.Narison, P.Pascual and R.Tarrach, Phys. Lett. 147B, 169 (1984).

[14] J.I.Latorre, P.Pascual and S.Narison, Z. Phys. C34, 347 (1987).

[15] A.Donnachie and Yu.S.Kalashnikova, "Four quark and hybrid mixing in the light quark vector sector." Manchester University preprint M-C-TH-93-02 (March 1993); S.Ishida et al., Prog. Theor. Phys. 88, 89 (1992).

[16] A.LeYaouanc, L.Oliver, O.Pène, J.-C.Raynal and S.Ono, Z. Phys. C28, 309 (1985); F.Iddir, A.LeYaouanc, L.Oliver, O.Pène, J.-C.Raynal and S.Ono, Phys. Lett. B205, 564 (1988).

[17] S.U.Chung, personal communication.

[18] A.Dzierba, personal communication.

[19] G.T.Condo et al., Phys. Rev. D43, 2787 (1991).

[20] T.Ferbel, Rochester University report UR-1129 (1989); Proceedings of the Third International Conference on Hadron Spectroscopy (Ajaccio, France, 23-27 Sept. 1989), pp.157-164.

[21] See F.E.Close, RAL-93-053 and T.Barnes, ORNL-CCIP-93-11 / RAL-93-065 for discussions of the possibility of detecting charm hybrids at a Tau-Charm Factory. (Both references to appear in Proceedings of the Third Workshop on the Tau-Charm Factory, Marbella, Spain, 1-6 June 1993.)

[22] W.Apel et al., Nucl. Phys. B193, 269 (1981).

[23] M.Boutemeur et al., in Proceedings of HADRON89, pp.119-126 (Ajaccio, France, 23-27 Sept. 1989); D.Alde et al., Phys. Lett. 205B, 397 (1988).

[24] E179 Collaboration, contribution to the Proceedings of HADRON93, Como, Italy, 21-25 June 1993.

[25] N.Isgur, personal communication.

[26] T.Barnes and F.E.Close, Phys. Lett. 123B, 89 (1983).

[27] E.Golowich, E.Haqq and G.Karl, Phys. Rev. D28, 160 (1983).

[28] C.E.Carlson and T.H.Hansson, Phys. Lett. 128B, 95 (1983).

[29] Z.Li and G.Karl, "Spin Structure of Nucleons in a Quark Model with Constituent Gluons", Carnegie-Mellon / Guelph University report (Sept. 1993).

[30] T.Barnes and F.E.Close, Phys. Lett. 128B, 277 (1983).

[31] R.G.Moorhouse, Phys. Rev. Lett. 16, 772 (1966).

[32] Z.Li, V.Burkert and Z.Li, Phys. Rev. D46, 70 (1992)

[33] F.E.Close and Z.P.Li, Phys. Rev. D42, 2194 (1990).

[34] See P.Stoler, these proceedings, for a summary of experimental electroproduction amplitudes.

[35] M.Warns, W.Pfeil and H.Rollnik, Phys. Rev. D42, 2215 (1990).

[36] S.Capstick, Phys. Rev. D46, 1965 (1992); Phys. Rev. D46, 2864 (1992).

[37] S.Capstick and B.Keister, in preparation.

[38] C.E.Carlson and N.C.Mukhopadhyay, Phys. Rev. Lett. 67, 3745 (1991).

SEMI-INCLUSIVE DEEP-INELASTIC SCATTERING ON $A = 1, 2$

A.E.L. Dieperink

and

G.D. Bosveld

Kernfysisch Versneller Instituut
Groningen, NL-9747 AA, The Netherlands

ABSTRACT

Semi-inclusive deep inelastic neutrino scattering with the detection of a slow proton
in coincidence with the scattered muon on $A = 1, 2$ is discussed.

1. Introduction

Inclusive deep inelastic lepton scattering (dils) on nucleons is a well established
tool for the investigation of the quark-parton model. In the Bjorken scaling limit the
cross section is proportional to deep inelastic structure functions which depend only
on one variable x, the light-cone momentum fraction carried by the hit quark.

In more exclusive experiments, in which also mesons and baryons are observed
in coincidence with the scattered lepton one can study the hadronization process. In
general the hadronization process is rather complicated; however, in two limiting cases
the situation becomes simpler, namely the detection of high energy hadrons coming
from fragmentation of the leading struck quark, and that of slow hadrons coming from
the spectator target quarks. Motivated by recent analysis of Cern (WA25) data of
deep inelastic neutrino scattering[1] I will discuss semi-inclusive deep inelastic scattering (sidis) in which at least one low energy proton (characterized by l.c. momentum
fraction $z = (p^0 + p^3)/m$) is present, i.e. $\nu(\bar{\nu}) + A \rightarrow \mu + p + X$.

2. The free proton

On a free nucleon slow protons originate from hadronization of the spectator
quarks in the struck nucleon (so-called debris fragmentation). To compute the cross
section I will treat the interaction of the lepton with a valence quark (leading to the
fragmentation of the spectator valence quarks) and that with a sea-quark separately,
i.e. $\sigma(x, z) = \sigma_{\text{val}} + \sigma_{\text{sea}}$.[2,3] The former is treated in a factorized form of a structure function times a di-quark fragmentation function $D^p_{\{qq\}}(z)$, (neglecting transverse
momentum), e.g.

$$\frac{d^2\sigma^{\nu p}_{\text{val}}(x, z)}{dx dz} = \frac{G^2 mE}{\pi} \frac{2x}{1 - x} d_{\text{val}}(x) D^p_{\{uu\}}\left(\frac{z}{1 - x}\right).$$

The sea-quark contribution, which dominates at small x values, has been simulated
by a pion cloud model, which has been shown to be able to simulate the effect of the

sea in inclusive dis very well,[4,5]

$$\frac{d^3 \sigma_{sea}^{\nu p}(x,z)}{dx\,dz\,dt} = \frac{G^2 s}{2\pi}(1-z)\frac{g_{\pi NN}^2}{4\pi^2}\frac{t}{(t+m_\pi^2)^2}F_2^\pi(\frac{x}{1-z})F_{\pi NN}^2(t),$$

where t is the four momentum transferred, $F_2^\pi(x)$ the pion structure function, and $F_{\pi NN}(t)$ the pion-nucleon form factor.

The observation[6] that the ratio $P(x,z)$ of the sidis cross section $\sigma(x,z)$ and the inclusive cross section $\sigma(x)$ has a strong bias for small x values can be explained[2,3] qualitatively by pure kinematics: a proton at rest can only be produced at $x = 0$. However, the detailed behavior of $P(x,z)$ depends on the choice of the fragmentation function.

3. The Deuteron

In a nuclear medium there are several processes that can give rise to slow protons.[2,7] First in analogy with sidis on a free nucleon target slow protons can originate from fragmentation of the spectator quarks (debris fragmentation) in the struck nucleon with the other nucleons acting as spectators. This process will be referred to as the *direct* process. A new effect, which only occurs for $A > 1$, is the deep inelastic scattering off a bound nucleon and the observation of a spectator proton (called the *spectator* process). Since in a mean field description the $A - 1$ nucleus is bound, this process provides evidence for the presence of correlations in the ground state (for $A > 2$). The important point is that in the independent pair approximation neglecting the centre-of-mass motion of the pair and final state interactions one has a direct relation between the l.c. momentum fractions of the detected (z) and struck (y) nucleons, $y + z = 2$.

In the following I consider the sidis cross section on the deuteron with a slow proton present assuming that the above processes contribute incoherently.

3.1 Classification of events

The data to be used as a comparison were obtained from an analysis of deep inelastic (anti-)neutrino scattering experiment on the deuteron in a bubble chamber in which all (charged) particles were detected. A special feature of the analysis was the ability to classify the events according to the various processes mentioned above.[1] First one can characterize the scattering process by considering the kinematical variable $\epsilon = \sum_f (E_f - p_f^3) - m$, where the sum runs over all secondary particles. From the conservation of light-cone momentum it follows that in a *direct* event, in which only one nucleon is involved in the interaction, one has $\epsilon = 0$ (assuming no events with neutral particles). All other cases (*spectator* and *rescattering*) correspond to $\epsilon > 0$.

A further refinement is possible by distinguishing between 'even' and 'odd' events, corresponding to an *even* or *odd* number of charged hadrons in the final state, respectively. In particular from charge conservation it follows that *odd* events correspond to the *direct process on a neutron*. In the category even events the protons observed in backward directions can be identified as spectators, since protons originating from the direct process only go forward (when corrected for Fermi motion.)

Furthermore, the forward spectator spectrum can be constructed from the observed backward spectrum. All backward spectator and constructed forward spectator events belong to the class of lepton-neutron interactions.

3.2 Direct process

The cross section for the direct process can be expressed as an integral over the light-cone momentum fraction carried by the nucleon

$$\frac{d^4\sigma_{\text{dir}}^{\nu d}}{dx\,dz\,d^2\vec{p}_\perp} = \sum_{\tau=p,n} \int_{x+z}^{m_d/m} dy\,d^2\vec{k}_\perp\, f_d(y,\vec{k}_\perp)\sigma^{\nu\tau}\left(\frac{x}{y},\frac{z}{y},\vec{p}_\perp - \frac{z}{y}\vec{k}_\perp\right) \tag{1}$$

where the convolution function is given by[2]

$$f(y,\vec{p}_\perp) = my \int \frac{d^3k}{E_N} n(k)\delta\left(y - \frac{m_d - \sqrt{k^2 + m^2} - k^3}{m}\right)\delta^{(2)}(\vec{k}_\perp - \vec{p}_\perp). \tag{2}$$

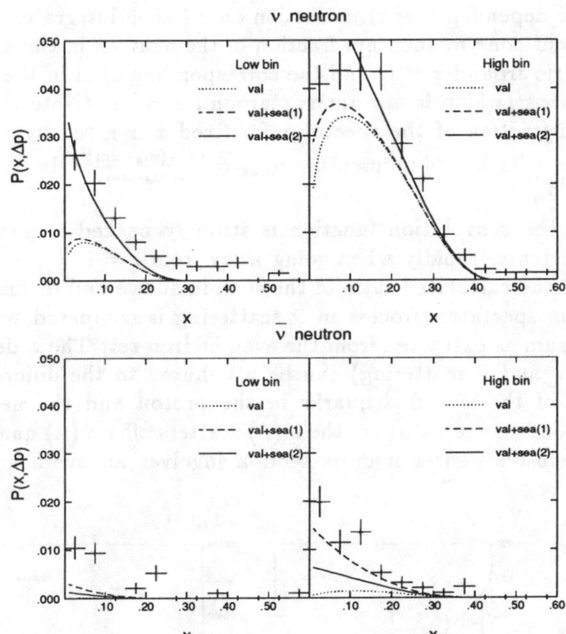

Figure 1: The ratio of the semi-inclusive and inclusive cross section $P(x, \Delta p)$ for the direct contribution to ν and $\bar{\nu}$ scattering on the neutron in the deuteron for two different momentum bins, $150 < p < 350$ MeV/c and $350 < p < 600$ MeV/c.

In the present work the momentum density $n(k)$ from the Bonn potential. For the deuteron the convolution function $f_d(y)$ is sharply peaked at $y = 1$, corresponding to a nucleon at rest, and falls of rapidly for $y \neq 1$ and as a result the cross section for the direct process differs very little from the one for the free nucleon case. The main difference is that, due to the fermi-motion, the calculated cross section extends to larger values of x.

The ratio $P(x, z)$ of semi-inclusive scattering on the neutron in the deuteron and the inclusive cross section is compared with experiment in Fig. 1. One interesting aspect is that the $\bar{\nu}$ cross section is much smaller than that for ν because it involves an unfavoured valence quark fragmentation function .

3.3 Spectator process

In the absence of final state interactions the cross section for the spectator process is given by

$$\frac{d^4\sigma^{\nu d}_{spec}}{dx\,dz\,d^2\vec{p}_\perp} = \int_x^{m_d/m} dy \frac{d\sigma^{\nu n \to X}(x/y)}{dx} f_d(y, \vec{p}_\perp) \delta(z + y - m_d/m) \tag{3}$$

Thus the dependence of cross section on z (when integrated over $p_\perp$) is determined by the light-cone momentum fraction of the neutron in the deuteron (approximately symmetric around $z = 1$) and the corresponding shift in the arguments of the structure functions (which is asymmetric around $z = 1$). (Note that the maximum l.c. momentum fraction of the spectator for fixed x is given by $z_{\max} = m_d/m - x$ corresponding to a backward momentum $p_{\max} = \frac{m^2 - (m_d - xm)^2}{2(m_d - xm)}$. In particular, for $x = 0$ one has $p_{\max} = \frac{3}{4}m$.)

Because the convolution function is strongly peaked at $y = 1$ the spectator cross section decreases rapidly when going away from $z = 1$.

In Fig. 2 the calculated ratio of the semi-inclusive and inclusive cross sections $P(x, \Delta p)$ for the spectator process in $\bar{\nu}$ scattering is compared with the backward spectator spectrum as extracted from the even hadron set. The x dependence (which is different for ν and $\bar{\nu}$ scattering) can be attributed to the difference between the x distributions of the u and d quarks in the proton and the neutron: spectator protons correspond to events where the ν ($\bar{\nu}$) scatters off a d (u) quark in the neutron whereas the inclusive deuteron cross section involves an average over proton and neutron events.

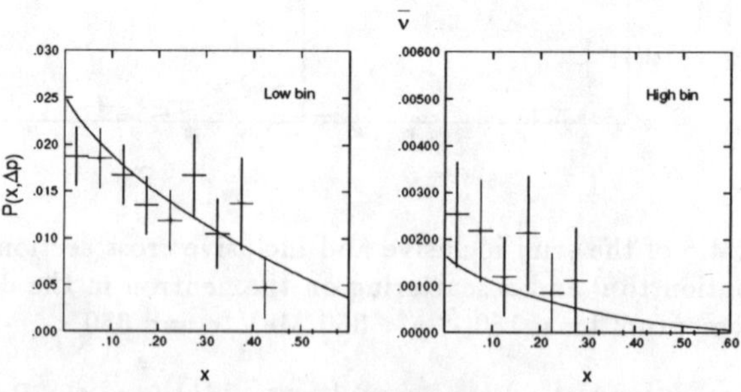

Figure 2: The ratio $P(x, \Delta)$ for the spectator process in $\bar{\nu} + d \to \mu + p + X$.

3.4 Rescattering

Above we have restricted ourselves to the PWIA, i.e. neglected effects from final state interactions between the spectator and the hadrons produced by the fragmentation. Here I briefly discuss the consequences of final state interactions (rescattering). The simplest approach to rescattering is to assume that the hit nucleon hadronizes instantaneously into a number of color singlets which can rescatter incoherently on the spectator. In the eikonal approximation the rescattering probability is

$$P_{\text{resc}}(x) = \frac{N(x)\sigma}{4\pi}\left\langle\frac{1}{r^2}\right\rangle_d,\tag{4}$$

where N is the multiplicity of the produced hadrons. Using for the fragment-nucleon cross section σ a value of 45mb one finds $P_{\text{resc}}(x) \approx 0.12N(x)$, to be compared with the averaged experimental value of 0.12, which suggests that effectively only one of the produced colorless fragments will interact.

The most plausible explanation of the small rescattering probability is the finite formation time for the created particles. This effect has been estimated by means of a Monte Carlo calculation and use the VENUS hadronization code (which is based on string fragmentation) to generate a space-time spectrum of fragments.

References

1. A. G. Tenner and N. N. Nikolaev, *Nu. Cim.* **A105** (1992) 1001.
2. G. D. Bosveld, A. E. L. Dieperink and O. Scholten, *Phys. Rev.* **C45** (1992) 2616
3. W. Melnitchouk, A. W. Thomas and N. N. Nikolaev, *Z. Phys.* **A342** (1992) 215
4. C. Korpa, A. E. L. Dieperink and O. Scholten, *Z. Phys.* **A343** (1992) 461
5. A. Scszurek and J. Speth, *Nucl. Phys.* **A555** (1993) 249
6. J. Guy et al., *Phys. Lett.* **B229** (1989) 421
7. L. L. Frankfurt and M. I. Strikman, *Phys. Rep.* **76** (1981) 215

QUARK CORRELATION FUNCTIONS IN
SEMI-INCLUSIVE HARD PROCESSES

P.J. MULDERS[1] and J. LEVELT
National Institute for Nuclear Physics and High Energy Physics
(NIKHEF-K), P.O. box 41882, NL-1009 DB Amsterdam, the Netherlands

ABSTRACT

We discuss the analysis of semi-inclusive hard scattering processes, such as hadroproduction in lepton-hadron scattering and electron-positron annihilation and Drell-Yan processes in terms of quark and gluon correlation functions. We discuss the dependence on perpendicular momenta of produced particles and polarization, with emphasis on lepton-hadron scattering.

1. Introduction

Semi-inclusive hard scattering processes offer opportunities to probe the quark and gluon structure of hadrons. The cross sections factorize into a calculable hard scattering part and soft parts which provide the connection between the quark and gluon degrees of freedom and the hadrons.

Processes that we study are lepton-hadron scattering proceeding through the exchange of a highly virtual $(-q^2 = Q^2 \geq 0)$ photon,

$$\gamma^*(q) + H(P) \longrightarrow h(p_h) + X, \tag{1}$$

electron-positron annihilation proceeding via a high mass photon $(q^2 = Q^2 \geq 0)$,

$$\gamma^*(q) \longrightarrow h_1(p_1) + h_2(p_2) + X, \tag{2}$$

and Drell-Yan processes with the production of a lepton pair originating from a high mass photon $(q^2 = Q^2 \geq 0))$,

$$H_A(P_A) + H_B(P_B) \longrightarrow \gamma^*(q) + X. \tag{3}$$

In all cases of course photons can be replaced by weak bosons.

At leading twist the cross section represented as the imaginary part of the forward $\gamma^* H$ amplitude in Fig. (1a) factorizes into quark distribution function $f(x_B)$ and quark fragmentation function $D(z)$, depending on the usual scaling variables $x_B = Q^2/2P \cdot q$ and $z = P \cdot p_h/P \cdot q$. Similarly the cross sections for electron-positron annihilation and the Drell-Yan process factorize as shown in Figs (1b) and (1c). For inclusive lepton-hadron scattering the factorization follows from the operator product expansion (OPE). In case of production of hadrons in the final state, however, one cannot use the OPE but we can use the hard scattering factorization procedure developed by Ellis, Furmanski and Petronzio[1].

---

[1] also at the Department of Physics and Astronomy, Free University, Amsterdam

Figure 1: *The amplitudes relevant for hadroproduction in lepton-hadron scattering (a), electron-positron annihilation into two hadrons (b) and Drell-Yan scattering (c)*

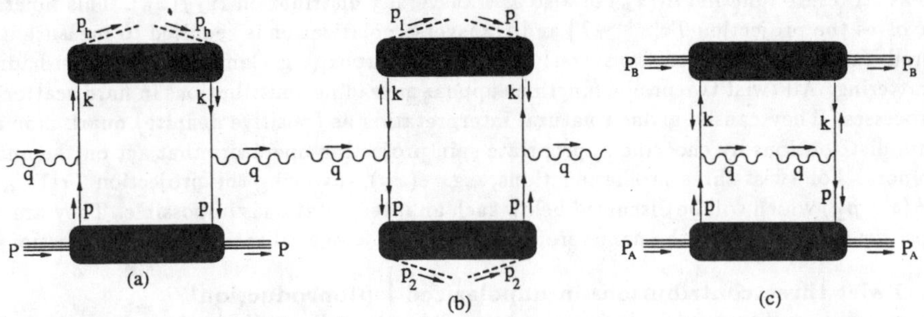

The soft parts, shown in Fig. (2), are the Fourier transforms of matrix elements of the nonlocal product of two quark fields,

$$T_{ij}(p) = \frac{1}{2(2\pi)^4} \int d^4x \ e^{-ip\cdot x} \langle P|\overline{\psi}_j(x)\psi_i(0)|P\rangle, \tag{4}$$

$$D_{ji}(k) = \frac{1}{4z(2\pi)^4} \int d^4x \ e^{ik\cdot x} \langle 0|\psi_j(x)a_h^\dagger a_h \overline{\psi}_i(0)|0\rangle. \tag{5}$$

Starting with these correlation functions a systematic classification of the contributions in hard scattering processes can be made according to twist[2] (the expansion in inverse powers of Q). Depending on the kinematical conditions of the process and the polarization of leptons and hadrons involved one can determine certain structure functions, that can be expressed in *profile functions* which are specific projections of the correlation functions. E.g. in unpolarized inclusive lepton-hadron scattering the structure function $2\,F_1(x_B, Q^2)$ can in leading twist be expressed as a sum over quark distribution functions $f(x_B)$ weighted by the quark charges squared. This quark distribution is an example of a profile function,

$$f(x_B) = \int dp^- d^2\boldsymbol{p}_\perp \text{Tr}\left[\gamma^+ T(p)\right]_{p^+ = x_B P^+} \tag{6}$$

$$= \frac{1}{4\pi} \int dx^- e^{iq^+ x^-} \langle P|\overline{\psi}(x)\gamma^+\psi(0)|P\rangle\Big|_{x^- = \boldsymbol{x}_\perp = 0}, \tag{7}$$

where $x^\pm = (x^0 \pm x^3)/\sqrt{2}$ and the +-component of q in the deep inelastic process is related to the quark momentum, $p^+ = -q^+ = x_B P^+$. The structure function $g_1(x_B)$ measured in polarized inclusive lepton-hadron scattering can be expressed in polarized quark distributions $\Delta f(x_B)$ which is given by Eq. (7), but with the projection $Tr[\gamma^+\gamma_5 T]$.

Figure 2: *The quark-quark correlation functions*

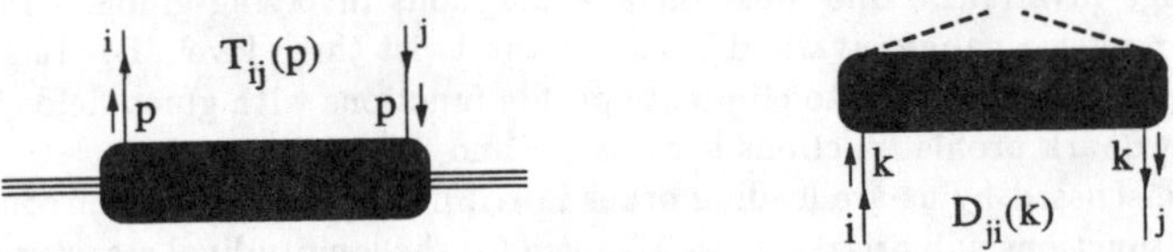

Jaffe and Ji[2] have discussed the twist expansion for the profile functions in Drell-Yan processes and lepton-hadron scattering. E.g., there is one more twist two distribution[3], referred to as structure function $h_1(x_B)$ or also as transversity distribution $\Delta_\perp f(x_B)$. This function involves the projection $Tr[\sigma^{+i}\gamma_5 T]$ and transverse polarization is required to measure it in Drell-Yan or detection of transversely polarized hadrons (e.g. lambdas) in lepton-hadron scattering. All twist two profile functions appear as leading contributions in hard scattering processes. They can be given a natural interpretation as (positive definite) quark momentum distributions by choosing appropriate spin projection operators that act on the quark spinors. For twist three profile functions, e.g. $e(x_B)$, involving the projection $Tr[1\,T]$, or $f^\perp(x_B, p_\perp)$ which will be discussed below such an interpretation is impossible. They are not positive definite, hence the name profile functions as a generalization for the projections.

2. Twist three contributions in unpolarized leptoproduction[4]

We will consider semi-inclusive processes of the type $eH \to e'hX$, where H is a hadronic target with mass M and h is a hadron with mass m_h detected in coincidence with the scattered electron. At leading twist the cross section factorizes into the distribution function $f(x_B)$ in the previous section and a fragmentation function $D(z)$ that represents the multiplicity of hadrons produced from a quark. This function is one of the profile functions that involve the quark correlation function $D_{ji}(k)$ in Eq. (5). We have

$$D(z) = \int dk^+ d^2k_\perp \, \mathrm{Tr}\left[\gamma^- D(k)\right]_{k^-=p_h^-/z} \tag{8}$$

$$= \frac{z}{8\pi} \int dx^+ e^{iq^-x^+} Tr \, \langle P|\psi(x) a_h^\dagger a_h \overline{\psi}(0)\gamma^-|P\rangle \Big|_{x^+=x_\perp=0}, \tag{9}$$

where q^- is related to the momentum of the struck quark, $k^- = q^- = p_h^-/z$.

In the unpolarized semi-inclusive process including twist three contributions four profile functions depending on x_B, z and perpendicular quark momenta $p_\perp$ and $k_\perp$ contribute

$$f(x_B, p_\perp) \equiv \frac{1}{2(2\pi)^3} \int dx^- \, d^2x_\perp e^{i(q^+x^- + p_\perp \cdot x_\perp)} \, \langle P|\overline{\psi}(x)\gamma^+\psi(0)|P\rangle \Big|_{x^+=0}, \tag{10}$$

$$\frac{p_\perp}{P^+} f^\perp(x_B, p_\perp) \equiv \frac{1}{2(2\pi)^3} \int dx^- \, d^2x_\perp e^{i(q^+x^- + p_\perp \cdot x_\perp)} \, \langle P|\overline{\psi}(x)\gamma_\perp\psi(0)|P\rangle \Big|_{x^+=0}, \tag{11}$$

$$D(z, -zk'_\perp) \equiv \frac{1}{4z(2\pi)^3} \int dx^+ \, d^2x_\perp e^{i(q^-x^+ - k'_\perp \cdot x_\perp)} \, \mathrm{Tr}\, \langle 0|\psi(x) a_h^\dagger a_h \overline{\psi}(0)\gamma^-|0\rangle \Big|_{x^-=0} \tag{12}$$

$$\frac{k'_\perp}{p_h^-} D^\perp(z, -zk'_\perp) \equiv \frac{1}{4z(2\pi)^3} \int dx^+ \, d^2x_\perp e^{i(q^-x^+ - k'_\perp \cdot x_\perp)} \, \mathrm{Tr}\, \langle 0|\psi(x) a_h^\dagger a_h \overline{\psi}(0)\gamma_\perp|0\rangle \Big|_{x^-=0} \tag{13}$$

where the last two are defined in the frame where the transverse component of the momentum of the produced hadron (p_h) is zero. In that frame $-zk'_\perp = -zk_\perp + p_{h\perp}$.

At the twist three level the contribution of the leading diagram, shown in fig. (1a) is not color gauge invariant. One must include diagrams involving gluons. Inclusion of these diagrams render a gauge invariant result at the twist three level. For this one must employ QCD equations of motion to eliminate profile functions with gluon fields in favor of nonleading twist quark profile functions such as $f^\perp$ and $D^\perp$.

The results discussed by us are leading order in α_s, i.e. α_s^0. In next order contributions to the structure functions will arise, e.g. well-known for the longitudinal structure function.

Also explicit gluon jet events appear in this order as calculated by König and Kroll[5]. Contributions with gluons in the final state and virtual gluon diagrams give rise to the QCD evolution of the leading twist profile functions f and D as has been proven by Ellis, Georgi, Machacek, Politzer and Ross[6]. These equations get slightly altered by the observation of transverse momentum[7].

Diagrams which give rise to logarithmic contributions in the transverse profile functions have not been calculated by us. It is plausible that the mass factorization properties still hold, i.e. that all logarithmic terms can be absorbed in the profile functions themselves and do not give rise to extra terms in the cross sections[8]. It is then possible to determine the twist three profile functions in a limited range of Q^2 where one neglects the logarithmic effects. Note that we are looking at Q^{-1}-effects, so Q^2 must be limited anyway.

The most general result is expressed as a cross section for the process were one detects transverse momentum of the produced jet ($\equiv$ the parton), the transverse momentum of the produced hadron and the invariants x_B, z and $y = (P \cdot q)/(P \cdot l)$, where l is the lepton momentum,

$$\frac{d\sigma}{dx_B dy dz d^2\,\boldsymbol{p}_{h\perp} d^2\boldsymbol{p}_\perp} = \frac{8\pi\alpha^2 M E}{Q^4}\left\{ \left(\frac{y^2}{2}+1-y\right) x_B f D \right.$$

$$+ 2(2-y)\sqrt{1-y}\cos\phi_h \frac{|\,\boldsymbol{p}_{h\perp}|}{Q}\frac{x_B}{z}\left[f\left(\frac{1}{z}D^\perp - D\right)\right]$$

$$\left. + 2(2-y)\sqrt{1-y}\cos\phi_j \frac{|\boldsymbol{p}_\perp|}{Q}\frac{x_B}{z}\left[-fD^\perp + z\left(f - x_B f^\perp\right) D\right]\right\}, \quad (14)$$

where f and $f^\perp$ are functions of x_B and $\boldsymbol{p}_\perp$ and D and $D^\perp$ depend on z and $\boldsymbol{p}_{h\perp} - z\boldsymbol{p}_\perp$. Integrating over all transverse momenta the $\cos\phi$ terms vanish and we are left with the wellknown leading twist parton model result.

A simple example of an observable where a transverse profile function enters is the azimuthal asymmetry in the process $e + H \to e' + jet + X$. In that case the fragmentation functions are integrated out. For the angle ϕ_{jet}, which is the angle between the scattering plane (defined by the momenta of e and e') and the plane formed by q and the transverse momentum $\boldsymbol{p}_\perp$ of the jet, one has

$$\langle \cos\phi_{\text{jet}}\rangle = -\left(\frac{2|\boldsymbol{p}_\perp|}{Q}\right)\frac{(2-y)\sqrt{1-y}}{2(1-y)+y^2}\left(\frac{x_B f^\perp(x_B, \boldsymbol{p}_\perp)}{f(x_B, \boldsymbol{p}_\perp)}\right). \quad (15)$$

The last factor is a ratio of profile functions. An earlier calculation by Cahn[9] folding a (on-mass-shell) parton cross section with a parton momentum distribution appears as a special case of our result. In that case $x_B f^\perp = f$.

Finally we show the result of a calculation in the bag model[10]. Expanding $\psi(x)$ in bag modes the matrix elements for f and $f^\perp$ in Eqs (10) and (11) can be calculated for bag states (considered as a $P = 0$ target). The results are shown in Fig. (3). As we have not corrected for violation of translation invariance in the bag model the structure functions extend beyond $0 \le x_B \le 1$. for the function $f^\perp(x_B)$, obtained after integration over $d^2\boldsymbol{p}_\perp$, the first two moments read for a bag with mass M and radius R

$$\int dx\; x f^\perp(x) = \frac{\omega}{MR} \int dx\; f^\perp(x) = \frac{2\omega - 3}{2(\omega - 1)} \approx 0.52. \quad (16)$$

Figure 3: *The profile functions $f(x_B)$ and $f^\perp(x_B)$ in the bag model for $MR = 4\omega$.*

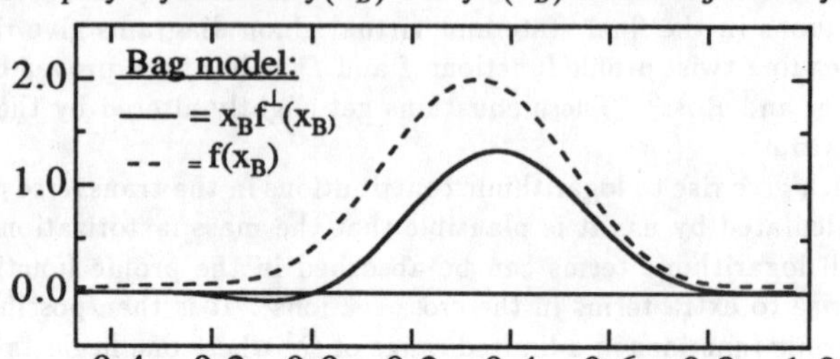

where ω/R is the energy of the (lowest) occupied mode with $\omega \approx 2.043$. The fact that $f^\perp$ and f are not simply related is analogous to to the situation of the vector current evaluated for a nucleon. In that case the nontrivial structure of the nucleon leads to independent form factors G_E and G_M for the different components of the current $\overline{\psi}\gamma^\mu\psi$. We note that for $f^\perp$ in the bag model the first moment is related to the magnetic moment, $\int dx_B\, f^\perp_{bag}(x_B) = 1.28\mu_{bag}$, independent of M and R.

Concluding, for a proper analysis of the data on leptoproduction it is important to realize that new independent profile functions can be measured that lead to a more complex transverse momentum dependence than that expected from a folding of partonic cross sections.

This work is supported by the foundation for Fundamental Research of Matter (FOM) and the National Organization for Scientific Research (NWO).

References

1. R.K. Ellis, W. Furmanski and R. Petronzio, Nucl. Phys. **B212** (1983) 29.
2. R.L. Jaffe and X. Ji, Nucl. Phys. **B375** (1992) 527.
3. J. Ralston and D.E. Soper, Nucl. Phys. **B152** (1979) 109;
   X. Artru and Z. Mekhfi, Z. Phys. **C45** (1990) 669.
4. J. Levelt and P.J. Mulders, *Quark correlation functions in deep inelastic semi-inclusive processes*, NIKHEF report 92-P9 (R).
5. A. König and P. Kroll, Z. Phys. **C16** (1982) 89.
6. R.K. Ellis et al., Nucl. Phys. **B162** (1979) 285.
7. H. Georgi, Phys. Rev. Lett. **42** (1979) 294.
8. J.C. Collins, D.E. Soper and G. Sterman in *Perturbative Quantum Chromodynamics*, ed. A. Mueller, World Scientific (1989).
9. R. Cahn, Phys. Lett. **78B** (1978) 269.
10. J. Levelt and P.J. Mulders, *Twist three contributions in unpolarized leptoproduction*, NIKHEF report 93-P7.

Exclusive charmonium decays in perturbative QCD

Mauro Anselmino

*Dipartimento di Fisica Teorica, Università di Torino
and Istituto Nazionale di Fisica Nucleare, Sezione di Torino
Via P. Giuria 1, I–10125 Torino, Italy*

Abstract

*Several exclusive charmonium decays are considered in the framework of pertur-
bative QCD. Many of these decays are forbidden by the helicity conservation of
massless quarks which couple to gluons; however, they are experimentally observed.
After reviewing some of these cases, possible non perturbative corrections or alternative decay mechanisms are considered and discussed.*

Exclusive decays of charmonium states into meson or baryon pairs supply an
interesting phenomenological ground for testing many of our current ideas about
the hadron structure. The initial $(c\bar{c})$ states can be safely treated in the non
relativistic approximation and we expect the decays to proceed through the $c\bar{c}$
annihilation into hard gluons with the subsequent creation of $q\bar{q}$ pairs, process
which can be computed in perturbative QCD. The final quark hadronization into
the observed particles is then described by a convolution of the hard part with the
soft, non perturbative, hadronic wave functions; in principle, QCD allows a good
description of both the hard and soft parts, at least at very large energies 1-3.

The above statements translate into the following schematic expression for
the helicity amplitudes describing the decay of a heavy quark bound state into a
hadronic pair $h\bar{h}$ (say two mesons):

$$A_{\lambda_h \lambda_{\bar{h}}} \sim \sum \int \psi^*_{\lambda_h} \psi^*_{\lambda_{\bar{h}}} \psi_{c\bar{c}}\, T(c\bar{c} \to q\bar{q}q\bar{q}) \tag{1}$$

where $\lambda_h, \lambda_{\bar{h}}$ are the final hadron helicities and the ψ's are the hadron wave functions; the T's are the helicity amplitudes describing the elementary interactions
among the constituents and we sum and integrate over all allowed quantum numbers.

So far only data on charmonium decays are available and the comparison
with the theoretical computations shows both successes 4-9 and failures 10-12.
The reason of all failures can be traced down to spin effects; the vector coupling
of quarks and gluons conserves, in the massless limit, the quark helicities, leading
to the "helicity conservation rule" in exclusive processes 4, which forbids many
two-body heavy meson decays 5. We consider here several examples of these
"forbidden" decays, which have nevertheless been observed, trying to understand

if corrections to the perturbative QCD scheme or alternative decay mechanisms might be able to give an explanation. We indeed expect modifications to the theoretical framework outlined in Eq. (1): the modest values of the Q^2 involved in charmonium decays (Q^2 up to few (GeV)^2) make non perturbative or higher order corrections non negligible.

Let us consider the helicity conservation rule. According to the picture outlined above, the initial heavy quarks annihilate into hard gluons, which then create $q\bar{q}$ pairs with the q helicity always *opposite* to the $\bar{q}$ one. The final quarks q_i then hadronize collinearly into a final hadron with helicity $\lambda_h = \sum_i \lambda_{q_i}$; similarly do the antiquarks, $\lambda_{\bar{h}} = \sum_i \lambda_{\bar{q}_i}$. We are thus led to final states with only opposite helicity particles, $\lambda_h = -\lambda_{\bar{h}}$, which immediately forbids many charmonium decays. For example, the decays η_c, $\chi_0 \to p\bar{p}$ should not occur, because a spin zero particle cannot decay into two opposite helicity fermions; similarly forbidden decays are $\eta_c \to VV$ (V = vector meson) and $J/\psi \to \pi\rho$, $K^*\bar{K}$ (in general $J/\psi \to$ any pseudoscalar-vector meson pair).

With the only possible exception of $\chi_0 \to p\bar{p}$ all of the above decays have been observed to occur with relative large branching ratios 13

$$B(\eta_c \to \rho\rho) = (2.6 \pm 0.9) \times 10^{-2} \tag{2a}$$

$$B(\eta_c \to K^*\bar{K}^*) = (8.5 \pm 3.1) \times 10^{-3} \tag{2b}$$

$$B(\eta_c \to \phi\phi) = (7.1 \pm 2.8) \times 10^{-3} \tag{2c}$$

$$B(\eta_c \to p\bar{p}) = (1.2 \pm 0.4) \times 10^{-3} \tag{2d}$$

$$B(J/\psi \to \rho\pi) = (1.28 \pm 0.10) \times 10^{-2} \tag{2e}$$

$$B(J/\psi \to K^*\bar{K}^*) = (4.2 \pm 0.4) \times 10^{-3} \tag{2f}$$

$$B(\chi_0 \to p\bar{p}) < 9.0 \times 10^{-4} \tag{2g}$$

Notice that some "forbidden" decays have branching ratios as large as "non forbidden" ones ($\chi_0 \to \pi\pi$, $\chi_{1,2} \to p\bar{p}$ 13). The experimental data (2) show beyond any doubt the limits of the perturbative QCD approach in describing exclusive hadronic charmonium decays. One should then conclude that, in this energy region, non perturbative, higher order effects or even new decay mechanisms can still be at work and cannot be neglected. Of course, the same consideration should not hold for, *e.g.*, ($b\bar{b}$) decays, which would involve much higher values of Q^2; the same discrepancies between theory and experiment in case of bottomonium decays would be a much more serious challenge to perturbative QCD. We shall now briefly discuss some possible corrections to the massless perturbative QCD calculations, which we expect to play a role in the few (GeV)^2 region.

a) - Higher Fock states in mesons

In the perturbative scheme of Eq. (1) only the leading Fock states (qqq for a baryon and $q\bar{q}$ for a meson) are included in the hadronic wave functions; higher order states are suppressed by additional powers of α_s/Q^2. However, they might still play a role for Q^2 values up to few $(\text{GeV})^2$. Such contributions have been computed for only few processes: $J/\psi \to \pi\rho$ 3 and $\eta_c, \chi_{0,2} \to \omega\phi$ 14, in which cases they seem to work. The same contributions should also allow a large value of the decay rate for $\psi' \to \pi\rho$, which, instead, is experimentally observed to be strongly suppressed 13, in agreement with perturbative QCD. A consistent explanation of both the J/ψ and $\psi' \to \pi\rho$, $K^*\bar{K}$ decays seems to be possible only via a totally different mechanism (see point **d**)).

b) - Mass corrections

In perturbative QCD calculations the light quarks are assigned their *current* masses, m_q, of few MeV, which makes any helicity flip of a quark of energy E_q, which couples to a gluon, proportional to m_q/E_q, in practice negligible; hence, the helicity conservation rule with all the problematic consequences discussed above. One might, however, assume that, at the Q^2 values involved in charmonium decays, the *constituent* quarks, that is the current quarks surrounded by their cloud of $q\bar{q}$ pairs and gluons, still act as single particles; the elementary interactions then involve, rather than the (almost) massless current quarks, the massive constituent ones. It is then natural, in the small Q^2 region, to assign the quarks an effective mass xm_h, like in the naïve parton model, where x is the fraction of the four-momentum of the hadron h (with mass m_h) carried by the quark. The different values of x are weighted by the hadron wave function. The massive quarks thus allow helicity flips in the elementary amplitudes proportional to m_h/m_c, where m_c is the charm quark mass. The "forbidden" $\eta_c \to VV$, $p\bar{p}$ and $\chi_0 \to p\bar{p}$ decays have been studied in such a scheme 11,12. Mass corrections do not help at all with $\eta_c \to VV$ and very little with $\eta_c \to p\bar{p}$; in the former case the model keeps predicting, even with massive quarks, $\Gamma(\eta_c \to VV) = 0$, and, in the latter, the actual value obtained for $\Gamma(\eta_c \to p\bar{p})$, although different from zero, is several orders of magnitude smaller than the data. Mass contributions to $\Gamma(\chi_0 \to p\bar{p})$ are more promising: they turn out to be sizeable, leading to results a factor ≈ 2 to 10 smaller than data on the analogous decays $\chi_{1,2} \to p\bar{p}$. Unfortunately, no good data on $\chi_0 \to p\bar{p}$ exist, apart from the large upper bound (2g), leading to $\Gamma(\chi_0 \to p\bar{p}) < 12\,\text{KeV}$.

c) - Two quark correlations in nucleons

In another attempt to overcome the difficulties encountered by the massless perturbative QCD scheme in describing many charmonium decays, quark-diquark models of the nucleon have been proposed and applied to many physical processes 15-19. Two quark correlations, induced by colour forces, are indeed present inside baryons 20; in intermediate energy regions these correlations might behave as actual particles, scalar or vector *diquarks*, participating as single entities to the underlying dynamics. Vector diquarks, in particular, might help with the spin problems: the coupling of gluons to spin 1 diquarks may change their helicity, thus avoiding the troublesome helicity conservation rule.

The quark-diquark model of the nucleon has been consistently applied to the description of several charmonium decays, in order to fix the parameters of the model and the properties of diquarks 15-17: it gives a correct description of $\chi_{1,2} \to p\bar{p}$ decays; it also yields a value of $\Gamma(\chi_0 \to p\bar{p})$ as big as, or even bigger than, the values measured for $\Gamma(\chi_{1,2} \to p\bar{p})$; such value is significantly larger than the value supplied by mass corrections (see point **b)**) and is in agreement with the generous experimental upper bound previously mentioned. Definite data on $\Gamma(\chi_0 \to p\bar{p})$ would be much helpful in clarifying the situation. Similar diquark models have been used in the description of the $p\bar{p}$ invariant mass distribution in $J/\psi \to \gamma p\bar{p}$ 18 and the $J/\psi \to B\bar{B}$ angular distribution 19.

The η_c decays, however, keep defeating any attempt of explanation; diquarks, although leading to values of $\Gamma(\eta_c \to p\bar{p})$ non zero and similar to those obtained with massive quarks, still give results much smaller than data. The η_c decays appear then to be somewhat special and we turn now to a brief discussion of a different decay mechanism which might explain some of them.

d) - Gluonic components in J/ψ and η_c

According to the helicity conservation rule, both the $J/\psi \to \pi\rho, K^*\bar{K}$ and the $\psi' \to \pi\rho, K^*\bar{K}$ decays should be forbidden and their experimental decay widths should be either zero or very small. However, this holds true only for the ψ', but not for the J/ψ, the so called $J/\psi\,(\psi') \to \pi\rho, K^*\bar{K}$ puzzle 10. This might be explained by assuming that the J/ψ couples directly to a glueball with its same quantum numbers and a similar mass value, so that the decay proceeds via the sequence $J/\psi \to$ glueball $\to \pi\rho, K^*\bar{K}$; for the ψ' this contribution would be much smaller, due to a larger mass difference with the glueball (mass difference which appears in the denominator of the glueball propagator) 10.

A similar explanation might hold for the η_c, which has so far caused so many troubles; a $0-+$ trigluonium state, with a mass close to that of the η_c, could explain the otherwise misterious η_c decays 21. The expectation of such a glueball in the η_c mass region is indeed plausible, if one assumes the existence a $1--$ trigluonium state in the J/ψ mass region. For this explanation to hold, the analogous decays of the η_c', similarly to what happens for the ψ', should be strongly suppressed, according to perturbative QCD. Data on η_c' decays should be available in the near future.

e) - Intrinsic quark transverse momentum

Another possible way of defeating the helicity conservation rule and the two-body decay selection rule, is that of properly taking into account the intrinsic transverse momentum, k_T, of quarks inside the hadrons. In such a case the quark helicities do not necessarily sum up to the hadron helicity. These corrections should be proportional to k_T/m_c; their evaluation immediately leads to a complication of the usual collinear kinenatics and has never been performed.

Charmonium decays probe a Q^2 transition region where both perturbative and non perturbative effects are important; a careful analysis, both experimental and theoretical, might then provide precious information on the subtelties of the internal structure of hadrons.

REFERENCES

1 G.P. Lepage and S.J. Brodsky, *Phys. Rev.* D **22**, 2157 (1980); S.J. Brodsky and G.P. Lepage, *Perturbative Quantum Chromodynamics*, A.H. Mueller Editor, World Scientific (1989)

2 A.H. Mueller, *Phys. Rep.* **73**, 237 (1981)

3 V.L. Chernyak and A.R. Zhitnitsky, *Phys. Rep.* **112**, 173 (1984)

4 S.J. Brodsky and G.P. Lepage, *Phys. Rev.* D **24**, 2848 (1981)

5 V.L. Chernyak and A.R. Zhitnitsky, *Nucl. Phys.* **B201**, 492 (1982)

6 A. Andrikopolou, *Z. Phys.* C **22**, 63 (1984)

7 V.L. Chernyak and I.R. Zhitnitsky, *Nucl. Phys.* **B246**, 52 (1984)

8 P.H. Damgaard, K. Tsokos and E. Berger, *Nucl. Phys.* **B259**, 285 (1985)

9 V.L. Chernyak, A.A. Ogloblin and I.R. Zhitnitsky, *Z. Phys.* C **42**, 569 (1989); **42**, 583 (1989)

10 S.J. Brodsky, G.P. Lepage and S.F. Tuan, *Phys. Rev. Lett.* **59**, 621 (1987)

11 M. Anselmino, F. Caruso and F. Murgia, *Phys. Rev.* D **42**, 3218 (1990)

12 M. Anselmino, R. Cancelliere and F. Murgia, *Phys. Rev.* D **46**, 5049 (1992)

13 Particle Data Group, *Phys. Rev.* **D45**, S1 (1992)

14 M. Benayoun, V.L. Chernyak and I.R. Zhitnitsky, *Nucl. Phys.* **B348**, 327 (1991)

15 M. Anselmino, F. Caruso and S. Forte, *Phys. Rev.* D **44**, 1438 (1991)

16 M. Anselmino, F. Caruso, S. Joffily and J. Soares, *Mod. Phys. Lett.* **6A**, 1415 (1991)

17 M. Anselmino and F. Murgia, *Z. Phys.* C **58**, 429 (1993)

18 C. Carimalo and S. Ong, *Z. Phys.* C **52**, 487 (1991)

19 E.-H. Kada and J. Parisi, Collège de France *preprint* LPC 9140 (1991)

20 See, e.g., M. Anselmino and E. Predazzi, Editors, *Proceedings of the Workshop on Diquarks*, World Scientific (1989); M. Szczekowski, *Int. J. Mod. Phys.* A **4**, 3985 (1989); M. Anselmino, S. Ekelin, S. Fredriksson, D. Lichtenberg and E. Predazzi, *Luleå preprint*, TULEA 1992:05, to appear in *Rev. Mod. Phys.*

21 M. Anselmino, M. Genovese and E. Predazzi, *Phys. Rev.* D **44**, 1597 (1991)

SPIN-1 FORM FACTORS

CHUENG-RYONG JI
Department of Physics, North Carolina State University
Raleigh, North Carolina 27695-8202, USA

ABSTRACT

We review the universality of spin-1 form factors and the high-Q^2 predictions of the deuteron form factors. We compare the two predictions from the standard light-cone $q^+=0$ frame and the Breit ($q^+\neq0$) frame and discuss their differences. We also discuss the application of universality to the heavy vector meson systems and comment on the constraints to the phase space of hard scattering amplitudes in the heavy meson form factors.

1.   Introduction

The deuteron is an interesting and important laboratory for the application of QCD to nuclear physics. At large distances the deuteron is evidently well described as a spin-one composite of two nucleon clusters with binding energy ~2.2 MeV, together with small admixtures of $\Delta\Delta$ and virtual meson components. However, at short distances, in the region where all six quarks overlap with a distance R ~ 1/Q, one can show rigorously that the deuteron state in QCD necessarily has "fractional parentage" 1/9np, 4/45$\Delta\Delta$, and 4/5 "hidden color" (nonnuclear) components[1]. In fact, at any momentum scale the deuteron cannot be described solely in terms of standard nuclear physics degrees of freedom, and in principle, any physical or dynamical property of the deuteron is modified by the presence of such non-Abelian components[2]. The universality of spin-one systems discussed by Brodsky and Hiller[3] recently made the deuteron and the other spin-one systems such as the ρ-meson and $W^\pm$ more interesting and important. According to the universality, the fundamental constraints on the magnetic and quadrupole moments of hadronic and nuclear states imposed by Compton-scattering sum rules[4] and the behavior of the electromagnetic form factors of composite spin-one systems[5] at large momentum transfer are the same as those of a corresponding elementary particle of the same spin and charge. At $Q^2=0$, the charge ($G_c(Q^2)$) magnetic ($G_M(Q^2)$) and quadrupole ($G_Q(Q^2)$) form factors define the charge e, the magnetic moment μ_1 and the quadrupole moment Q_1, respectively. In the limit of zero radius of the bound states or large excitation energies, μ_1 and Q_1 approach cannonical values[3]:

$$\mu_1 = \frac{e}{M}, \ Q_1 = -\frac{e}{M^2}, \tag{1}$$

where M is the mass of the spin-one system. These are the same values obtained for the intermediate vector bosons $W^\pm$ in the tree approximation to the standard model. Also at large Q^2, these form factors approach the universal ratios given by[3]

$$G_c(Q^2) : G_M(Q^2) : G_Q(Q^2) \xrightarrow{\ Q>>\sqrt{2M\Lambda_{QCD}}\ } \left(1 - \frac{Q^2}{6M^2}\right) : 2 : -1, \tag{2}$$

These ratios were obtained in the standard light-cone frame ($q^+=0$) and hold at large spacelike or timelike momentum transfer in the case of composite systems such as the ρ or deuteron in QCD with corrections of order Λ_{QCD}/Q and $\Lambda_{QCD}/M_{\rho,d}$. They are also the ratios predicted for the electromagnetic couplings of the $W^{\pm}$ for all Q^2 in the standard model at the tree level. However, the corresponding ratios were predicted a little less informatively in the Breit frame $(q^+\neq 0)$[6]:

$$G_c(Q^2) : G_Q(Q^2) \xrightarrow{Q>>2M} \frac{Q^2}{6M^2} : 1. \tag{3}$$

In this talk, we review the universality of spin-one form factors, especially the high-Q^2 predictions of the deuteron form factors and compare the two different predictions at the $q^+=0$ standard light-cone frame[3] and the Breit $(q^+\neq 0)$ frame[6]. We also discuss the application of universality in the heavy vector meson systems[3,7] and comment on the constraints from the Lorentz and Gauge invariance to the phase space of hard scattering amplitudes in the heavy meson form factors. We will proceed in the following orders. In the next section, section 2, we define the spin-one form factors and relate them to the experimental observables[5]. In section 3, we relate the helicity amplitudes at $q^+=0$ frame[3] and the Breit frame[6] to the three form factors and present the high Q^2 predictions of the deuteron form factors[8]. In section 4, we discuss the application to the heavy vector meson systems[7]. The summary and conclusions are followed in section 5.

2.   Spin-1 Form Factors and Relation to Experimental Observables

The Lorentz and gauge invariant form factors for the spin-one systems are defined[3,5] by the electromagnetic coupling of the spin-one systems:

$$ie\Gamma^{\mu}_{\lambda'\lambda} = <p', \lambda'|J^{\mu}|p, \lambda>, \tag{4}$$

where $|p, \lambda>$ is an eigenstate of momentum p and helicity λ and J^{μ} is the electromagnetic current. These helicity amplitudes $\Gamma^{\mu}_{\lambda'\lambda}$ can be written in terms of three form factors $G_i(Q^2)$ $(i=1, 2, 3)$:

$$\Gamma^{\mu}_{\lambda'\lambda} = \epsilon'^*(\lambda')\cdot\epsilon(\lambda)(p^{\mu}+p'^{\mu})G_1(Q^2)$$
$$+ \left\{\epsilon'^*(\lambda')\cdot q\epsilon^{\mu}(\lambda) - \epsilon(\lambda)\cdot q\epsilon'^*(\lambda')\right\}G_2(Q^2) \tag{5}$$
$$+ \frac{1}{2M^2}\epsilon(\lambda)\cdot q\epsilon'^*(\lambda')\cdot q(p^{\mu}+p'^{\mu})G_3(Q^2),$$

where $Q^2=-q^2$, $q=p'-p$ and ϵ and ϵ' are the initial and final polarization vectors, respectively. The Lorentz and gauge form factors $G_i(Q^2)$ are related to the charge, magnetic and quadrupole form factors:

$$G_c = G_1 + \frac{2}{3}\eta G_Q$$
$$G_M = G_2 \tag{6}$$
$$G_Q = G_1 - G_2 + (1 + \eta)G_3,$$

where $\eta = Q^2/4M^2$ is a kinematic factor. At $Q^2=0$, these form factors define the charge e, the magnetic moment μ_1 and the quadrupole moment Q_1:

$$e\,G_c(0) = e$$
$$e\,G_M(0) = 2M\mu_1 \tag{7}$$
$$e\,G_Q(0) = M^2Q_1.$$

The standard Rosenbluth cross section[9] for elastic electron scattering on a spin-one particle in the laboratory frame is given by these form factors

$$\frac{d\sigma}{d\Omega} = \frac{\alpha^2\cos^2\left(\frac{\theta}{2}\right)}{4E^2\sin^4\left(\frac{\theta}{2}\right)}\frac{E'}{E}\left\{A(Q^2) + B(Q^2)\tan^2\left(\frac{\theta}{2}\right)\right\}, \tag{8}$$

where

$$A = G_c^2 + \frac{2}{3}\eta G_M^2 + \frac{8}{9}\eta^2 G_Q^2, \tag{9}$$

$$B = \frac{4}{3}\eta(1 + \eta)G_M^2.$$

However, in order to determine three form factors, one needs another observable beyond A and B. The commonly used observable is the tensor polarization T_{20} which is defined[10] by

$$T_{20}(Q^2, \theta) = \frac{1}{\sqrt{2}}\frac{\left(\frac{d\sigma}{d\Omega}\right)_+ + \left(\frac{d\sigma}{d\Omega}\right)_- - 2\left(\frac{d\sigma}{d\Omega}\right)_0}{\left(\frac{d\sigma}{d\Omega}\right)_+ + \left(\frac{d\sigma}{d\Omega}\right)_- + \left(\frac{d\sigma}{d\Omega}\right)_0}, \tag{10}$$

where $\left(\frac{d\sigma}{d\Omega}\right)_\lambda$ is the cross section for an unpolarized target deuteron giving a deuteron with final polarization λ (or vice versa). The tensor polarization T_{20} can also be written in terms of the three form factors as[3]:

$$T_{20} = -\frac{\frac{8}{9}\eta^2 + \frac{8}{3}\eta G_c G_Q + \frac{2}{3}\eta G_M^2\left\{\frac{1}{2} + (1+\eta)\tan^2\left(\frac{\theta}{2}\right)\right\}}{\sqrt{2}\left\{A + B\tan^2\left(\frac{\theta}{2}\right)\right\}}. \tag{11}$$

In principle, measurements of A, B and T_{20} are sufficient to determine G_c, G_M and G_Q. However, other components of the tensor polarization, $T_{2\pm1}$ and $T_{2\pm2}$, are also measurable and they can provide the complimentary informations on the three form factors[11]. For example, since experimental errors are finite, it is desirable to investigate numerically the sensitivity of T_{20} and $T_{2\pm1}$ to the changes in G_c and G_Q individually. These other tensor polarizations $T_{2\pm1}$ and $T_{2\pm2}$ are given by[11]

$$T_{2\pm1} = \frac{\pm2\eta(|\vec{p}|-\eta)\tan\frac{\theta}{2}G_M G_Q}{\sqrt{2}\left\{A + B\tan^2\left(\frac{\theta}{2}\right)\right\}}, \tag{12}$$

$$T_{2\pm2} = -\frac{\eta G_M^2}{2\sqrt{3}\left\{A + B\tan^2\left(\frac{\theta}{2}\right)\right\}}.$$

3. Helicity Amplitudes at $q^+=0$ Frame and Breit Frame: The High Q^2 Predictions of Deuteron Form Factors

Because of the advantages[12] in the quantization[13] of fields at equal light-cone time $\tau=t+z/c$ over the ordinary equal t quantization, we use the light-cone quantum field theory for the following discussions. To construct the helicity amplitudes given by Eq. (4), we consider the absorption of the virtual photon which has the four-momentum q^μ to the spin-one target which has the initial and final four-momenta p^μ and p'^μ and the corresponding polarizations $\varepsilon^\mu(\lambda)$ with the helicity λ and $\varepsilon'^\mu(\lambda')$ with the helicity λ', respectively. Each of these four-vectors can be represented differently depending on the choice of the reference frame. In Table 1, we present the explicit representations of these four-vectors in $q^+=0$ frame and Breit frame comparatively. In $q^+=0$ frame, the virtual photon is coming in the direction $\vec{q} = Q\hat{x} - (Q^2/2P^+)\hat{z} = Q\hat{x} - (Q^2/2M^2)(\sqrt{M^2+P^2}-P)\hat{z}$ and the initial target is moving in the direction $\vec{p} = P\hat{z}$, while in the Breit frame $\vec{q} = -Q\hat{z}$ and $\vec{p} = \frac{Q}{2}\hat{z}$.

Table 1: Comparison of Explicit Representations in Two Reference Frames.

Reference Frames (vector components)	$q^+=0$ frame (t+z, t-z, x, y)	Breit frame ($q^+\neq0$) (t, x, y, z)
q	$\left(0, \dfrac{Q^2}{p^+}, Q, 0\right)$	$(0, 0, 0, -Q)$
p	$\left(p^+, \dfrac{M^2}{p^+}, 0, 0\right)$	$\left(\sqrt{\dfrac{Q^2}{4}+M^2}, 0, 0, \dfrac{Q}{2}\right)$
$\varepsilon(0)$	$\dfrac{1}{M}\left(p^+, -\dfrac{M^2}{p^+}, 0, 0\right)$	$\dfrac{1}{M}\left(\dfrac{Q}{2}, 0, 0, \sqrt{\dfrac{Q^2}{4}+M^2}\right)$
$\varepsilon(+1)$	$-\dfrac{1}{\sqrt{2}}(0, 0, 1, i)$	$-\dfrac{1}{\sqrt{2}}(0, 1, i, 0)$
$\varepsilon(-1)$	$\dfrac{1}{\sqrt{2}}(0, 0, 1, -i)$	$\dfrac{1}{\sqrt{2}}(0, 1, -i, 0)$
p'	$\left(p^+, \dfrac{M^2+Q^2}{p^+}, Q, 0\right)$	$\left(\sqrt{\dfrac{Q^2}{4}+M^2}, 0, 0, -\dfrac{Q}{2}\right)$
$\varepsilon'(0)$	$\dfrac{1}{M}\left(p^+, \dfrac{-M^2+Q^2}{p^+}, Q, 0\right)$	$\dfrac{1}{M}\left(-\dfrac{Q}{2}, 0, 0, \sqrt{\dfrac{Q^2}{4}+M^2}\right)$
$\varepsilon'(+1)$	$-\dfrac{1}{\sqrt{2}}\left(0, \dfrac{2Q}{p^+}, 1, i\right)$	$-\dfrac{1}{\sqrt{2}}(0, 1, i, 0)$
$\varepsilon'(+1)$	$\dfrac{1}{\sqrt{2}}\left(0, \dfrac{2Q}{p^+}, 1, -i\right)$	$\dfrac{1}{\sqrt{2}}(0, 1, -i, 0)$

Using the Table 1 and Eq. (5), one can derive the relations[14] between the helicity amplitudes and the three form factors. In $q^+=0$ frame, one obtains[3]

$$G_c = \frac{1}{2p^+(2\eta+1)}\left\{\frac{2\eta-3}{3}\Gamma^+_{00} - \frac{16}{3}\eta\frac{\Gamma^+_{+0}}{\sqrt{2\eta}} - \frac{2}{3}(2\eta-1)\Gamma^+_{+-}\right\}$$

$$G_M = \frac{2}{2p^+(2\eta+1)}\left\{-\Gamma^+_{00} - (2\eta-1)\frac{\Gamma^+_{+0}}{\sqrt{2\eta}} + \Gamma^+_{+-}\right\} \tag{13}$$

$$G_Q = \frac{1}{2p^+(2\eta+1)} \left\{ \Gamma_{00}^+ - 2\frac{\Gamma_{+0}^+}{\sqrt{2\eta}} + \frac{\eta+1}{\eta}\Gamma_{+-}^+ \right\}$$

and also in the Breit frame[6],

$$G_c = \frac{-1}{2M\sqrt{1+\eta}} \left(\frac{1}{3}\Gamma_{00}^0 - \frac{2}{3}\Gamma_{+-}^0 \right)$$

$$G_M = \frac{2}{2M\sqrt{1+\eta}} \frac{\Gamma_{+0}^x}{\sqrt{2\eta}} \tag{14}$$

$$G_Q = \frac{-1}{2M\sqrt{1+\eta}} \frac{\Gamma_{00}^0 + \Gamma_{+-}^0}{2\eta}.$$

Note that, in the Breit frame, $\Gamma_{00}^0 = \Gamma_{00}^+$ and $\Gamma_{+-}^0 = \Gamma_{+-}^+$. In the $q^+=0$ frame one needs only $\mu=+$ component to determine three form factors while in the Breit frame one needs both $\mu=+$ (or 0) and x components. In both frames, one can show that helicity zero to zero matrix element Γ_{00}^+ is the dominant helicity amplitude at large Q^2 for lepton scattering on a spin-one bound state. However, in the Breit frame, the dominance of Γ_{00}^+ is not sufficient to determine G_M because Γ_{00}^+ does not contribute to G_M. This makes the Breit frame less informative in predicting the asymptotic form of three form factors than the $q^+=0$ frame. In principle, the same results should be obtained no matter what reference frames are used. On the other hand, the light-cone perturbation theory analyses in the Breit frame must take into account nondiagonal Z-graph contributions[15] or the surface terms[16] to the electromagnetic current. Since these quantities are not yet known, some informations are lost in the Breit frame. In any case, Brodsky and Hiller used the $q^+=0$ frame and obtained the ratio[3] from Eq. (13)

$$G_c : G_M : G_Q = (1 - \frac{2}{3}\eta) : 2 : -1 \tag{15}$$

modulo $\frac{\Lambda_{QCD}}{M}$ under the assumption $\Gamma_{+0}^+ \sim \frac{\Lambda_{QCD}}{Q}\Gamma_{00}^+$ and $\Gamma_{+-}^+ \sim \left(\frac{\Lambda_{QCD}}{Q}\right)^2 \Gamma_{00}^+$. One should note that the condition for this ratio to be valid is $Q \gg \sqrt{2M\Lambda_{QCD}}$ not $Q \gg \Lambda_{QCD}$. This can be shown by examining all the coefficients of the helicity ampltiudes in Eq. (13). For the deuteron, the difference between $\sqrt{2M\Lambda_{QCD}}$ and Λ_{QCD} is significant and postpones the applicability of perturbative QCD (PQCD) to the moderate momentum transfer square, e.g.

$Q^2 \gg 2M\Lambda_{QCD} \sim 0.8$ GeV^2. However, since $\frac{\Lambda_{QCD}}{M} \sim \frac{1}{10}$ for the deuteron, $\eta = \frac{Q^2}{4M^2} \gg$

$\frac{\Lambda_{QCD}}{2M} \sim 0.05$ meaning that η does not need to be larger than 1. In order for the Breit frame to yield the same asymptotic ratio given by Eq. (15), one needs the assumption

$\Gamma^+_{+-} = \frac{3\Gamma^+_{00}}{4\eta}$ and $\Gamma^x_{+0} = \frac{\Gamma^+_{00}}{\sqrt{2\eta}}$. The ratio given by Eq. (15) is identical to the ratio of the tree level three electromagnetic form factors for the elementary particles $W^\pm$. Since the electromagnetic vertex for $W^\pm$ is given by

$$\Gamma^\mu_{\lambda'\lambda} = ige\left[g^{\alpha\beta}(p+p')^\mu + g^{\beta\mu}(-p'-q)^\nu + g^{\mu\alpha}(q-p)^\beta\right]\varepsilon_\alpha(\lambda)\varepsilon'^{*}_\beta(\lambda'). \tag{16}$$

One can easily show $G_1=1$, $G_2=2$, $G_3=0$, and thus $G_c=1-\frac{2}{3}\eta$, $G_M=2$, $G_Q=-1$ leading to the ratio given by Eq. (15). Using Eq. (15), one can predict the asymptotic behavior of the observables given by Eqs. (9) and (11). In the domain $Q^2 >> 2M_d\Lambda_{QCD} \sim 0.8$ GeV^2, Brodsky and Hiller[3] derived the expressions for $\frac{B}{A}$ and T_{20} and compared with the experimental data. The present available data do not seem to come close to the prediction for $\frac{B}{A}$. However, for T_{20} the trend of the data is not inconsistent with the prediction. Data at larger momentum transfers are clearly needed. The predictions from the Breit frame produce completely different results in the regime currently accessible to experiment because of the large difference in the magnitudes of M_d and Λ_{QCD}.

The asymptotic high Q^2 behavior of the Γ^+_{00} itself has also been analyzed[8] for the deuteron system by considering the six-quark hard scattering amplitude and solving the evolution of the quark distribution amplitude[17] of the helicity zero component of the deuteron. The reduced form factor of the deuteron is defined by[2]

$$f_d(Q^2) \equiv F_d(Q^2)/F^2_N\left(\frac{Q^2}{4}\right), \tag{17}$$

where F_d and F_N are the asymptotically dominant form factors (e.g. magnetic form factors) of the deuteron and the nucleon, respectively. F_d is proportional to Γ^+_{00} and the prediction for f_d is given by[8,17]

$$f_d(Q^2) \sim \frac{\alpha_s(Q^2)}{Q^2}\left(\ell n \frac{Q^2}{\Lambda^2_{QCD}}\right)^{C_F/2\beta}, \tag{18}$$

where $C_F=\frac{4}{3}$ and $\beta=11-\frac{2}{3}n_f$ (n_f is the number of flavors). The present data[18] seems to be consistent with the prediction. Thus, optimistically, one could expect that the dominance of Γ^+_{00} begins at $Q^2 \sim 1$ GeV^2. However, the validity of the universal ratio given by Eq. (15) and the asymptotic behaviors of the observables $\frac{B}{A}$ and T_{20} should start at moderate momentum transfer $Q^2 >> 0.8$ GeV^2.

4. Universality in the Heavy Vector Meson Systems

The form factors of the spin-one heavy meson systems can be predicted[7] by PQCD rigorously in the time-like region above the meson threshold $q^2>4M^2>>\Lambda^2_{QCD}$. The result is given by

$$G_1 : G_2 : G_3 = \frac{e_1 m_1^4 + e_2 m_2^4}{(m_1 + m_2)^4} : \frac{e_1 m_1^3 + e_2 m_2^3}{(m_1 + m_2)^3} : 0, \tag{19}$$

where $e_1(e_2)$ and $m_1(m_2)$ are the charge and the mass of the quark (antiquark), respectively. Interestingly, this ratio becomes

$$G_1 : G_2 : G_3 = 1 : 2 : 0 \tag{20}$$

if $|m_1 - m_2| < \dfrac{\Lambda_{QCD}}{m_1 + m_2}$. Thus, the ratio given by Eq. (15) is also obtained in the heavy vector meson systems which have the equal mass for the quark and the antiquark. Further investigation on the reason why it requires the equal mass for the constituents is needed. In the following, we comment on the PQCD calculation of the heavy meson form factors. The hard scattering amplitude T_H is usually computed by replacing each hadron by collinear constituents $p_i^\mu = x_i P_H^\mu$. This implies $x_i = \dfrac{m_i}{M_H}$ due to the on-shell conditions for the external particles in T_H, i.e. $p_i^2 = m_i^2 = x_i^2 P_H^2 = x_i M_H^2$. Thus, the only allowed quark distribution amplitude which can be taken in the consistent PQCD calculation of the heavy meson form factors is the delta function

$$\phi(x, q^2) \sim f_M \, \delta\!\left(x - \frac{m_1}{M_H}\right), \tag{21}$$

where f_M is the meson decay constant. Even though, the collinear energy condition is removed; i.e. $p_i^- \neq x_i P_H^-$, the constraint in the momentum fraction space still remains. Consequently, we find a certain constraint in taking the quark distribution amplitude for the PQCD calculation of the heavy meson form factors. The details on this investigation will be presented elsewhere[19].

5. Summary and Conclusions

The QCD predicts the dominance of the helicity conserving amplitude Γ_{00}^+ over the helicity flip amplitudes Γ_{+0}^+ and Γ_{+-}^+ at high Q^2. If $\Gamma_{+0}^+ \sim \dfrac{\Lambda_{QCD}}{Q} \Gamma_{00}^+$ and $\Gamma_{+-}^+ \sim \left(\dfrac{\Lambda_{QCD}}{Q}\right)^2 \Gamma_{00}^+$, then the dominance[8] of Γ_{00}^+ begins at $Q^2 \sim 1$ GeV^2. However, in the high-Q^2 prediction of spin-1 form factors, the contained kinematical factor $\eta = \dfrac{Q^2}{4M^2}$ introduces another scale M in addition to Λ_{QCD} and the PQCD prediction of the ratio, $G_c : G_M : G_Q = \left(1 - \frac{2}{3}\eta\right) : 2 : -1$, is valid when $Q \gg \sqrt{2M\Lambda_{QCD}}$, not Λ_{QCD}. For the deuteron, this difference is significant and does postpone the applicability of PQCD ($Q^2 \gg 0.8$ GeV^2).

While PQCD predictions should be independent from the choice of reference frames, the standard light-cone $q^+ = 0$ frame gives more detailed predictions over the Breit ($q^+ \neq 0$) frame[6] due to the absence of Z-graphs[15,16] and no need of $\mu \neq +$ components. The universality to the point-like spin-one objects adds more confidence on $q^+ = 0$ frame.

The universality can also be applied in the PQCD calculations of heavy vector meson form factors[7]. When the quark and antiquark masses in the meson are same

($m_1=m_2$), then we show that the universality holds. We also find a certain constraint in taking the quark distribution amplitude for the consistent PQCD calculation of the heavy meson form factors.

This work was supported by the Department of Energy under contract DE-FG05-90ER40589 and the author thanks Stan Brodsky, Carl Carlson and Adam Szczepaniak for useful discussions and Jennifer Tuten for typing the manuscript.

References
1. M. Harvey, *Nucl. Phys.* **A352**, 301 (1981); **A352**, 326 (1981); **A424**, 428 (1984); For a more comprehensive review, see S. J. Brodsky and C.-R. Ji, Lecture Notes in Physics, Vol. **248**, pp. 153-245, Quarks and Leptons, Springer-Verlag, Berlin, 1985.
2. S. J. Brodsky and C.-R. Ji, *Phys. Rev.* **D33**, 2653 (1986).
3. S. J. Brodsky and J. R. Hiller, *Phys. Rev.* **D46**, 2141 (1992).
4. S. D. Drell and A. C. Hearn, *Phys. Rev. Lett.* **16**, 908 (1966); S. B. Gerasimov, *Sov. J. Nucl. Phys.* **2**, 430 (1966); S. J. Brodsky and J. R. Primack, *Ann. Phys.* (N.Y.) **52**, 315 (1969); W.-K. Tung, *Phys. Rev.* **176**, 2127 (1968).
5. R. G. Arnold, C. E. Carlson and F. Gross, *Phys. Rev.* **C21**, 1426 (1980).
6. C. E. Carlson and F. Gross, *Phys. Rev. Lett.* **53**, 127 (1984).
7. S. J. Brodsky and C.-R. Ji, *Phys. Rev. Lett.* **55**, 2251 (1986).
8. S. J. Brodsky, C.-R. Ji and G. P. Lepage, *Phys. Rev. Lett.* **51**, 83 (1983).
9. R. Hofstadter, *Annu. Rev. Nucl. Sci.* **7**, 231 (1957).
10. C. E. Carlson, *Nucl. Phys.* **A508**, 481C (1990).
11. M. I. Haftel, L. Mathelitsch and H. F. K. Zingl, *Phys. Rev.* **C22**, 1285 (1980).
12. C.-R. Ji, in Dirkfest '92, CEBAF, 1992, edited by W. Buck, K. Maung and B. Serot (World Scientific Publishing, Singapore, pp. 147-163).
13. For a review, see C.-R. Ji, in LAMPF Workshop; Nuclear and Particle Physics on the Light Cone, Los Alamos, 1988, edited by M. B. Johnson and L. S. Kisslinger (World Scientific Publishing, Singapore, pp. 291-308).
14. P. L. Chung, F. Coester, B. D. Keister and W. N. Polyzou, *Phys. Rev.* **C37**, 2000 (1988); F. Coester, in Proceedings of the International Conference on Medium and High-Energy Nuclear Physics, Taipei, Taiwan, 1988, edited by W.-Y. P. Hwang, K. F. Liu and Y. Tzeng (World Scientific, Singapore, 1989).
15. M. Sawicki, *Phys. Rev.* **D46**, 474 (1992).
16. M. Burkardt and A. Langnau, *Phys. Rev.* **D44**, 3857 (1991).
17. C.-R. Ji and S. J. Brodsky, *Phys. Rev.* **D34**, 1460 (1986).
18. R. G. Arnold et al., *Phys. Rev. Lett.* **35**, 776 (1975); R. G. Arnold, SLAC Report No. SLAC-PUB-2373, 1979 (unpublished).
19. C.-R. Ji and C.-Y. Pang, in preparation.

DEUTERON PHOTODISINTEGRATION: DATA AND PLANS

J. NAPOLITANO[a], P. ANTHONY[b], R. ARNOLD[c], J. ARRINGTON[e], D. BECK[d], E. BEISE[e],
E. BELZ[e], P. E. BOSTED[c], H. BULTEN[g], M. CHAPMAN[g], K. COULTER[f], F. DIETRICH[b],
R. ENT[g], M. EPSTEIN[h], B. W. FILIPPONE[e], H.-Y. GAO[e], D. F. GEESAMAN[f], J. O. HANSEN[g],
R. J. HOLT[f], H. E. JACKSON[f], C. E. JONES[i], C. E. KEPPEL[c], E. R. KINNEY[j], S. KUHN[k],
K. LEE[g], W. LORENZON[e], A. LUNG[c], N. MAKINS[g], D. MARGAZIOTIS[h], R. D. MCKEOWN[e],
Z. E. MEZIANI[k], R. MILNER[g], B. MUELLER[e], J. NELSON[g], T. O'NEILL[e], V. PAPAVASSILIOU[f],
G. PETRATOS[l], D. POTTERVELD[f], S. ROCK[c], R. E. SEGEL[m], M. SPENGOS[c],
Z. SZALATA[c], L. TAO[c], K. VAN BIBBER[b], J. VAN DEN BRAND[i], J. WHITE[c], B. ZEIDMAN[f]

[a]*Department of Physics, Rensselaer Polytechnic Institute, Troy, NY 12180, USA*
[b]*Lawrence Radiation Laboratory, Livermore CA 94550, USA*
[c]*American University, Washington DC 20016, USA*
[d]*Department of Physics, University of Illinois, Urbana IL 61801, USA*
[e]*California Institute of Technology, Pasadena CA 91125, USA*
[f]*Physics Division, Argonne National Laboratory, Argonne IL 60439, USA*
[g]*Massachussetts Institute of Technology, Cambridge MA 02139, USA*
[h]*Department of Physics, California State Univ., Los Angeles, CA 90032, USA*
[i]*Department of Physics, University of Wisconsin, Madison WI 53706, USA*
[j]*Department of Physics, University of Colorado, Boulder CO 80309, USA*
[k]*Stanford University, Stanford CA 94305, USA*
[l]*Stanford Linear Accelerator Center, Stanford CA 94305, USA*
[m]*Northwestern University, Evanston IL 60201, USA*

ABSTRACT

We review existing data for the exclusive reaction $\gamma d \to pn$ at photon energies above ≈ 1 GeV. We compare these data to various models, most notably those which rely on fundamental quark and gluon degrees of freedom. We include a discussion of future plans for related measurements.

1. Introduction

Two-body photodisintegration of the deuteron, i.e. $\gamma d \to pn$, is a high momentum transfer exclusive process which should be described by perturbative QCD (pQCD) above some energy[1,2,3]. At low energies, effective field theories in terms of meson and nucleon degrees of freedom, have been applied[4,5] using various formulations of the deuteron wave function[6]. At high energies, the momentum transferred to the nucleon is quite large, and surpasses that for elastic electron scattering at comparable beam energies[7]. For momentum transfer much greater than ≈ 1 GeV^2, we should expect these effective field theories to break down and some other description, perhaps pQCD, to take over.

Unfortunately, the signature of pQCD is not entirely clear. In its most simple form[1,2], it predicts the differential cross section $d\sigma/dt$ should "scale" with energy like $1/s^{n-2}$ where s is the square of the center-of-mass energy and n is the total number of pointlike constituents involved in the reaction. This behavior has in fact been observed in many reactions, for example $\gamma p \to \pi^+ n$ at photon energies above ≈ 2 GeV. [3] For the reaction $\gamma d \to pn$, we expect that, above some energy,

$$\frac{d\sigma}{dt} = \frac{\hat{f}(\theta_{CM})}{s^{11}}$$

(1)

where the function $\hat{f}(\theta_{CM})$ is unknown. A different pQCD formulation[8], specific to $\gamma d \to pn$, is based on a successful description of elastic ed scattering[9]. This model aims to describe the data at particularly low energies by using the empirical proton and neutron form factors to remove higher order contributions. In this case, one has

$$\frac{d\sigma}{d\Omega}_{CM} = \frac{1}{[s(s - M_d^2)]^{1/2}} F_p^2(t_p) F_n^2(t_n) \frac{1}{p_T^2} f^2(\theta_{CM})$$

(2)

where $F_N(t)$ are the nucleon elastic form factors. The "reduced nuclear amplitude" $f^2(\theta_{CM})$ is, again, unknown. **Equations 1 and 2 BOTH represent predictions of pQCD, and give the same energy dependence at very high energy.** One might expect, however, that Eq.(2) better approximates the data at lower energies. It has been argued, however, that no exclusive reaction should be describable by pQCD at experimentally accessible energies[10].

Several effective meson/nucleon field theories [11,12,13] have been developed to describe the data for photon energies above ≈ 1 GeV, and they show some success in describing the differential cross section and its energy dependence. In addition, a picture based on Regge phenomenology has also been developed[14]. We will not review these here, but only point out that a measurement of the energy dependence alone cannot conclusively rule out one picture or another.

Perhaps the best way to discriminate models will be to measure the proton polarization in the reaction $\gamma d \to \vec{p}n$, as a function of θ_{CM}. If pQCD describes the process, then the polarization must be zero[15]. Furthermore, we note that since this reaction involves a pointlike particle (the photon), Landshoff diagrams[16] should not contribute at high energy and this prediction is sound. On the other hand, predictions of effective field theories depend on the specific amplitudes included, and the result is unlikely to tend to zero.

2. Data and Comparisons

Until recently, the highest energy data[17,18,19] on $\gamma d \to pn$ was for photon energies just reaching 1 GeV. Since then, two experiments at SLAC have extended the data up

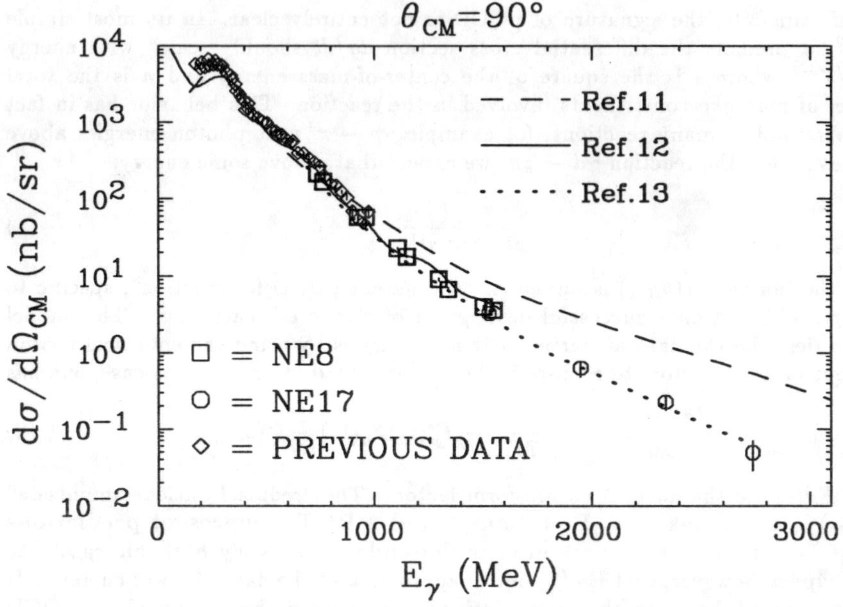

Fig. 1. Cross section $d\sigma/d\Omega$ vs. E_γ at $\theta_{CM} = 90°$ for $\gamma d \to pn$.

to ≈ 4 GeV. One of these, experiment NE8[20,21], took data up to 1.8 GeV at $\theta_{CM} = 90°$ and predominantly backward angles. The second, experiment NE17[22], took data at $\theta_{CM} = 90°$ and forward angles, with a single data point at $E_\gamma = 4.2$ GeV and $\theta_{CM} = 37°$. Data analysis for NE17 is ongoing.

Both NE8 and NE17 used bremsstrahlung beams and single arm focussing spectrometers to detect the proton from $\gamma d \to pn$. Untagged bremsstrahlung is quite suitable for this reaction since the endpoint can be kinematically isolated up to the threshold for π^0 production. In addition, the good resolution ($\sim 10^{-3}$) of focussing spectrometers makes it possible to cleanly avoid the bremsstrahlung tip and the π^0 threshold, and divide the region up into more than one photon energy bin.

We used a removable Cu radiator that allowed us to subtract the electroproduction yield in the target, typically a 30% correction. The radiator was located just far enough upstream so that it was not viewed by the spectrometer, but still allowed the entire bremsstrahlung angular cone to be intercepted by the long cryogenic targets. This allowed us to integrate the entire bremsstrahlung spectrum, dramatically decreasing the sensitivity to systematic error in the photon flux calculation[4]. Proton backgrounds from the target windows were accurately subtracted by taking data with a liquid hydrogen target, again with and without the bremsstrahlung radiator. Systematic errors to the cross section, under 7%, dominated the uncertainties in the NE8

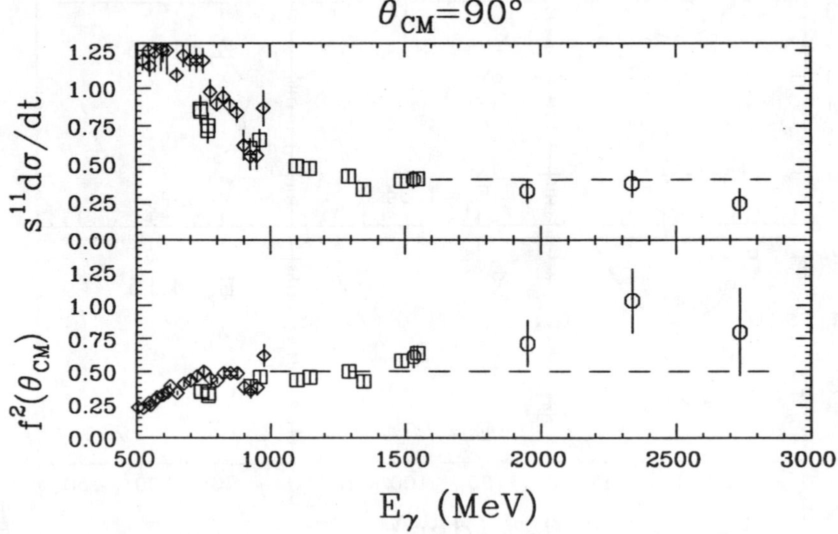

Fig. 2. Comparisons of $\gamma d \to pn$ at $\theta_{CM} = 90°$ with two formulations of pQCD.

data[21], and the energy dependence is not affected by them. The NE17 uncertainties are dominated by statistical error.

Figure 1 plots the measurements at $\theta_{CM} = 90°$, where the momentum transferred to both nucleons is maximized, and compares them to non-pQCD models. The cross section falls very rapidly, and measurements at large photon energy are limited by count rate. Models[11,12,13] based on conventional degrees of freedom do a more or less reasonable job of describing the data, but there are caveats. For example, there is considerable freedom as to what amplitudes are included and in the choice of parameters. In one case[13], the normalization itself is not determined. This has been described in more detail elsewhere. [21]

In order to test agreement with pQCD, that is *either* Eq.(1) or Eq.(2), it is easiest to take out the explicit dependence on photon energy and see if the data is consistent with a constant. We do this in Fig.(2) for the same data shown in Fig.(1). For convenience, we draw a horizontal line arbitrarily normalized to the data at 1.6 GeV for Eq.(1) and 1.0 GeV for Eq.(2). There is the suggestion that one or the other describes the data, but both cannot be true. In fact, the data below 1.6 GeV appear to be very consistent[20] with $d\sigma/dt \propto 1/s^{11}$, but this is not clearly supported by the higher energy data. With the existing data, one cannot claim that either asymptotic formulation is correct.

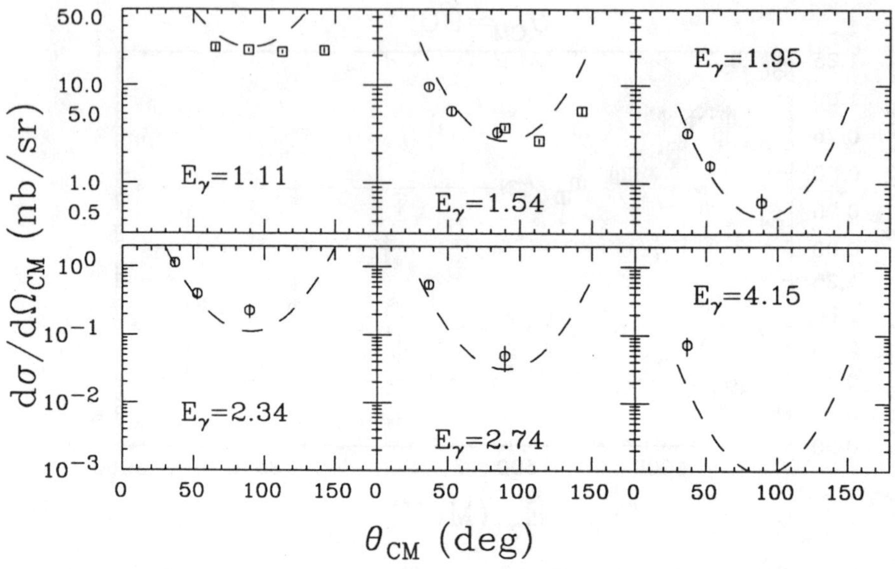

Fig. 3. Cross section $d\sigma/d\Omega$ vs. θ_{CM} for $\gamma d \to pn$. The photon energy is in GeV.

Figure 3 shows how the angular distribution evolves as the energy increases. The cross section is relatively independent of θ_{CM} for energies near 1 GeV, but at 1.6 GeV it is minimized near $\theta_{CM} = 90°$, with a somewhat symmetric enhancement for forward (experiment NE17) and backward (NE8) angles. As the energy increases further, the NE17 data show that this "forward peaking" becomes even more pronounced. That is, the data far from $\theta_{CM} = 90°$ fall much more slowly[22] than $1/s^{11}$, and there is not good evidence for a universal function $\hat{f}(\theta_{CM})$ in Eq.(1). For comparison, though, we also plot in Fig. 3 the angular distribution obtained from Eq.(2) **if we assume the reduced nuclear amplitude function $f^2(\theta_{CM}){\equiv}0.5$**. This is a strictly arbitrary choice, however, and the predictions of the model[8] are not firm. What's more, a similar behavior is predicted by at least one of the conventional calculations[13].

3. Plans

Clearly, the differential cross sections at $\theta_{CM} = 90°$ must be measured more precisely up to high energy so that we can establish whether or not some sort of asympotic behavior has been reached. (Recall that the momentum transferred to both nucleons is maximized at $\theta_{CM} = 90°$.) In addition, the angular distributions must be determined more carefully, particularly at high energy, so that we have additional constraints on the models, pQCD and otherwise. Finally, new observables should be studied that can help distinguish pQCD from conventional models, and the various conventional models from each other. In fact, experiments are planned for each of these areas.

An experiment to measure the inverse reaction, $np \to d\gamma$, is now in progress at KEK[23]. This is a collaboration between U. Kentucky, KEK, Kyoto U., Osaka U., Miyazaki U., and Nagoya U. A free neutron beam, with 10% energy resolution, is produced by an external deuterium beam on a Be target. The neutrons then impinge on a liquid hydrogen target. Neutron beam energies of 1.0, 2.0, 2.35, 2.7, and 3.0 GeV are used, corresponding to equivalent disintegration photon energies up to 1.5 GeV. The radiative photon and the recoil deuteron are detected in coincidence, covering an angular range $85° \leq \theta_{CM} \leq 165°$, where θ_{CM} is the angle between the photon and the incident neutron. Photons are detected in two shower counters consisting of converter, scintillators, drift chambers, and lead glass blocks. Deuterons are tracked through a 1.5 Tm dipole field using proportional wire chambers and drift chambers, and identified using time-of-flight. The cross section is normalized by taking concurrent data on np elastic scattering. Data was taken in April, 1993, and $d\gamma$ coincidences have been clearly identified. Present efforts are aimed at distinguishing $np \to d\gamma$ from backgrounds produced by associated π^0 and η production. These data will fill out the angular distributions over the range covered by NE8.

A particularly exciting possibility for the KEK experiment is the potential to use a polarized neutron beam, so that the analyzing power for $\vec{n}p \to d\gamma$ can be measured. This analyzing power is zero in pQCD[15].

Further data on $\gamma d \to pn$ up to 1.6 GeV or so will be taken at CEBAF using a tagged photon beam and the CLAS detector to identify the protons[24]. Very thorough data in fine angle and energy bins will be acquired over a large angular range. In addition, the reaction $\gamma d \to p\Delta^0$ will be measured so that Regge-phenomenological models[14] can be tested.

In order to push the cross sections to the highest possible energies, we must resort to bremsstrahlung and single-arm spectrometers (as in NE8 and NE17), so that the prerequisite luminosity, resolution, and particle identification can be achieved. To this end, a CEBAF experiment[25] will be carried out using this technique and the High Momentum Spectrometer (for forward angles) and the Short Orbit Spectrometer (for backward angles). A relatively complete angular distribution should be achieved for photon energies up to 4 GeV.

Finally, a CEBAF experiment will be done to measure the proton polarization in the reaction $\gamma d \to \vec{p}n$ using the bremsstrahlung technique and a polarimeter located near the focal plane of the spectrometer[26]. In this way, the proton polarization can be measured for photon energies approaching 2 GeV. This polarization should be zero if pQCD describes the reaction, and there are indications that for photon energies near 1 GeV that the polarization is indeed quite large[27].

4. Summary

The differential cross section for $\gamma d \to pn$ shows dramatic effects as the photon energy increases beyond ≈ 1 GeV. The cross section for $\theta_{CM} = 90°$ falls very rapidly, in a way consistent with pQCD, although the highest energy data is not precise enough to distinguish between two such models. The angular distribution also changes dramatically, showing a change from relative isotropy at $E_\gamma \approx 1$ GeV, to a strong forward peaking at the highest energy. A strong backward peaking is also suggested, but the data only go up to ≈ 1.6 GeV. This can be understood in at least some models using conventional meson and nucleon degrees of freedom, but the predictions of pQCD are not clear.

An experiment in progress at KEK and one to be done at CEBAF will fill out and verify the angular distribution up to photon energies near 1.6 GeV. A different experiment at CEBAF will measure the angular distributions at limited angles up to 4 GeV. Finally, an experiment at CEBAF (and possibly one at KEK) will measure the analyzing power in $\gamma d \to \vec{p}n$ (or $\vec{n}p \to d\gamma$), where pQCD predicts one should measure zero.

5. Acknowledgements

I am pleased to thank the organizers of this conference on Exclusive Reactions at High Momentum Transfer for their hospitality and support. I also thank those additional members of the NE8 collaboration, namely G. Boyd, D. Collins, S. Freedman, R. Gilman, M. Green, J. Jourdan, R. Kowalczyk, C. Marchand, R. Minehart, J. Nelson, T.-Y. Tung, and R. Walker. I also wish to acknowledge M. Kovash for informing me of the scope and status of the KEK experiments. This work was partially supported by the National Science Foundation under Grant No. PHY-9208119.

6. References

1. V. A. Matveev, R. M. Muradyan, A. N. Tavkhelidze, Lett. Nuovo Cimento **7**, 719 (1973).
2. S. Brodsky and G. Farrar, Phys. Rev. Lett. **31**, 1153 (1973).

3. R. L. Anderson, *et.al.*, Phys. Rev. D **14**, 679 (1976).
4. H. Arenhovel and M. Sanzone, "Photodisintegration of the Deuteron – A Review of Theory and Experiment," Few Body Systems Supplement 3, Springer-Verlag (1991)
5. J. M. Laget, Nucl. Phys. **A312**, 265 (1978).
6. R. B. Wiringa, R. A. Smith, and T. L. Ainsworth, Phys. Rev. C **29**, 1207 (1984); R. Machleidt, K. Holinde, and Ch. Elster, Phys. Rep. **149**, 1 (1987); M. Lacombe, *et.al.*, Phys. Rev. C **21**, 861 (1980); M. M. Nagels, T. A. Rijken, and J. J. DeSwart, Phys. Rev. D **17**, 768 (1978).
7. R. J. Holt, Phys. Rev. C **41**, 2400 (1990).
8. S. J. Brodsky and J. R. Hiller, Phys. Rev. C **28**, 475 (1983).
9. S. J. Brodsky and B. T. Chertok, Phys. Rev. D **14**, 3003 (1976).
10. N. Isgur and C. H. Llewellyn-Smith, Nucl. Phys. **B117**, 526 (1989).
11. T.-S. H. Lee, Argonne National Laboratory report PHY-5253-TH-88; T.-S. H. Lee, Proceedings of the International Conference on Medium and High Energy Nuclear Physics, May 23-27, 1988, Taipai, Taiwan, World Scientific (1988) pg.563
12. Y. Kang, P. Erbs, W. Pfeil, H. Rollnik, Abstracts of the Particle and Nuclear Intersections Conference, MIT, Cambridge MA, I-40 (1990); W. Pfeil, private communication.
13. S. I. Nagornyi, Yu. A. Kasatkin, and I. K. Kirichenko, Sov. J. Nucl. Phys. **55**, 189 (1992).
14. P. Rossi, these proceedings; E. DeSanctis, *et. al.*, Proceedings of the 13th European Conference on Few Body Problems in Physics, Elba, Sept. 9-14 (1991); E. DeSanctis, A. B. Kaidalov, and L. A. Kondrachuk, Phys. Rev. C **42**, 1764 (1990).
15. Carl E. Carlson and M. Chachkhunashvili, Phys. Rev. D **45**,2555 (1992).
16. C E. Carlson, M. Chachkhunashvili, and F. Myhrer, Phys. Rev. D **46**,2891 (1992); J. Botts and G. Sterman, Nucl. Phys. **B325**, 62 (1989); J. Ralston and B. Pire, Phys. Rev. Lett. **57**, 2330 (1986).
17. P. Dougan, et.al., Z. Physik **A276**, 55 (1976).
18. H. Myers, R. Gomez, D. Guinier, and A. V. Tollestrup, Phys. Rev. **121**, 630 (1961).
19. R. Ching and C. Schaerf, Phys. Rev. **141**, 1320 (1966).
20. J. Napolitano, *et al.*, Phys. Rev. Lett. **61**, 2530 (1988).
21. S. J. Freedman, *et al.*, Phys. Rev. C, in press.
22. David H. Potterveld, in the CEBAF 1992 Summer Workshop, AIP Conference Proceedings No.269, Particles and Fields Series 51 (1993); David H. Potterveld, proceedings of PANIC 1993 Conference, to be published.
23. M. Kovash, U. Kentucky, private communication.
24. P. Rossi, INFN/Frascati, CEBAF proposal PR-93-017.
25. R. J. Holt, Argonne National Laboratory, CEBAF proposal PR-89-012.
26. R. Gilman, Rutgers University, CEBAF proposal PR-89-019.
27. A. S. Bratashevskii, *et.al.*, JETP Lett. **36**, 216 (1982)

Deuteron photodisintegration at small t or u

L.A. Kondratyuk,[*] E. De Sanctis, P. Rossi, N. Bianchi, A.B.
Kaidalov,[*] M.I. Krivoruchenko[*], P. Levi Sandri, V. Muccifora, E.
Polli, A.R. Reolon

INFN-*Laboratori Nazionali di Frascati, P.O. Box 13, I-00044 Frascati, Italy*

Abstract

We have derived an expression for the cross section for the $\gamma d \to pn$ reaction at small momentum transfers t and u in the framework of the quark-gluon model and Regge phenomenology. Our predictions reproduce very well the few data available at energies between 1.5 and 4.2 GeV, where our approach can be applied. Moreover, we found a good agreement also with the forward and backward angle data in the intermediate energy region.

We have examined the data of the reaction $\gamma d \to pn$ available at high and intermediate energies with the aim of determining whether this process is more economically described in terms of the quark degrees of freedom, rather than nucleon and meson degrees of freedom. In particular, we tried to apply the quark-gluon string (QGS) model[1] and Regge phenomenology to the reaction $\gamma d \to pn$ in the region of small momentum transfers, t or u. At high energy, it is convenient to distinguish two different kinematical regions: *i*) high momentum transfers [$t \gg 1$ $(GeV/c)^2$], and *ii*) small momentum transfers [$t \lesssim 1$ $(GeV/c)^2$]. In the region *i*) according to quark-dimensional counting rules, the cross section of the reaction at constant centre-of-mass angle should be very fast decreasing function of energy: [2]

$$\frac{d\sigma}{dt} (\vartheta_{c.m.} = const.) \sim s^{-11} \qquad (1)$$

In the region *ii*), that is at sufficiently high energy and small t or u, the photodisintegration amplitude is dominated by the exchange of 3 valence quarks in t- or u-channels (see Fig. 1a) with any number of gluons exchanged between them. In the Regge language, the dominant contribution of 3-quark-exchange corresponds to the fermion

---

[*] Permanent address: Institute of Theoretical and Experimental Physics, Moscow 117259, RUSSIA

Regge pole [see Fig. 1b where the line $\alpha_N(t)$ describes the exchange of a Reggeon, which is an assembly of 3 quarks plus many gluons with angular momentum $\alpha_N(t)$].

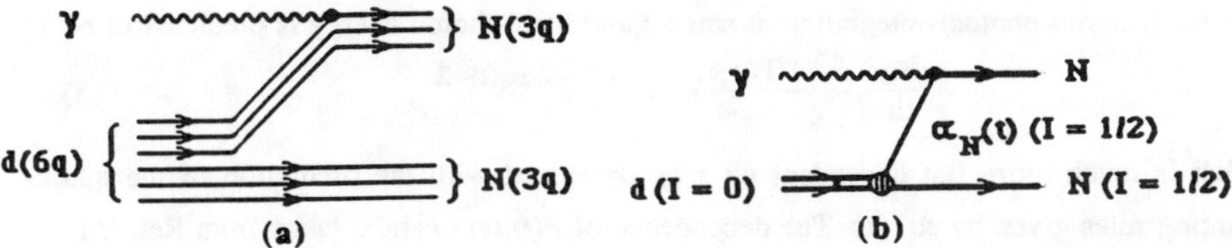

Fig.1 The 3-quark-exchange diagram which dominates the deuteron photodisintegration amplitude at sufficiently high energy and small t or u (a). In the Regge language this contribution correponds to the exchange of the Reggeon $\alpha_N(t)$ (b).

The analysis of binary hadronic reactions[1,3,4] showed that this picture works very well at high energies and $|t|$ or $|u| \leq 1(\text{GeV/c})^2$. However, due to the duality property of scattering amplitudes[5], this approach can work also in the intermediate energy region[3,6]. Recently, Kaidalov[6] has shown that this approach can describe the reactions $pp \to \pi^+ d$ and $\bar{p}d \to \pi^- p$ (which are also dominated by the 3-quark-exchange diagrams similar to one shown in Fig.1a) in the full energy range, starting almost from the threshold (this is clearly shown in Fig. 2 of Ref. [7]).

According to the Regge phenomenology and the QGS model, we have parametrizedthe cross section for the photodisintegration of the deuteron in the form:

$$\frac{d\sigma_R}{dt} = \frac{1}{64\pi s} \frac{1}{P^2_{c.m.}} (\,|\,T(s,t)\,|^2 + \frac{1}{R}\,|\,T(s,u)\,|^2\,), \qquad (2)$$

where $P_{c.m.}$ is the photon momentum in the center-of-mass system, $T(s,t)$ and $T(s,u)$ are the photodisintegration amplitudes, and R is the forward-to-backward ratio of the cross section values. The energy behaviour of $T(s,t)$ for fixed t, which corresponds to the fermion Regge pole exchange, can be written as:[1,3]

$$T(s,t) \approx F(t) \left[\frac{s}{s_0}\right]^{\alpha_N(t)} exp\{-i\frac{\pi}{2}(\alpha_N(t) - \frac{1}{2})\}, \qquad (3)$$

where $\alpha_N(t)$ is the trajectory of the N Regge-pole, $F(t)$ is the residue of the pole, s_0 is equal to the square of the deuteron mass, and the factor in the brackets is the phase factor . The baryon Regge trajectory deduced from the data on πN backward scattering is :[4]

$$\alpha_N(t) = \alpha_N(0) + \alpha_N'(0)\,t + \frac{1}{2}\alpha_N''t^2 , \qquad (4)$$

224

where $\alpha_N'(0) = 0.9$ GeV^{-2}, $\alpha_N'' = 0.25$ GeV^{-4}, and the intercept for the nucleon Regge trajectory N_α is $\alpha_N(0) = -0.5$ GeV^{-2}. Therefore, the energy behaviour of the cross section for the deuteron photodisintegration at small t and high photon energy is predicted to be:

$$\frac{d\sigma_R}{dt} \sim \frac{|T(s,t)|^2}{s^2} \sim [\frac{s}{s_0}]^{2\alpha_N(t)-2} , \qquad (5)$$

which is much more flat dependent on s as compared with the prediction of the quark-counting rules given by eq. (1). The dependence of $F(t)$ on t can be taken from Ref. [6]:

$$F(t) = B \left[\frac{1}{m_N^2 - t} \, exp(R_1^2 \, t) + C \, exp(R_2^2 \, t) \right] , \qquad (6)$$

We used the following values of the parameters: $R_1^2 = 1$ GeV^{-2}, $R_2^2 = -0.1$ GeV^{-2}, and $C = 0.7$ GeV^{-2} (Ref.[7]). The constants C_1 and C_2 were determined from a fit to existing low energy data at forward and backward angles. We found $C_1 = 6.99$ $\mu b^{1/2} GeV^3$ and $C_2 = 7.12$ $\mu b^{1/2} GeV^3$.

At fixed c.m. angle ϑ and in a limited energy region, the behaviour of the cross section as a function of energy can be parametrized as a simple power law, $d\sigma/dt \sim s^{-n}$. In distinction from the quark counting rules, the above expression for the cross section predicts different powers at different angles, specifically: $n=-5.4$ for $\vartheta=0°$, $n=-9.1$ for $\vartheta=37°$, $n=-10.3$ for $\vartheta= 53°$, and $n=-11.8$ for $\vartheta=90°$.

In Fig 2 we compare the predictions of our calculations (solid line curve) with the preliminary data (circles) of the new NE17 experiment carried out at SLAC.[8] The good agreement between our predictions and the experimental points is encouraging.

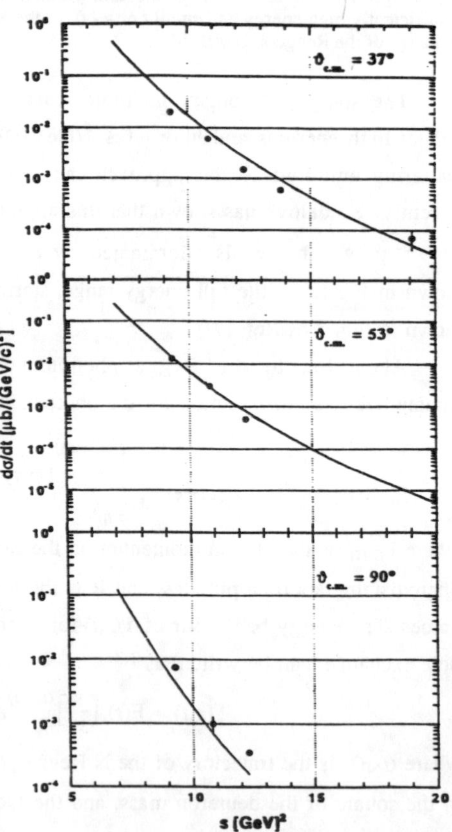

Fig.2 Differential cross section for deuteron photodisintegration at $\vartheta=0°$, 37°, and 53° as a function of s compared with the data of the NE17 experiment.[8]

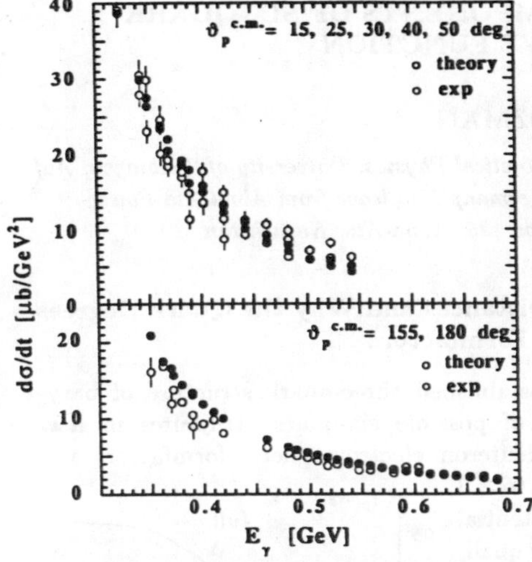

Fig. 3 Comparison of forward and backward data[9-11] $\gamma d \rightarrow pn$ above 375 MeV to our prediction.

Moreover, we compared our calculations with the scarce data, available at forward and backward angles and in the energy interval (375-700) MeV, used to determine with low uncertainties the parameters C_1 and C_2. In order to describe these data at low energies we had to take into account the tail of the Δ-resonance. Then, we used the expression :

$$\frac{d\sigma}{dt} = \frac{d\sigma_R}{dt} + \frac{d\sigma_\Delta}{dt} , \qquad (8)$$

where $d\sigma_R/dt$ is the Regge prediction as given by Eq. (2), and $d\sigma_\Delta/dt$ is the Δ-resonance tail contribution (for the expression of $d\sigma_\Delta/dt$ see Ref [7]).

In Fig.3 we compare all the experimental data available at (375-700) MeV and at forward and backward angles with the results obtained with Eq. (8). As it is seen, our calculation reproduces rather well the experimental values.

REFERENCES

[1]   A.B.Kaidalov, Z. Phys., **C 12**, 63 (1982) .

[2]   S. Brodsky and G. Farrar, Phys. Rev. **D 11**, 1309 (1975); and S. Brodsky and J. Hiller Phys. Rev. **C 28**, 475 (1985).

[3]   P.D.B. Collins and E.J. Squires "Regge poles in particle physics" Springer tracts in modern physics, v. **45**, Springer-Verlag, 1968;A.B. Kaidalov, Sov. Physics Uspekhi **14**, 600 (1972); and E.M. Levin, Sov. Physics Uspekhy **16**, 600 (1973).

[4]   V.A. Lyubimov. Sov. Phys. Uspekhy **20**, 691 (1975).

[5]   G.C. Rossi and G. Veneziano, Nucl. Phys. **B 123**, 507 (1977).

[6]   A.B.Kaidalov, Sov. J. Nucl. Phys., **53**, 872 (1991).

[7]   L.A.Kondratyuk et al., LNF-93/015 (P).

[8]   D.H. Potterweld, CEBAF 1992 Summer Workshop, Newport News WA 1992, AIP Conference Proc. n. **269**, Eds. F. Gross and R. Holt, p. 291

[9]   K.Baba et al. Phys. Rev. **C28**, 286 (1983).

[10] J. Arends et al., Nucl. Phys. **A412**, 509 (1984).

[11] K.H. Althoff et al., Z. Phys. **C21**, 149 (1983)].

NUCLEON AND BARYON COMPONENTS OF SIX-QUARK DEUTERON WAVE FUNCTION

L.Ya.GLOZMAN

Alexander von Humboldt Fellow, Institute of Theoretical Physics, University of Tuebingen, Auf der Morgenstelle 14, D–72076 Tuebingen, Germany ; on leave from Alma-Ata Power Engineering Institute, Kosmonavtov 126, Alma-Ata, Kazakhstan

1. The NN-System Structure at Small Distances and Why the Quark Degrees of Freedom are not Visible in Deuteron Formfactors

One would think that due to the well established three-quark structure of baryon one should expect some manifestations of possible six-quark structures in few-nucleon systems. But comparison of the deuteron electromagnetic formfactors and tensor polarization at momentum transfer up to 2-3 GeV^2 with the quark[1,2] and the usual NN potential description shows that there is no room for new qualitative quark effects which are not reduced to the usual NN picture in these observables.

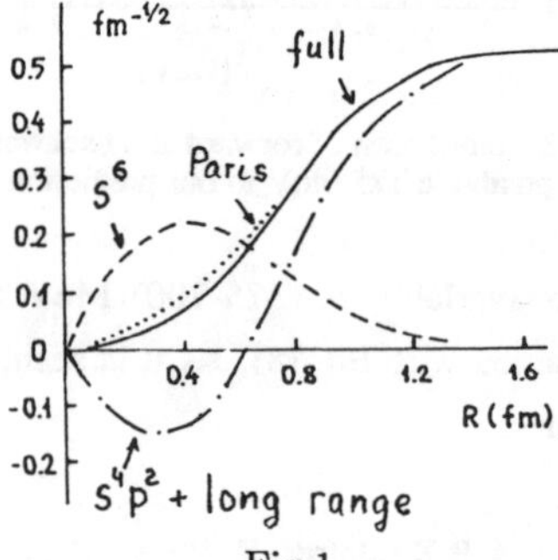

Fig.1

Let us trace back shortly the origin of this situation. The small amplitude of the deuteron wave function at small distances $r \leq 0.5 - 0.7$ fm (Fig.1), which is a result of the repulsion in the usual NN phenomenology, was easily explained through the destructive interference of the quark configuration s^6 and the superposition of all excited configurations with different spatial and color-spin (or spin-isospin) symmetries in the s^4p^2 quark shell[1,3].

This destructive interference can be clearly demonstrated with the magnetic structure function of the deuteron on Fig.2 (which is proportional to square of the magnetic form-factor). Namely this destructive interference leads to a typical minimum at $q^2 \sim 50 fm^{-2}$.

The mixing of the s^6 and s^4p^2 configurations is provided by the color-magnetic $\sim \lambda_i \lambda_j \sigma_i \sigma_j$ and color-electric $\sim \lambda_i \lambda_j$ parts of the effective one-gluon-exchange potential and by the pion-exchange (chiral phase) $\sim \tau_i \tau_j \sigma_i \sigma_j$ quark-quark interaction[3]. Due to this mixing the six-quark wave function at small distances is similar to the resonating group method (RGM) wave function[2,5]

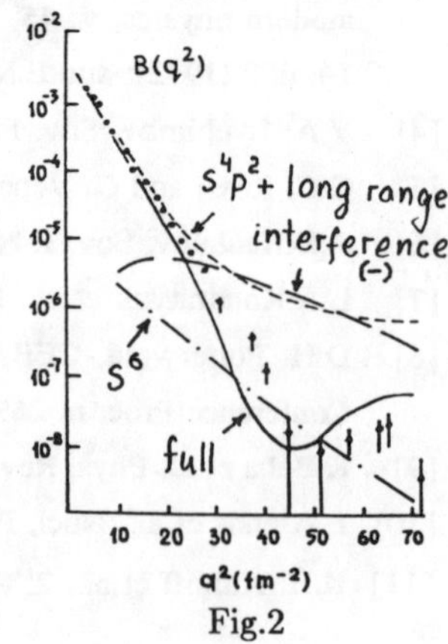

Fig.2

$$\psi^d(1,...,6) = \hat{A}\{\varphi_N(1,2,3)\varphi_N(4,5,6)\chi_{NN}(\mathbf{r})\} \quad (1)$$

The difference between the microscopic quark description of the 2N system at small energies and the usual phenomenological ones lies practically only in the presence of the quark antisymmetrizer $\hat{A}$ in (1) (of course, up to some quantitative distinction of the trial function $\chi_{NN}(\mathbf{r})$ from

the phenomenological NN wave functions at small range; but now we discuss only qualitative effects).

The quark antisymmetrizer leads to the quark-exchange current (QEC) contributions of Fig.3 in the deuteron electromagnetic formfactors, tensor polarization, etc.But the QEC contribution is small compared to the conventional terms.

This smallness is quite understand-
able. First, the exchange terms are important at small distances, but the NN
pair in the deuteron is dominantly at
intermediate and large distances. Sec-
ond, the $\chi_{NN}(\mathbf{r})$ in (1) dies out when
penetrating into the "core zone". The
third reason is the small value of the
spin-

Fig.3   Fig.4

isospin quark-interchange contribution in the $NN \to NN$ channel (see below).

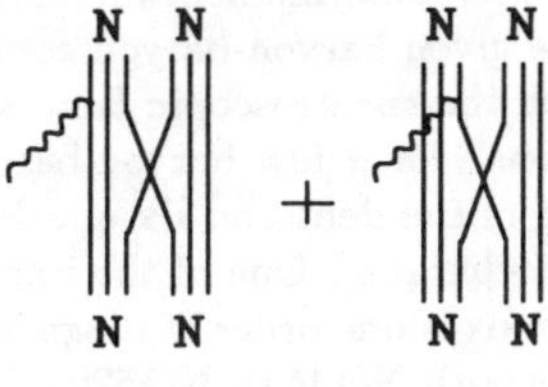

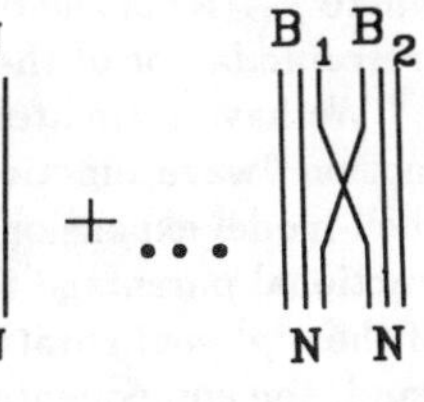

Thus, the achievement of the constituent quark concepts in the simplest nuclear system - NN - is mainly the explanation of the soft-core-like behaviour of this system at
small distances, nucleon-nucleon phase shifts, deuteron properties, etc. This achievement
is very essential. But nevertheless any serious concept must not only explain the existing
phenomena but also predict new ones. This is a main subject of my talk.

2. The Baryon-Baryon Content of the Deuteron and $d(e, e'p)R$ Processes

The six-quark wave function of the deuteron shows very rich cluster properties[5]. Namely, together with the NN, $\Delta\Delta$ and CC components as the well known characteristics of
the s^6 configuration, due to the essential contribution of the s^4p^2 configurations there
are many different components NN^*, N^*N^*, $\Delta\Delta^*$, CC^*,... where one or two oscillator
quanta of excitation are concentrated on the internal Jacobi coordinates of the 3q cluster. Here one-quantum excitations N^* correspond to negative parity low-laying baryons
N(1520), N(1535),... while the lowest of the two-quantum states N^{**} is the Roper resonance N(1440), etc.

The semi-exclusive process $d(e, e'p)R$ at CEBAF and ELSA energies, where various
states of the resonance-spectator R can be identified by coincidence measurement of fast
final e' and p particles, offers the opportunity to reveal the baryon-baryon composition of
the deuteron and, by this, to reconstruct the deuteron six-quark wave function [5]. The final
proton must have a kinetic energy above 1 GeV to avoid the admixture with the process
where the resonance R is excited off a nucleon by the incident photon. At the same time
the resonance-spectator R must be quite slow (or, better, moving to a backward direction)
to avoid the influence of the final state meson exchange diagram with an excitation of a
resonance R on one of the nucleons.

The electromagnetic current for these reactions with this kinematics can be presented
as a sum over all possible 3q clusters B $(N, N^*, \Delta, \Delta^*, ...)$in the intermediate state:

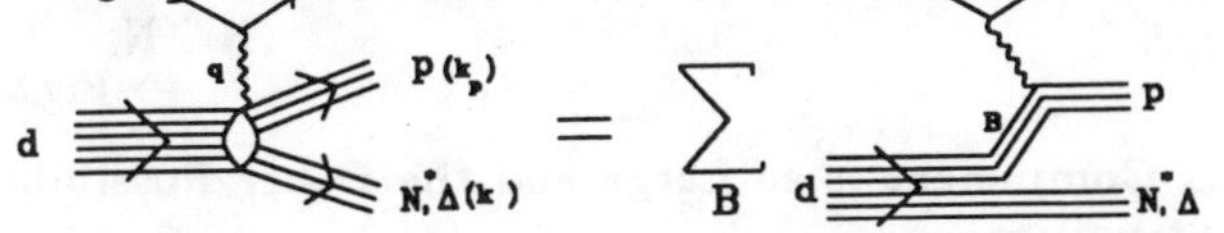

228

The most simple situation takes place when only one state B is possible or one of them is dominant. This is typical, for example, for the $d(e,e'p)\Delta$ and $d(e,e'p)N^*(1440)$ reactions with Δ and p as B, respectively. In this case the cross section is

$$\frac{d^2\sigma}{dk'\,dk_p} = \frac{\sigma_{eB\to e'p}(\Theta_e, q_\mu, k_p)}{k'^2}\rho_{BR}(\mathbf{k})\,\delta(E_f - E_i - \omega)$$

where $\rho_{BR}(\mathbf{k})$ is the observable momentum distribution which is normalized to the spectroscopic factor of the given baryon-baryon component in the deuteron.

We have estimated the spectroscopic factors (the effective numbers) and the relative motion "wave functions" for a few baryon-baryon components $\Delta\Delta$, *, N^*N^* using the shell-model expansion of the deuteron six-quark wave function at small distances and the fractional parentage technique . One of the important results is that the $\Delta\Delta$ component of the s^4p^2 configurations is one order of magnitude smaller than that of s^6. On the other hand, the components with N(1440), N(1520), N(1535) are connected only to the excited quark configurations. Thus, with the $d(e,e'p)\Delta$ reaction ($e + \Delta \to e' + p$ as elementary process) we could separate the s^6 configuration in the deuteron while the channels with N^* would display the configuration s^4p^2. This is important, as far as any reaction like $d(e,e'p)n$ and $d(\gamma,np)$ will be characterized at large momentum transfer by the destructive interference mentioned above.

There is an alternative interpretation of all these baryon-baryon components[6]. Namely, as far as the six-quark deuteron wave-function can be approximately presented in form (1), all these non-nucleonic components are generated by the quark antisymmetrizer in (1) and are, in essence, the quark-exchange terms of Fig.4. We have estimated[6] the quark-exchange virtual disintegration amplitude $d \to B_1 B_2$. The corresponding effective numbers for the most essential components are:

$$N^d_{\Delta(1232)\Delta(1232)} \simeq (2.5 \div 4.3)10^{-2} \; ; \; N^d_{\Delta(1232)\Delta(1600)} \simeq 6*10^{-3} \; ; \; N^d_{NN(1710)} \simeq 2.6*10^{-3};$$
$$N^d_{N(1520)N(1520)} \simeq 2.4*10^{-3} \; ; \; N^d_{N(1520)N(1535)} \simeq 1.6*10^{-3} \; ; \; N^d_{NN(1520)} \simeq 1.4*10^{-3} \; ;$$
$$N^d_{N(1535)N(1700)} \simeq 1.3*10^{-3} \; ; \; N^d_{N(1535)N(1650)} \simeq 9.6*10^{-4} \; ; \; N^d_{N(1520)N(1700)} \simeq 8.4*10^{-4} \; ;$$
$$N^d_{NN(1700)} \simeq 8.6*10^{-4} \; ; \; N^d_{NN(1535)} \simeq 7.1*10^{-4} \; ;...$$

We should stress here that all these effective numbers are connected only to the six-quark region. But in addition there are baryon-baryon components at a little larger distances $r \leq 1.5 - 2fm$ which are generated by the usual meson-exchange dynamics between two "point-like" baryons. In principle both these mechanisms are coherent. To check the role of the second one we have taken into account both of them when describing the virtual $d \to \Delta\Delta$ disintegration amplitude[7]. It turned out that the "quark-exchange $\Delta\Delta$" in $^3S_1$ wave dominates one order over the "meson-exchange" one[7]. Thus in any process with one of the Δ as a spectator, for instance in $d(e,e'p)\Delta$, we shall see the quark-exchange contribution of Fig.5.

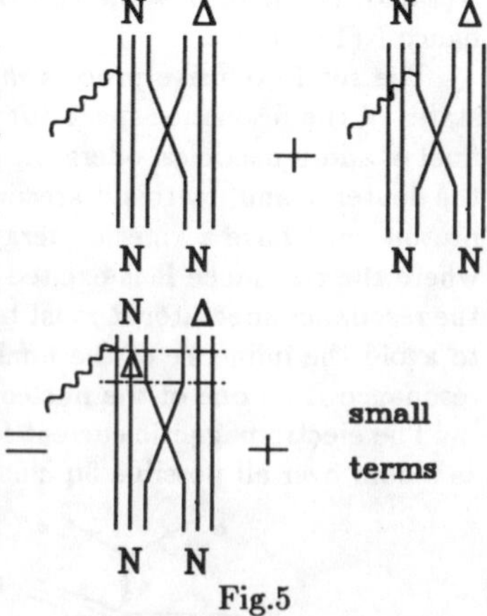

Fig.5

3. Why the $\Delta\Delta$ Component is so Large and the Other Possibility to Study the Six-Quark Structure

I think one should clarify here why the $d \to \Delta\Delta$ virtual disintegration amplitude is so large compared to the other ones. The reason is in the large value of the spin-isospin quark-interchange $NN \to \Delta\Delta$ transition

$$\langle([3]_{ST}ST = \frac{3}{2}\frac{3}{2}) \otimes ([3]_{ST}ST = \frac{3}{2}\frac{3}{2})|\hat{P}_{36}^{ST}|([3]_{ST}ST = \frac{1}{2}\frac{1}{2}) \otimes ([3]_{ST}ST = \frac{1}{2}\frac{1}{2})\rangle_{ST=10} = \frac{4\sqrt{5}}{27}$$

compared,for instance, to the $NN \to NN$ ST-exchange contribution

$$\langle([3]_{ST}ST = \frac{1}{2}\frac{1}{2}) \otimes ([3]_{ST}ST = \frac{1}{2}\frac{1}{2})|\hat{P}_{36}^{ST}|([3]_{ST}ST = \frac{1}{2}\frac{1}{2}) \otimes ([3]_{ST}ST = \frac{1}{2}\frac{1}{2})\rangle_{ST=10} = -\frac{1}{27}$$

It means, in particular, that if the baryon-baryon interaction in some kinematical region would be determined mainly by the color-exchange $f(r_{ij})\lambda_i\lambda_j$ forces followed by constituent quark interchange (that could be at quite large momentum transfer) we would have in this region $\sigma_{NN\to\Delta\Delta}/\sigma_{NN\to NN} \sim 80$. This ratio is a direct consequence of the identical spatial and color structure of N and Δ. But I should stress that it is valid only for the $f(r_{ij})\lambda_i\lambda_j$- type quark-quark forces (for example, color-electric and confinenent) and not valid for the hyperfine interaction of $\lambda_i\lambda_j\sigma_i\sigma_j$-type. In the latter case the $NN \to NN$ and $NN \to \Delta\Delta$ cross-sections would be comparable. Mention should be made of the fact that starting from the Oka and Yazaki works[4] it is traditionally believed that the hyperfine interaction plays a dominant role in the explanation of the short-range NN repulsion at small energies in the NN system. But an analysis by Obukhovsky et al[3] shows that the main role in the $s^6 - s^4p^2$ mixing, i.e. in the short-range repulsion in the NN system, is due to the color-electric forces. What is clear, that the $\sigma_{NN\to\Delta\Delta}/\sigma_{NN\to NN}$ ratio at large momentum transfer must be measured and by this we could clarify the question which type of the quark-quark forces is really responsible for the short range baryon-baryon phenomena.

1. L.Ya.Glozman, N.A.Burkova and E.I.Kuchina, *Z. Phys.* **A332**, 339(1989)

2. A.Faessler,A.Buchmann and Y.Yamauchi, *Int. J. Mod. Phys.* **E2**, 39(1993)

3. I.T.Obukhobsky and A.M.Kusainov, *Sov. J. Nucl. Phys.* **47**, 313(1988); *Phys. Lett.* **B238** ,142(1990); A.M.Kusainov, V.G.Neudatchin and I.T.Obukhovsky, *Phys. Rev.* **C44**, 2343(1991)

4. M.Oka and K.Yazaki, *Progr. Theor. Phys.* **66**, 556(1981); **66**, 572(1981)

5. L.Ya.Glozman, V.G.Neudatchin, I.T.Obukhovsky and A.A.Sakharuk, *Phys. Lett.* **B252**, 23(1990) L.Ya.Glozman, V.G.Neudatchin and I.T.Obukhovsky, *Phys. Rev.* **C48**, 389(1993)

6. L.Ya.Glozman and E.I.Kuchina, submitted

7. L.Ya.Glozman, U.Straub and A.Faessler, submitted

Relativistic light-cone approach
to the study of deuteron properties

L. Kondratyuk
Institute for Theoretical and Experimental Physics
Moscow - Russia (CSI)

M.M. Giannini
Dipartimento di Fisica and INFN - Sez. Genova
Via Dodecaneso, 33 - I16146 Genova (Italy)

P. Saracco*
INFN - Sez. Genova
Via Dodecaneso, 33 - I16146 Genova (Italy)

The deuteron is the simplest nuclear system and therefore it is a preferred test for studying the fundamental laws governing the dynamics of hadronic systems. Many of its properties can be satisfactorily described within the conventional non relativistic scheme [1], but various strong indications support the idea that other topics can be relevant to obtain a realistic description: relativistic effects, Meson Exchange Currents, quark degrees of freedom, etc. [2]

The effect of inelasticities in the deuteron wave function is small, at least up to energies of 1.5 GeV, as one can infer from the partial wave analysis of NN scattering data. Then relativistic effects should play a significant rôle in the description of the deuteron properties [3], like, e.g., the electromagnetic elastic form factors.

Moreover it is a general idea that the low Q^2 behavior of QCD should be well described by an effective lagrangian involving hadronic degrees of freedom only. In the intermediate Q^2 region, where one should look for finding a matching between low- and high-energy models of strong interactions, it is then preferable to deal with a relativistic, but fixed-number of particle model in which the nucleonic degrees of freedom can be present just from the beginning. Relativistic Hamiltonian Dynamics seems to be the best suited theoretical tool for this and it is here employed in the form it takes on the Light Front (LCD) [4], with the advantage of the possibility of interpreting the results with the help of Feynman diagrams evaluated in the Infinite Momentum Frame. There is not, however, a complete correspondence between field theory and LCD that, in this respect, can be interpreted as a sort of perturbative expansion in the effects of inelasticities and of retardation. The detailed study of this correspondence is the subject of recent investigations [5].

The matrix elements of the current taken between light-front deuteron states

$$_F < d'\lambda'|\hat{J}_\alpha(q)|d\lambda >_F = \frac{1}{2\sqrt{d'_+ d_+}} e^*_{F,\mu}(d',\lambda') J^{\mu\nu}_\alpha(q) e_{F,\nu}(d,\lambda) \tag{1}$$

define in the usual way the covariant form-factors; the matrix elements of the "good" component J_+ can be written also in the form

$$_F < d'\lambda'|\hat{J}_+(\vec{q}_\perp)|d\lambda >_F = \int \frac{d^3k}{\sqrt{\varepsilon_k \varepsilon_{k'}}} A_{\lambda'\lambda}(\vec{q}_\perp, \vec{k}) \tag{2}$$

where

$$A_{\lambda'\lambda}(\vec{q}_\perp, \vec{k}) = \frac{1}{2} \mathrm{Tr} \left\{ \psi^\dagger_{\lambda'}(\vec{q}_\perp + \vec{k}) \left[F^S_1(q_\perp^2)\hat{H} - \frac{F^S_2(q_\perp^2)}{2m}\hat{G} \right] \psi_\lambda(\vec{k})\hat{\tilde{H}} \right\} . \tag{3}$$

The matrices $\hat{H}$, $\hat{G}$ and $\hat{\tilde{H}}$ depend on the Wigner rotations of the proton and neutron spins and ψ_λ has the structure of the conventional non relativistic wave function

$$\psi_\lambda(k) = u(k)\sigma_\lambda - \frac{1}{\sqrt{2}} \left(\sigma_\lambda - \frac{3k_\lambda \vec{\sigma}\cdot\vec{k}}{k^2} \right) w(k) \tag{4}$$

---

*presented by P. Saracco

and obeys to the same equation. This is the central result of this approach, in that it directly connects the evaluation of the leading relativistic corrections to a well established phenomenological background. There is obviously some price to be paid for it: in fact the invariance of $J_+^{\lambda\lambda'}(\vec{q}_\perp)$ under P,T transformations and under rotations around the $\hat{z}$-axis restricts the number of independent matrix elements to 4, via:

$$J_+^{\lambda'\lambda}(\vec{q}_\perp) = J_+^{\lambda\lambda'}(-\vec{q}_\perp)$$
$$J_+^{\lambda'\lambda}(\vec{q}_\perp) = \left(J_+^{\lambda\lambda'}(\vec{q}_\perp)\right) ,$$

but only 3 physical form factors are needed. This means that between these matrix elements another relation should be fulfilled, the so-called "angular condition",

$$(1 + 2\eta)\, J_+^{+1,+1} - J_+^{0,0} - 2\sqrt{2}\eta J_+^{+1,0} + J_+^{+1,-1} = 0 \tag{5}$$

which can be derived in two ways: either directly from the relations existing between current the matrix elements and the covariant form factors, or transforming to the helicity basis and requiring that matrix elements connecting states with helicity different by two units to vanish. This relation is not analitically fulfilled by the previous expressions: its violation, is a measure of the inconsistencies, intrinsic in the approach, connected in particular with the difficulties in defining the states with longitudinal polarization $\lambda = 0$ in the IMF. In particular it results impossible to extract out of the integral over the relative IMF momentum the expression of polarization vectors in the case of λ and/or $\lambda' = 0$: this means that eq. (2) cannot be factorized in the form(1) for the case of longitudinal polarizations. It is useful to note that F_{ch} and F_{Q} can be derived without making use of longitudinally polarized states, while this it is not possible when looking for F_{mag}. Since the angular condition is not fulfilled, we shall not calculate directly the "most dangerous" matrix element $J_+^{0,0}$, but instead, we shall derive it from the other 3, when needed. Anyway, at least for $Q^2 < 4$ GeV^2, the angular condition is violated less than 15%.

The result of this calculation [see Fig. 1] can be summarized as follows:

- the relativistic effects are relevant, as expected, for $Q^2 \gtrsim .8$ GeV^2.

- the consistency of this approach is within 15% for $Q^2 \lesssim 4$ GeV^2.

- the different off-shell behaviors of the various phenomenological potentials reflect in different results for the form-factors. Paris, Urbana V14 and RSC potentials yields more or less the same results both in the charge and quadrupole form factors while the Bonn potential has a peculiarly different behavior; in the case of magnetic form factor all the potentials behave very similarly up to $Q^2 \simeq 1.5$ GeV^2, but the very different positions of the first minimum force discriminate strongly between them in the higher Q^2 region. Obviously the OPE potential is not expected to give a realistic result and it is here reported for the sake of comparison.

- other (model-dependent) effects must be included in the calculation (MEC, 6-quark bag, etc.) before drawing definite conclusions, but it seems difficult that they can override completely the differences between the various potential here examined.

- the results depend significantly, in the high Q^2 region, on the electromagnetic form factors assumed for nucleons, particularly on the neutron magnetic form factor.

It is not surprising that relativistic effects can play a relevant rôle starting from $Q^2 \sim .8$ GeV^2. Here however we are mainly interested in understanding to what extent a detailed knowledge of the short range part of the NN interaction can be relavant. So at least four points seems of higher priority at present:

- the evaluation of polarization observables for the deuteron [6].

- the evaluation of the effect of the MEC and of other relativistic corrections within a definite scheme.

- the extension of this treatment to inelastic processes, like, e.g., the electrodisintegration of deuteron, or to exclusive processes.

- the evaluation of 3-nucleon system elastic form factors.

232

[1] H. Arenhövel and M. Sanzone, *Few-body Syst. Suppl.* **3**(1991)1.

[2] S. Boffi, C. Giusti and F. D. Pacati. To be published on Phys. Rep. C.

[3] L.L. Frankfurt and M.I. Strikman, *Phys. Rep.* **C76**(1981)215.

[4] B.D. Keister and W.N. Polyzou, *Adv. Nucl. Phys.* **20**(1991)225.

[5] M.M. Giannini, L. Kondratyuk and P. Saracco, GEF-Th-9/1993, submitted to *Few-body Syst.*.

[6] M.M. Giannini, L. Kondratyuk and P. Saracco, Proc. of VI Workshop on "Perspectives in nuclear physics at intermediate energies", Trieste 1993, World Scientific Pub. Co. (Singapore) - in press.

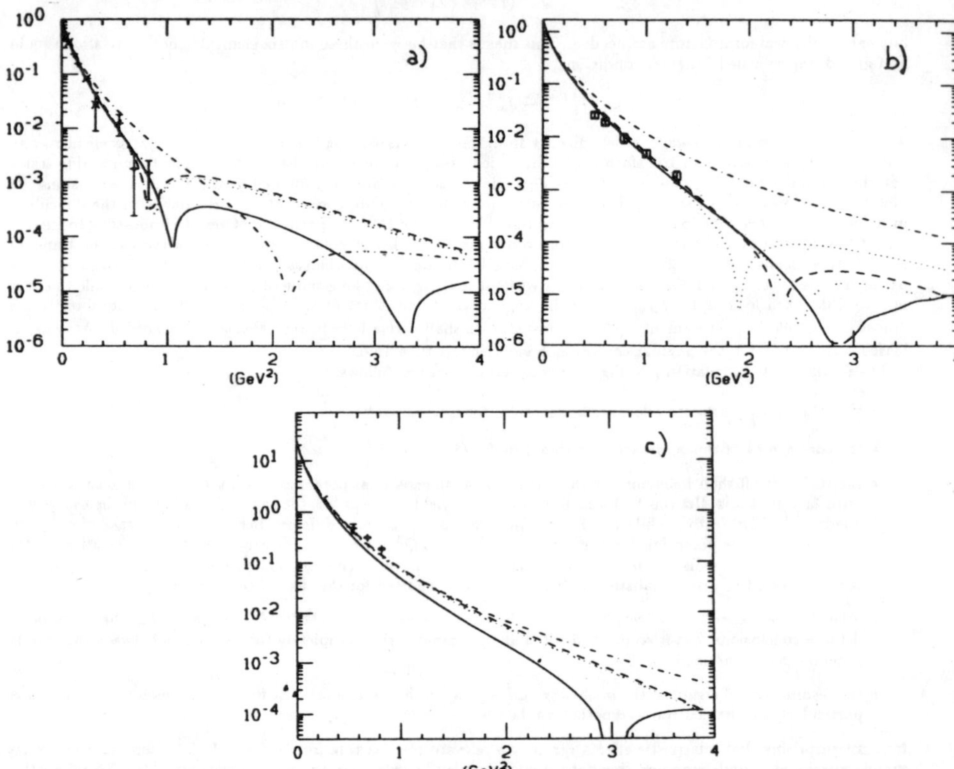

Fig. 1 - *Charge (a), magnetic (b) and quadupole (c) elastic electromagnetic form factors of deuteron: Bonn pot. - solid line, OPE - dotdashed line, Urbana V14 - double dotted, Paris, dotted and RSC - dashed line. The electromagnetic form factors of nucleons are the usual Dipole ones, but the neutron magnetic f.f., for which Saclay parametrization have beeen used.*

Color Transparency

B.K. Jennings

TRIUMF, 4004 Wesbrook Mall, Vancouver, B.C., V6T 2A3, Canada

and

G.A. Miller

Department of Physics, FM-15, University of Washington, Seattle, WA 98195, USA

Abstract

The anomously large transmission of nucleons through a nucleus following a hard collision is explored. This effect, known as color transparency, is believed to be a prediction of QCD. In this talk we discuss the necessary conditions for its occurrence and the effects that must be included in a realistic calculation.

1. Introduction

In this talk we consider hard exclusive reactions on a nuclear target. The idea is to use the nucleus to analyze the size (or interactions) of a nucleon immediately before or after a hard interaction. The expectation is that it may be different[1,2] then predicted by the naive Glauber theory. The original motivation behind this was to test a prediction of perturbative QCD and check its validity.

The expectation is that after a hard interaction, interactions with the nucleus will be reduced. This is based on three ideas:

1. A small object is produced in a hard reaction.

2. Small objects interact weakly with the nuclear medium.

3. The expansion time is sufficiently long that the object can exit the nucleus before it expands.

This reduction in the interactions with the nucleus is known as color transparency. It is the first of these three ideas that we really wish to check. However it turns out that is it the third that causes the most uncertainty and unfortunately has the least intrinsic interest. We will consider the three ideas in turn.

2. Hard interactions produce small objects

Perturbative QCD based arguments lead to the conclusion that small objects are produced in hard interactions. The idea is quite simple and we refer to Fig. 1a. For all the quarks in the hadron to go in the same direction after a hard interaction the momentum must be shared between the constituents (two in the diagram) of the hadron. The gluon responsible for the sharing is highly virtual and hence can not travel very far. Thus the constituents must be close together in the transverse direction. In fact the size distribution goes like

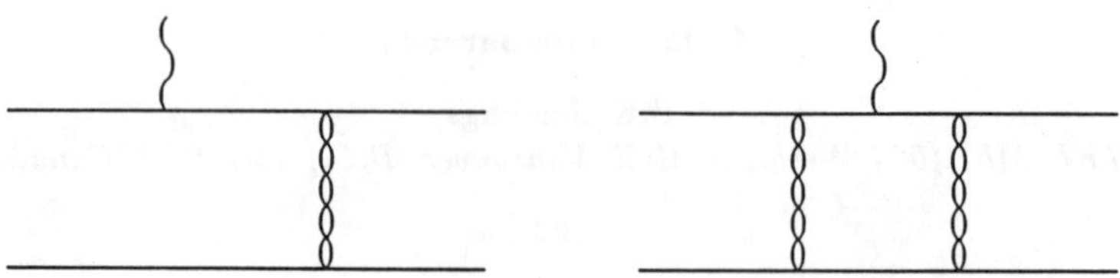

Fig. 1. Perturbative QCD (a) and end point singularity (b) diagrams

$\int \frac{d^2 q_\perp}{(2\pi)^2} \frac{e^{iq_\perp \cdot b}}{q_\perp^2 + (Q/2)^2} \sim \frac{1}{2\pi} \sqrt{\frac{\pi}{2}} \frac{\exp(-Qb/2)}{\sqrt{Qb/2}}$. In the longitudinal direction Lorentz contraction makes the size small.

There are alternate processes that can be imagined to contribute to hard interactions. For example in fig. 1b we show another diagram that can in principle contribute. In this diagram the momentum can be redistributed without any particle being far off shell.[3] This occurs when the spectator has zero momentum and is sometimes referred as an end point singularity. In this case since the particles are all almost on shell we can not use perturbative QCD to estimate this diagram but must use also the confining potential. Naively this diagram does not require the constituents to be close together.

Now the controversy starts. Is this second diagram important? At some energy it must be suppressed. If a charged particle under goes a hard interaction it will bremsstrahlung off photons. Similarly if a color charged object under goes a hard interaction it will bremsstrahlung off gluons. This leads to inelastic processes and not to the exclusive reaction under consideration. The net effect is that this diagram will be suppressed. Actually more is achieved. All processes where the colored constitutions are not close together will be suppressed. This is known as as Sudakov suppression and was discussed by Carlson at this meeting. The only question is at what energy does this suppression occur. There is no consensus on this question.[4,3] Thus this diagram may make a contribution that is not spatially small. If this is dominant we would not see color transparency. One of the interests in color transparency is to see if such diagrams are important. However in the numerical work described in this talk these diagrams will be ignored.

An additional contribution is shown in fig. 3. This is known as the Landshoff term and should contribute to proton-proton scattering. Like the end point singularity it should be suppressed by Sudakov effects. Again we do not know at what energy this suppression will occur. However one thing we do know is that the proton-proton 90° scattering data is not given by the S^{-10} predicted by perturbative QCD but rather oscillates around this value.[5] The interference between the Landshoff term and the perturbative QCD term will generate such an oscillation.[5] Other sources for the oscillation have been suggested. For example Brodsky[6] has suggest that the wiggles arise from the opening of the strangeness and charm thresholds. With this approach is he is able to fit both the cross-section and spin observables. It is also possible to fit the spin observables with the Landshoff term as shown by Carlson et al.[7] Both approaches to describing the oscillations have two interfering terms one of which is the perturbative QCD term and other which presumably does not have a small spatial size. We thus expect the second term to have normal distortions. This, as we will see later, has a significant impact on the observed color transparency. In principle the (p,2p) reaction on

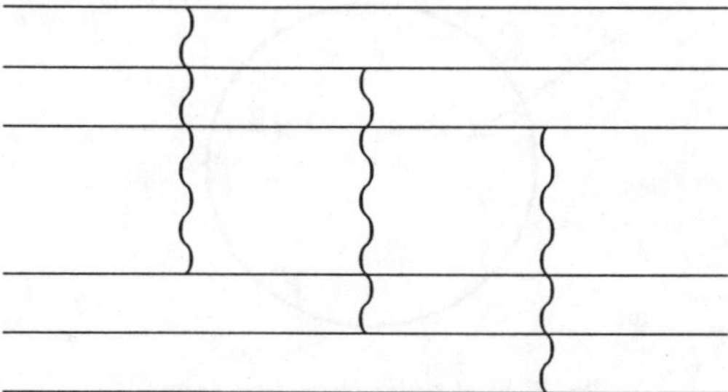

Fig. 2. The Landshoff diagram

a nucleus can distinguish between these two alternatives but currently neither the data nor theory are sufficiently precise.

3. Small Objects Interact Weakly

A gluon will only interact strongly with a colorless object only if its wavelength is less then or approximately equal to the color separation. Thus for small objects we expect the interactions to be small. Other processes, for example pion exchange, must also be considered. This has been done by Strikman et al[8] who argues that all interactions are suppressed for small objects. Even the Skyrme model predicts small objects interact weakly.[9] This is the best founded and least controversial of the three assumptions needed to get color transparency.

4. The Expansion Time

4.1. Small Objects Expand Rapidly!

Let us consider the form factor in the laboratory frame. The photon four momentum is $(q_0, \vec{q})$. The outgoing (on shell nucleon) has four momentum $(E_q, \vec{q})$ with $E_q^2 - \vec{q}^2 = m^2$. The on shell condition leads to $Q^2 = \vec{q}^2 - q_0^2 = \vec{q}^2 - E_q^2 - m + 2mE_q = 2m(E_q - m) \approx 2mq$ The small object produced has transverse size the order of $1/Q$ hence transverse momenta the order of Q and is off shell by amount $\sqrt{Q^2 + q^2} - \sqrt{m^2 + q^2} \approx Q^2/2q = m$. Thus it lives for a time $1/m$ and can travel a distance of $c/m \approx 0.2\text{fm}$. With this kinematics a small object will expand rapidly for any incident energy! This problem must be overcome if we are to have color transparency. As we will see shortly the solutions comes from considering the complete problem.

This example also gives a warning for other problems. Never assume that the frozen approximation is valid just because the incident energy is large. It must be checked in every case.

Even if the above estimate is off by an order of magnitude the expansion is still too fast. The argument relies on just the uncertainty principle so it should be quite robust. The important point is that the small object is far off shell so non-quantum treatments are of no use.

This rapid expansion is actually useful. It means that we can use factorization. The hard interaction is over and done with before the particle has time to move one nucleon radius

236

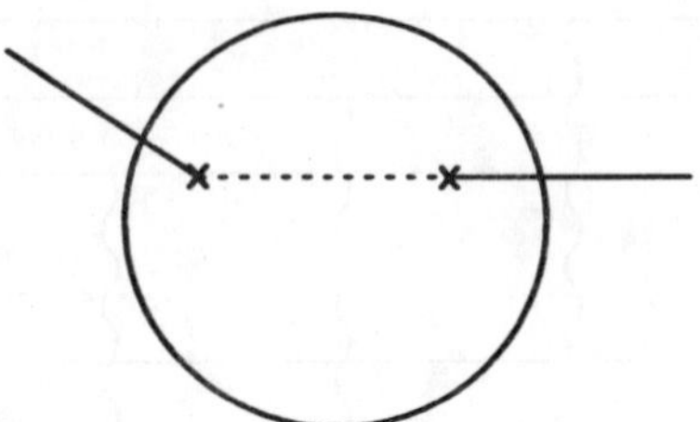

Fig. 3. First order correction term

and interact with the other nucleons.

4.2. Perturbative Treatment

To see how rapid the expansion time is we treat the interactions with the medium as a perturbation. In fig. 3 we show the first order correction to the amplitude. For color transparency this term should vanish.

Notice that we have two interactions. First we have the hard interaction that makes the small object. Second we have the interaction with nuclear medium. This second interaction give the final state interactions. It must also convert the small object to the nucleon that is observed in the detector.[10,11]

The contribution to the amplitude to second order can be written:

$$
\begin{aligned}
\mathcal{M}_\alpha &= B_\alpha + ST_\alpha. \\
B_\alpha &= \langle \vec{p}|T_H(Q)|\alpha\rangle = F(Q^2)\langle \vec{p} - \vec{q}|\alpha\rangle, \\
ST_\alpha &= \langle \vec{p}|U\,G\,T_H(Q)|\alpha\rangle,
\end{aligned}
$$

$$(1)$$

where B_α is the Born amplitude, $F(Q^2)$ the nucleon form factor, and α labels the nuclear state. Notice the form of the second term; it has U, G and T. If G were not there we would have UT which is zero (or at least small) since this is just the statement that a hard interaction produces a weakly interacting object. The expansion is in the propagator G. The propagator depends on both the internal degrees of freedom and center of mass coordinate. We use a spectral representation of $G = \sum_m \int d^3q'|n,\vec{q}\,> /(E_0(q) - E_n(q') + i\epsilon) < n,\vec{q}'|$ where n denotes the intrinsic excited states and q' is the center of mass momentum. The condition to have just UT is that we can use closure on the sum over intermediate states n. The hard interaction produces very high energy intermediate states n which by them self would not permit the use of closure. However U prefers low lying states and if we are lucky will cut off the sum at sufficiently low energies to permit the use of closure. Thus the expansion time is more a property of U then of T (although of course both are involved). This is a useful observation since there are other reactions such as total cross-sections that allow the study of U.[12] In fact[13] for the total cross-section calculation the frozen approximation is not good. This is quite worrying for color transparency.

An alternate way of looking at the problem is as follows: The energies can be written as $E^* = \sqrt{m_m^2 + q^2}$ and $E = \sqrt{m_0^2 + q^2}$. Thus the intermediate state is off shell by $\Delta E =$

$E^* - E \approx (m^{*2} - m^2)/2q$ and travels a distance $R = 2q/(m^{*2} - m^2)$. Transparency requires that the nuclear size must be less then R. As with the closure argument the whole question of expansion time comes down to which states are important.

If we take a simple harmonic oscillator model[10,11] and use $U \propto r^2$ only two states will appear in the sum since r^2 only connects the ground state to ground state and the second excited state. This is an interesting example since it shows that it possible to get the cancellation in the scattering amplitude with only two states in spite of the fact we need many states to get a small size. The second point is that it is U that cuts off the sum.

Unfortunately the oscillator model is not necessarily a good approximation. However there is some experimental data available. We need $\langle n|T_H(Q)|n = 0\rangle$ which is related to the transition form factor (or the nucleon form factor for n=0). We also need $U_{N,m}(\vec{B}, Z)$ which is related to diffractive disassociation. Unfortunately what we need are matrix elements for specific states while all we have experimentally are cross-section for fixed mass m. Assuming the matrix elements are just square roots of cross-sections and add coherently gives contributions at very large mass m. However we know that the hard interaction and diffractive dissassociation produce different states at the same energy (the pion multiplicities are different) so we introduce a cutoff[14] which we constrain by requiring that color transparency is exact in the closure limit. It is quite worrying that the rather add hoc cut off is all that makes color transparency work.

In some calculations the extreme frozen approximation is used where all effects of the expansion are ignored. In the hadron matrix element approach this does not happen and we can not imagine a reasonable scenario where it would be valid at the energies under discussion. There are simply not the necessary states at very low energy.

5. Results

So far we have considered just the results to first order in perturbation theory. We take the higher order terms approximately into account using an exponentiating technique.[11] The approach has been checked numerically and found quite accurate.[15]

The results are shown in fig. 4. There is little transparency in the Q^2 range of the SLAC experiment.[16] The results from that experiment, presented at his workshop, fall between the two limits for different cuts off given in this figure and are thus nicely in agreement with our calculations. The results are also in the same ball park as the Brookhaven results.[17]

So far we have considered only quasifree kinematics. This, of course, is optimized for nucleon production. Since color transparency arises by a cancellation between nucleon and resonant states we can enhance color transparency by shifting off the quasi-elastic peak in a direction to optimize for the production of excited states. This can be done by varying the Bjorken x_B or by transferring momentum to the residual nucleus as in the Brookhaven experiment. This should have been obvious from the beginning but we did not realize it until it was pointed out by Boris Kolepiovich[18] who discussed in the context of the role of Fermi motion (see also Bianconi[19]).

The x_B dependence is shown in fig. 5. Only the values of x_B near one should be taken seriously. For comparison with experiment it is probably better to do as in fig. 5b and just compare above the quasi-elastic peak with below the quasi-elastic beak.

So now we come to our final calculation.[20] We include in this calculation the basic color transparency, the distributed mass with a power law cut off, the Landshoff term as

238

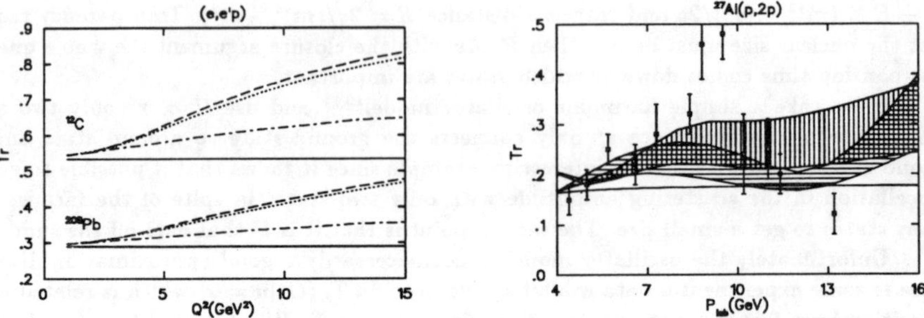

Fig. 4. (a) The transparency, Tr, for the $(e, e'p)$ reaction. The solid line represents the standard Glauber calculation $(\sigma_{eff} = \sigma_p)$. The lines are as follows dashed: sharp cutoff $g(M_X^2)$, dotted: eq. (5) with $M_1 = 1.44 GeV$, dash-dot: power law $g(M_X^2)$. (b) Energy dependence of the transparency Tr. The data points are from Carroll et al..[17] The area shaded vertically is obtained from the mechanism of Ref.[5] and amplitude of Ref..[7] The area shaded horizontally is obtained from the mechanism of Ref..[6] In each case the upper bound uses the sharp cutoff for $g(M_X^2)$ and the lower bound a power law. The solid curve assumes no color transparency.

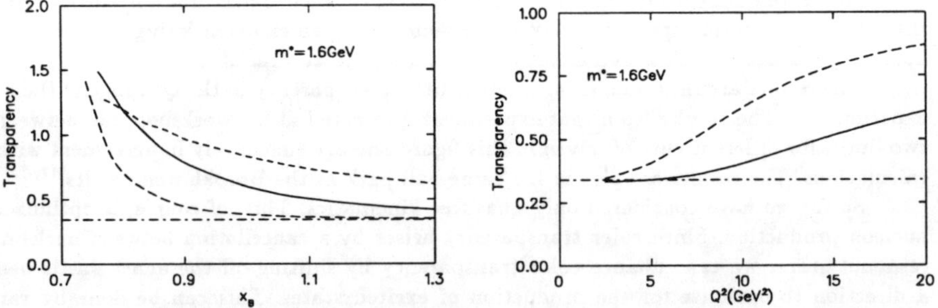

Fig. 5. (a) The nuclear transparency as a function of x_B. The curves correspond to Q^2 of 7 GeV^2 (long dashed curve), 15 GeV^2 and 30 GeV^2. The dash-dotted curve is the Glauber model. (b) The nuclear transparency as a function of Q^2. The solid curve is for $x_B < 1$ while the dashed curve is for $x_B > 1$. The dash-dotted curve is the Glauber model.

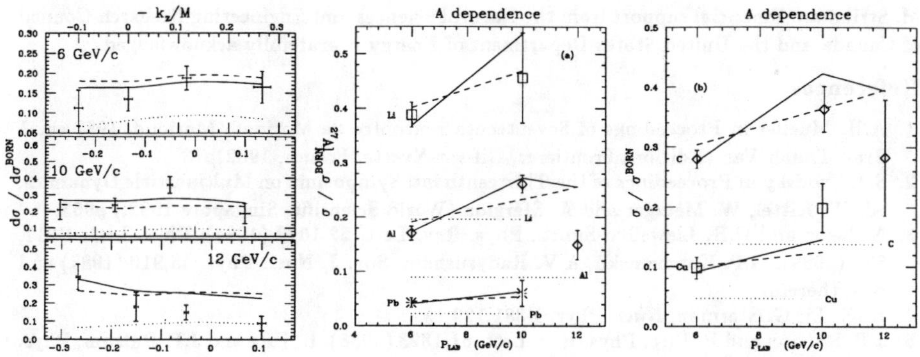

Fig. 6. Transparency for (p,2p). Solid-curve full calculation, dashed-curve without the Ralston-Pire effect, dotted-curve Glauber.

fit by Ralston and Pire,[5] and the kinematic effect relating to x_B. The results are shown in the next figures. In the calculations the energy dependence is more reliable then the normalization. Also there is an uncertainty in the normalization of the experimental data. In the figure we have therefore shifted the data down by about one standard deviation of the quota experimental normalization error. For the ^{27}Al data we have not used the conventional presentation employing a p_{lab} effective. Instead we show the data separately at each incident momenta. This is because the effective p_{lab} approach assumes that there is only one energy scale in the problem, namely S. Certainly when the expansion time is important this is not correct. Part of the downward slope in the 12 GeV data comes about precisely because of the breakdown of the concept of effective p_{lab}. We strongly discourage the use of this concept.

The down turn in the data in fig. 6b and c is due to the Landhoff term (or the threshold effects). We believe that both this term and the proper treatment of the kinematic effects are necessary in order to describe the data.

The agreement while not perfect is certainly at least qualitative. Thus it is possible to describe simultaneously the SLAC and Brookhaven data. That latter give an indication that color transparency has been seen and its main features understood.

In conclusion we think color transparency has indeed been seen at in the Brookhaven experiment. Although it will have to be confirmed by more precise experiments. The SLAC experiment is at too low an energy to see much effect but their results as presented at this conference are consistent with our calculations. It is thus possible to have qualitative agreement with both the SLAC and Brookhaven experiments. More data is however desperately needed.

6.   Acknowledgments

Part of this work was done in collaboration with Boris Kopeliovich. We have also benefited from discussions with C. Carlson, L. Frankfurt, W.R. Greenberg, J. Ralston, and

M. Strikman. Financial support from the Natural Sciences and Engineering Research Council of Canada and the United States Department of Energy is gratefully acknowledged.

References

1. A.H. Mueller in Proceedings of Seventeenth rencontre de Moriond, Moriond, 1982 ed. J Tran Thanh Van (Editions Frontieres, Gif-sur-Yvette, France, 1982)p13.
2. S.J. Brodsky in Proceedings of the Thirteenth intl Symposium on Multiparticle Dynamics, ed. W. Kittel, W. Metzger and A. Stergiou (World Scientific, Singapore 1982,) p963.
3. N. Isgur and C.H. Llewellyn-Smith, Phys. Rev. Lett. 52,1080 (1984); Phys. Lett. B217, 535 (1989); G.P. Korchemskii, A.V. Radysushkin, Sov. J. Nucl. Phys. 45,910 (1987) and refs. therein.
4. H.-N. Li, G. Sterman, Nucl. Phys. B381,129,1992
5. J.P. Ralston and B. Pire, Phys. Rev. Lett. 61,1823 (1988). B. Pire and J.P. Ralston, Phys. Lett. 117B,233 (1982) J.P Ralston and B. Pire, Phys. Rev. Lett. 49,1605 (1982).
6. S.J. Brodsky & G.F. De Teramond, Phys. Rev. Lett. 60,1924 (1988).
7. C.E. Carlson ,M.Chachkhunashvili, and F. Myhrer, Phys. Rev. D46,2891 (1992).
8. L. Frankurt and M. Strikman, Phys. Rep. 160, 235 (1988), L. Frankurt and M. Strikman, Nucl. Phys. B250,143 (1985).
9. G. Kalbermann and J.M. Eisenberg, Phys. Lett. B 286,24 (1992).
10. B.K. Jennings and G.A. Miller, Phys. Lett. B236,209 (1990).
11. B.K. Jennings and G.A. Miller, Phys. Rev. D 44,692 (1991)
12. P.V.R. Murthy et al, Nucl. Phys. B92,269 (1975).
13. B.K. Jennings and G.A. Miller, work in progress.
14. B.K. Jennings and G.A.Miller Phys. Rev. Lett. 69,3619 (1992).
15. W.R. Greenberg and G.A. Miller, Phys. Rev. D47,1865 (1993).
16. R. Milner et al., SLAC proposal NE18
17. A.S. Carroll et al. Phys. Rev. Lett. 61 (1988) 1698. M.B. Johnson and L.S. Kisslinger, World Scientific (Singapore, 1989)
18. B. Kopeliovich, priviate communication, B.K. Jennings and B. Kopeliovich, Phys. Rev. Lett. 70,3384 (1993).
19. A. Bianconi, S, Boffi, and D.E. Kharzeev, Phys. Lett. B205,1 (1993), Nucl. Phys. A, (in press).
20. B.K. Jennings and G.A. Miller, Phys. Lett. B (submitted).

HOW TO TEST EXPERIMENTALLY FOR COLOR TRANSPARENCY AND NUCLEAR FILTERING

PANKAJ JAIN

Department of Physics and Astronomy, University of Oklahoma
Norman OK 73019 USA

JOHN P. RALSTON

Department of Physics and Astronomy, University of Kansas
Lawrence, KS 66045 USA

ABSTRACT

We review the formalism for color transparency and nuclear filtering based on factorization in perturbative QCD. The theory leads to a new systematic data analysis procedure which can be used to extract signals of color transparency without excessively biasing the analysis with models. With minimal assumptions of filtering for $A>>1$, the method allows extraction of the hard scattering rate inside the nuclear target as well as the survival rate. We can minimize model dependence further by considering a scaling law, which predicts that the survival probability P depends on a special combination of the momentum transfer Q^2 and nuclear number A, namely, $P(Q^2, A) = P(Q^2/A^{1/3})$. Analysis of the BNL data on $pA \rightarrow p' p'' (A-1)$ scattering provides evidence for observation of color transparency by three independent tests.

1. The Interface of Nuclear and Particle Physics

Color transparency and nuclear filtering are subjects on the interface of nuclear and particle physics. Traditional nuclear physics studies the strong interactions using hadronic degrees of freedom; now both particle and nuclear physics use quarks. At first sight there seems to be more quark degrees of freedom than the hadronic one, because each hadron can be described not only by its "center of mass" variables, but also the internal configurations of the quarks. A natural question arises: can the interactions of quark degrees of freedom produce observable effects that are not expected from traditional hadron ones?

The fundamental theory, quantum chromodynamics (QCD), predicts that the effective strong interactions will indeed depend on the internal quark configurations. If, for example, a pair of quarks with opposite color quantum numbers is localized in a small spatial region, then the pair's strong interactions in the form of gluon exchange will cancel by destructive interference. If such a small color singlet configuration is a predominant amplitude in an experiment, then the system will look as if it has reduced strong interactions. Color transparency[1] is the name for this effect: it is as if the strong interactions could, under certain circumstances, be turned "off".

Nuclear filtering[2] is complementary to color transparency. It hinges on two things: knowledge from experiment that typical strong interaction cross sections are "big", and the perturbative QCD prediction that the forward scattering amplitude (total cross section) of a transversely small color singlet configuration is "small". Filtering is the phenomenon of relative enhancement of those components which do not produce inelastic events simply by the selection of the quasi-exclusive events. One expects to observe relatively more short distance effects in exclusive processes in large nuclear targets than in small ones.

Both color transparency and nuclear filtering are consequences of a systematic perturbative formalism[3] which contains non-perturbative ingredients. Even the perturbative part is controversial. The short distance model[4] of free space scattering in QCD predicts that quarks participating with large momentum transfer Q^2 are selected from regions of the quark wave function with rather small relative spatial separation b. Is this true?

Experimentally the short distance model has some successes, and some failures. The successes include observation of a consistent pattern of power law behavior in most fixed angle or "one-scale" processes. The failures include a consistent pattern of hadron helicity violation, (the total helicity of hadrons in reactions is not conserved), which violates a dynamical symmetry of the short-distance model, and *cannot be explained within that model to all orders of perturbation theory*. Another discrepancy is the observation of patterns of oscillations with energy in both pp and πp 90° cm scattering (Fig. (1), top). The oscillations contradict the known leading Q^2 dependence of the model, which to all orders of perturbation theory consists of powers of Q^2 and powers of $log(Q^2)$, and no oscillations.

In full view of the shortcomings of the free space application, we have explored the idea indicated by nuclear filtering that *for targets with A>>1, the survivors come from small configurations so that the short distance model should apply*. In this case, factorization can be used to separate the hard scattering from the nuclear target dependence. As we will see, both color transparency and nuclear filtering are consequences of factorization[3] along with very minimal assumptions about the target dependence.

A simple test of the idea of filtering is given by the BNL data[5]. In this experiment, $Q^2 = -t = s/2$. The s-dependence of the *ratio* of the BNL data to the free space data shows a bump (Fig 1, bottom). However, this bump is apparently being produced by dividing a smooth, monotonically varying quantity by an oscillating one.

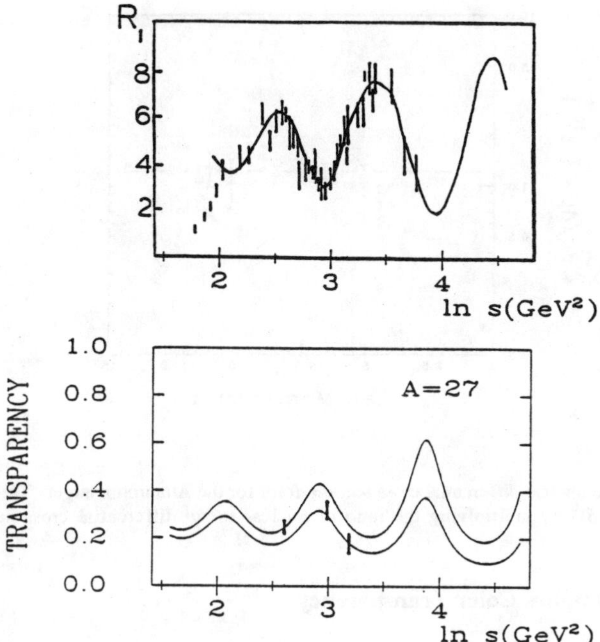

Fig 1. *top*: Oscillations of the scaled fixed angle *pp* scattering cross section $s^{10}d\sigma/dt$ with respect to *s*. *bottom*: The "transparency ratio" $1/Z\,d\sigma/dt\,(pA \to p'p''(A\text{-}1))\,/\,d\sigma/dt(pp \to p'p'')$ for Aluminum has a bump where the $pp \to p'p''$ scattering has a dip. Data from Ref. (5); curves from Ralston and Pire, Ref. (2).

However, as Fig. (2) shows, the oscillations are much reduced in the cross section taken on the nuclear target. This is rather direct evidence that components of the amplitudes that were interfering in the free-space $pp \to pp$ scattering have been depleted in the nuclear target.

For many years we interpreted the oscillations as a perturbative QCD "chromo-Coulomb phase" interference effect caused by the separation of color in independent scattering[6]. It is beautifully consistent with this picture that the independent scattering regions are filtered away in a large nuclear target. To test this interpretation further, we would like to see more experiments on both πp and *pp* scattering on nuclear targets, and also measurements of hadronic helicity conservation, which should be restored for $A \gg 1$ even at fixed, moderately large Q^2.

244

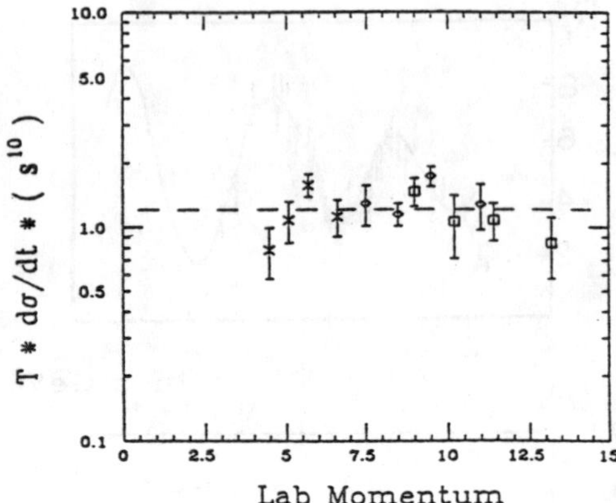

Fig. 2 The s^{10} - scaled nuclear differential cross section $d\sigma/dt$ for the Aluminum target. This was obtained (Heppelmann, Ref. (5)) by multiplying the ratio of nuclear to pp differential cross sections by the denominator.

2. Factorization Implies Color Transparency

Factorization, the field-theoretic expression of the impulse approximation, allows scattering to be expressed in terms of wave functions before and after the "fast" hard scattering event, along with a hard scattering kernel which is calculated perturbatively. The hard scattering kernel is the same for any target with the same number of quarks, inside a nucleus or in free space. However, the results of hard scattering are not always the same, because the wave functions the hard kernel couples to are modified by propagation and interaction inside the nucleus. Factorization makes few assumptions about the nature of the hadronic wave functions, which are determined by the long-time evolution of the strong interactions.

The formalism for the nuclear target is as follows[3]. A scattering amplitude $M(Q^2, A)$ can be written in an appropriate gauge as

$$M(Q^2, A) = \prod_{i,f} \int [dx\, d^2 k_T] \left[\psi_A^{(f)*}(x_f, k_{Tf}) H(x_i, x_f, k_{Ti}, k_{Tf}, Q^2) \psi_A^{(i)}(x_i, k_{Ti}) \right] \tag{1}$$

where $H(x, k_T, Q^2)$ represents the hard scattering kernel and $\psi_A(x, k_T)$ is the wave function to find quarks with relative transverse momentum k_T and momentum fraction x. Here we are emphasizing the internal quark momentum variables and suppressing integration over hadron center of mass variables, indicated by curly brackets. The wave function ψ_A

represents the overlap of the hadronic wave function with the process of propagation through the nuclear medium, and can be expressed as

$$\psi_A(x', k_T'^2) = \int dx\, d^2 k_T\, \delta(x - x')\delta^2(k_T - k_T') - F_A\left[s, \left(\frac{x}{x'}(k_T - k_T')\right)^2\right]\psi_0(x, k_T^2) \quad (2)$$

where $\psi_0(x, k_T)$ is the free space hadronic wave function and F_A is the nuclear inelastic scattering amplitude. The nuclear filtering of the internal quark coordinates depends mainly on the transverse variables; we suppress indices i,f and the x dependence, which is integrated over. Going to the quark transverse separation b space by a Fourier transform, we get $\psi_A(x, b) = f_A(s, b)\psi_0(x, b)$, where $f_A(s, b) = 1 - F_A(s, b)$ can be called the *survival amplitude*. The distribution amplitude ϕ_A used in the nuclear target can then be written

$$\phi_A(x, Q^2) = (2\pi)^2 Q \int_0^\infty db\, J_1(Qb)\tilde{f}_A(s, b^2)\tilde{\psi}_0(x, b). \quad (3)$$

The Bessel function tells us that the above integral gets most of its contribution from the region $b^2 < 1/Q^2$; this a theoretical output, not an input. With the perturbative QCD result that the forward survival amplitude $f_A(s, b)$ goes like $1 - const\; b^2 A^{1/3}$ as $b^2 \to 0$, one readily obtains color transparency at large Q^2 as a prediction. ($A^{1/3}$ is the target length.)

Controversy over the so-called "expansion time" problem amounts to disagreements on the non-perturbative survival amplitude $f_A(s, b)$. Our assumption that it is fairly independent of s is based on the perturbative prediction, and data from conventional strong interaction processes, showing weak energy (s) dependence in the region of $s = 2 - 20$ GeV^2 for hadronic scatttering. However, models[7] inserting baryon resonances in a "dilute gas" approximation predict extra energy dependence.

Our assumptions so far have been kept to a minimum: for example, the x-and b dependence of the hadronic wave functions have been left unspecified. To extract more information, we can tell the mathematics that the wave functions are smooth near the origin. (Large k_T^2 perturbative tails are removed; that part goes into the hard scattering). With an expansion of the wave functions about $b = 0$, the distribution amplitudes take the form

$$\phi_A(Q^2, x) \approx 2\pi\tilde{\psi}(b = 0, x) \int_0^\infty db\, Q J_1(bQ)\tilde{f}(s, b^2 A^{1/3})$$

$$(4)$$

which exhibits[8] a "scaling dependence" on two large dimensionful variables: Q^2 and $nA^{1/3}$, where n is the nuclear number density.

From Eqs. (2-4) we see that the amplitude $M(Q^2, A)$ has a factorized form:

$$M(Q^2, A) \equiv < \left\{ < \tilde{f}_A(b^i \approx 1/Q) \dots \tilde{f}_A(b^j \approx 1/Q) \right\} H(x, 0, Q^2) >,$$

(5)

where the pointed and curly brackets indicate the integrations over x variables and hadron cm coordinates. Note that $H(x, k_T=0, Q^2)$ comes out of the curly brackets once the value of A is large enough to deplete large configurations such as independent scattering. In general we do not expect H to be the same kernel which dominates the free space scattering, since filtering should remove most of the independent scattering. It would be a mistake, and in any case it is unnecessary, to assume that the hard scattering simply "divides out" in forming transparency ratios. The transparency ratio $T(Q^2, A)$ can, however, be written

$$T(Q^2, A) = d\sigma/dt_A / (Zd\sigma/dt_{free\ space}) \rightarrow P(Q^2, A)\, R(Q^2)$$
$$\rightarrow P(Q^2/A^{1/3})\, R(Q) \qquad (6)$$

indicating the factorization into an effective survival probability $P(Q^2, A)$ and the ratio of the hard scatterings $R(Q^2)$; the last step follows from the scaling relation for the distribution amplitudes.

3. Experimental Tests

Our methodology is strongly tied to experiment and we do not go too far into formalism or excessive modeling without guidance from experiment. To this end we present three tests which we believe strongly confirm that we are on the right track:

test : depletion of large configurations: We have already discussed this test, namely the decrease of the oscillations for A>>1. We eagerly await further data to confirm this pattern on other targets, including information on helicity violation and with higher experimental precision. A transparency ratio oscillating with energy has been predicted[3,4].

test : increasing survival with Q^2. The BNL experiment was a pioneering one, but since it is the only data currently available, we used it to illustrate a new data analysis procedure[9] based on Eq. (6). The first step is to calculate the survival probability P as a function of an attenuation cross section σ_{eff}; we used exponential attenuation and realistic nuclear densities. The functional form of Eq. (6) then suggests fitting the data at fixed $Q^2=Q_o^2$ as a function of A. By taking the logarithm of the cross section (or transparency ratio), we can determine the value of the attenuation cross section $\sigma_{eff}(Q_o^2)$ from the *shape* of the A-dependence, treating the logarithm of $R(Q_o^2)$ as an unknown additive constant. The results of this procedure are twofold[9]: (1) The BNL data shows a statistically significant signal for a decreasing attenuation cross section, fit by $\sigma_{eff}(Q^2) = 40\,mb\,(2.2\,GeV^2/Q^2) \pm 2\,mb$. (2) The hard scattering rate in the nuclear target is then extracted by examining the ratio $R(Q^2)$. In the case of the BNL data, we found[9] to our surprise that the nuclear hard scattering was not only smooth but also falling faster than s^{-10}. After some thought we realized that perturbative QCD predicts not s^{-10} but $\alpha_s^{10}(\mu^2)s^{-10}$, where μ^2 is a scale of order $-t = s/2$. We indeed find that the nuclear hard scattering rate agrees well with this. The interplay of two effects, which in retrospect are generic and not unexpected, also explains a puzzle[9]: how the nuclear data shown in Fig. (2), which is the product of a hard scattering rate and a survival rate, could be flat compared to the s^{-10} rule and at the same time show evidence for increasing survival with increasing energy.

test : transparency scaling. The scaling law that survival probabilities should be functions of the ratio $Q^2/A^{1/3}$ was predicted in Ref. (8). In that paper the prediction was applied to the transparency ratio using the assumption that in electroproduction the free space and nuclear hard scatterings would be identical. We no longer believe any assumption about the nuclear target hard scattering is necessary or desirable, and we treat all experiments on an equal footing in our procedure. The scaling prediction can now be examined with the BNL data. There is a way to do this by fitting the data to an unknown function of Q^2 times an unknown function of $Q^2/A^{1/3}$, but the range of the current BNL data is too limited to check this. However, we can create a working model of a transparency ratio by simply dividing the nuclear cross section by $\alpha_s^{10}(\mu^2)s^{-10}$. Since this is done with no modeling of the survival probability at all, this constitutes a logically independent test. After doing the division, we binned all the data as a function of Q^2/A^α, and examined the results of plots for $\alpha = 0.2$, 1/3, and 0.5. These results, presented elsewhere,[10] show that the value $\alpha = 1/3$ is considerably favored, while $\alpha = 0.2$ and 0.5 are ruled out. In Fig. (3) we display the results only for the case $\alpha = 1/3$. One universal function of $Q^2/A^{1/3}$ could fit all the BNL data independent of the model for that function.

It remains to be seen whether these results can be accommodated by "rapid-expansion" models[7], which tend to overshoot the bump in the s-dependence in any case, after taking into account the oscillations in the denominator of the transparency ratio.

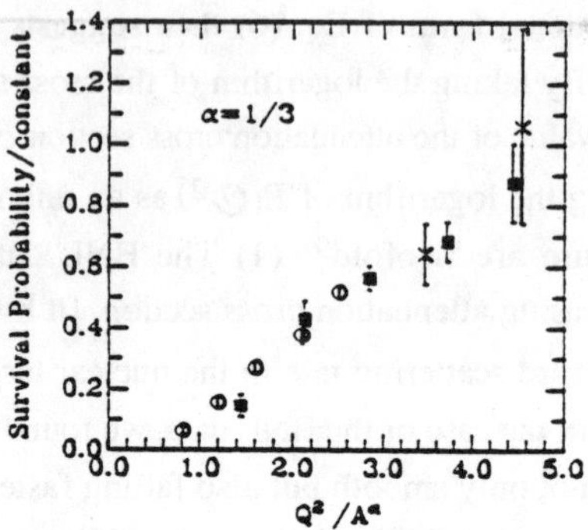

Fig. (3). Scaling dependence as a function of $Q^2/A^{1/3}$. BNL Data points are circles ($Q^2 = 4.8\ GeV^2$), squares ($Q^2 = 8.5\ GeV^2$) and crosses ($Q^2 = 10.4\ GeV^2$).

Acknowledgments: This work was sponsored in part by DOE Grants number DE-FG02-85ER40214 and DE-FG05-91ER40636 and the *Kansas Institute for Theoretical and Computational Science.*

References

1. S. J. Brodsky and A. H. Mueller, *Phys. Lett.* **B 206**, 685 (1988).

2. G. Bertsch, S. J. Brodsky, A. Goldhaber and J. Gunion, *Phys. Rev. Lett.* **47**, 297 (1981); J. P. Ralston and B. Pire, *Phys. Rev. Lett.* **61**, 1823 (1988).

3. J. P. Ralston and B. Pire, *Phys. Rev. Lett.* **65**, 2343 (1990).

4. P. LePage and S. J. Brodsky, *Phys. Rev.* **D22**, 2157 (1980); *ibid* **D24**, 2848 (1981).

5. A. S. Carroll *et. al.*, *Phys. Rev. Lett.* **61**, 1698 (1988). See also S. Heppelmann, in *Nuc. Phys.* **B** *(Proceedings Supplement)* **12**, 159 (1990).

6. B. Pire and J. P. Ralston, *Phys. Lett.* **117B**, 233 (1982); Kansas preprint, 1992.

7. See, e.g. G. R. Farrar, *et al. Phys. Rev. Lett.* **61**, 686 (1988); B. Jennings and G. Miller, *Phys. Lett.* **B274**, 442 (1992); *Phys. Rev.* **D44**, 692 (1992).

8. B. Pire and J. P. Ralston, *Phys. Lett.* **B256**, 523 (1991).

9. P. Jain and J. P. Ralston, *Phys. Rev.* **D48**, 1104 (1993); see also *ibid* **D46**, 3807 (1992).

10. P. Jain and J. P. Ralston, in *Proceedings of the XXVIII International Rencontre de Moriond* (Les Arcs, 1993), J. Tranh Van Than, editor, (Editions Frontiers, in press).

EXCLUSIVE HADRONIC INTERACTION MECHANISM
AND COLOR TRANSPARENCY

Steven Heppelmann

*Department of Physics, Pennsylvania State University,
University Park, Pa. 16802, USA*

ABSTRACT

In a series of experiments at Brookhaven, exclusive scattering at large angles (near $90°$ CM) has been studied. The purpose of these experiments is to search for evidence of simple quark configurations or the "minimal fock states" which have been predicted to play an important role in hard exclusive hadronic interactions. In this paper, an analysis of 20 exclusive reactions at beam energy of 5.9 GeV/c ($q^2 \simeq (6 \, GeV/c)^2$) is presented. Evidence for the dominance of quark interchange amplitudes which connect valence quark initial and final states is considered. Also discussed is the apparatus for a new generation of color transparency measurements. The observation of patterns in the flavor flow dependence of exclusive scattering or the observation of color transparency phenomena, can serve as an evidence for the simple valence quark dominated and spatially compact picture of hard scattering, the picture inspired by perturbative QCD.

1. Exclusive Hard Processes

From the point of view of quarks and gluons, the two body exclusive scattering between hadrons seem to be a complex reaction. The initial and final state hadrons can only be expressed as a superposition of a large number of quark-gluon states, some containing very many partons. Furthermore, it is only when the interacting currents of QCD are localized in space that the quark-gluon and gluon-gluon interaction is predicted to be weak enough to validate the intuitive-perturbative picture with well defined particle states interacting via exchange of virtual quanta.

It is often the case in physics that one approaches a very complex problem by identifying circumstances where simplification occurs. The simplification in the case of exclusive hadron scattering is expected in those relatively rare interactions with large p_T (transverse momentum). It is argued, from the rules of quantum mechanics and QCD, that these interactions are associated with intermediate states which are simpler than those used to describe the asymptotic incident or outgoing hadrons.[1,2] These intermediate states are expected to be in the valence fock configuration, the minimal quark content necessary to carry the quantum numbers of the corresponding hadrons. Furthermore, and by closely related reasoning, it is expected that the spatial transverse size of these intermediate states must be small, reflecting the scale $\frac{1}{p_T}$.[3,4]

This picture of interaction, via this special intermediate state, is not only simple but also provides the prerequisite condition in which perturbation theory should be valid for calculating cross sections. It is known that an expression for the cross section factors into two parts. The first is a projection of these intermediate states (with minimal parton content and with small spatial extent) onto a hadron wave function. The second part is the scattering amplitude between the incoming and outgoinging compact "minimal fock states". The first factor involves hadron wave functions which are not currently calculable. The nature of the wave functions depends upon the full complexity of the strong interaction and is perhaps outside the scope of current perturbative techniques. The second part, the interaction of the small size configuration should be calculable in perturbative QCD.

A series of experiments at Brookhaven National Lab has focused on exploring phenomena which reflect the consequences of this picture of hard hadronic interactions. The most recent emphasis has been directed toward a new series of experiments to study color transparency effects in hard hadronic exclusive interactions.[5] But even before the color transparency experiments were considered, a program was started at Brookhaven to explore the consequences of the perturbative QCD inspired picture of hard interactions.

The most striking consequence of this picture, which remains a compelling success, is the dimensional scaling law.[6] The signature of entry into the energy and p_T regime where this picture may be applicable should be the onset of power law s dependence of the fixed angle cross sections. The variable "s" is the square of the center of mass energy. The dimensional scaling relation is

$$\frac{d\sigma}{dt} \propto s^{2-n} \tag{1}$$

where n is the sum of the number of initial state and final state partons in the valence or minimal fock representation. This functional form is seen in many exclusive processes at values of s not too far above $1\ GeV^2$. While the scaling laws are specifically predicted by a detailed analysis of the perturbative hard scattering picture, the possibility that this success is a coincidence still fuels an active debate.[7]

A different approach to understanding these processes could be to attempt to classify the patterns seen in exclusive scattering. For example, if a formula like Eq. 1 can be written, which relates the observed cross section to the number of partons in the minimal fock state, it is reasonable to expect that these minimal representations of the hadrons must play an important role in a correct phenomenological picture of hard scattering. This point would be apparent even if details of the perturbative QCD scattering model failed to predict the form of Eq. 1. The fact that the various exclusive cross sections can be sorted according by values of n and reactions with the same n have similar energy dependence should be a signal about the importance of valence configurations. In complex systems, finding new patterns can be the key to understanding. We try to observe that a particular set of variables establishes a clear pattern in the experiments. Such an outcome is then evidence to support a phenomenological picture that has important degree's of freedom related to these variables.

The principle direction of Brookhaven experiments E755,E834,E838 and E850 has been to look for patterns in hard exclusive hadron scattering. The goal has not been limited to searching for phenomena that can be predicted by the present state of theory, like color transparency. Also we search for patterns that have not been specifically anticipated.

2. Flavor Flow In Hard Exclusive Scattering

There are very many feynman graphs that contribute to the hard exclusive hadronic scattering amplitude in Perturbative QCD, even when only leading twist graphs (those involving valence quark representation of hadrons) are considered. The evaluation of these graphs is technically difficult, however, it is possible to sort these amplitudes in terms of four basic topologies of flavor flow.[8]

To sort interactions by allowed flavor flow topologies, one starts with the assumption that only the valence or leading quark configurations of each hadron contribute to the scattering amplitudes. In this picture, the proton is characterized as (uud) with states like $(uud\overline{u}u)$ playing no role. From this point of view, we can identify 4 different categories of scattering amplitudes which are possible and may contribute to the cross section.

1. <u>Gluon Exchange</u>: These amplitudes involve no exchange of flavor and can be present whenever each final state hadron has the same quark content as its corresponding initial state hadron. These amplitudes are always present in elastic scattering processes.

2. <u>Quark Interchange</u>: These amplitudes involve pairs of quarks, one in each of the initial state hadrons, exchanging in the final state.

3. <u>Quark-Antiquark Annihilation or Creation</u>: These amplitudes are present if there are flavors in the final state which are not present in the initial state.

4. <u>Combination</u>: This is a class of amplitudes which satisfy both 2 and 3.

Each exclusive process may proceed via some or all of the above amplitude classes.

We have measured 20 exclusive reactions at 90° in the center of mass frame at a beam momentum of 5.9 GeV/c with q^2 near 6.0 $(GeV/c)^2$. The analysis of this experiment involves the classification of each reaction according to the subset of the four flavor flow topologies which can contribute to the scattering amplitude. It is interesting to remark that this division is only meaningful if valence hadron configurations are dominant. For example, if sea quarks are included in the hadron representation, then all interactions would allow interchange of quarks as well as annihilation. To the extent that nonvalence configurations of the hadrons are important, we would expect the dependence on this sorting scheme to wash out and we would not expect to see patterns in an analysis of cross section vs flavor flow topology.

In Figure 1, 20 exclusive processes are tabulated. A detailed description of the experimental technique used in these measurements is described elsewhere.[9] The measured cross sections vary from 10^{-9} to 10^{-6} barns.

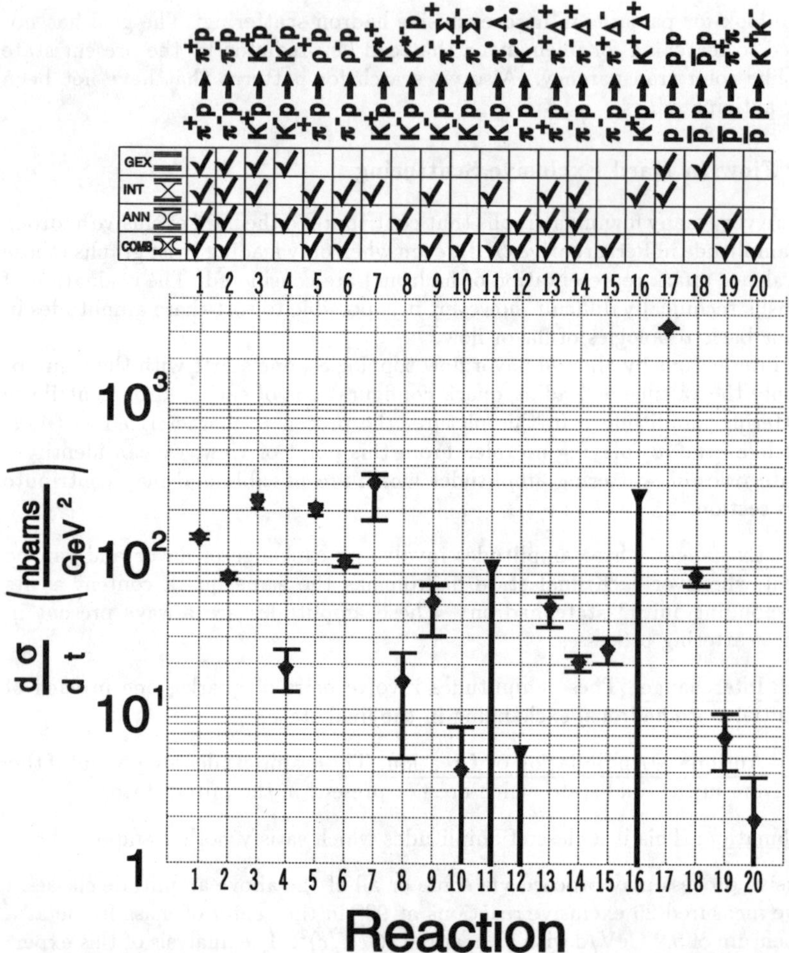

Fig. 1. The cross sections for 20 exclusive interations at center of mass angle of $90°$ are plotted vs reaction number ranging from 1 to 20. The classes of quark flow amplitudes which can contribute are indicated for each reaction.

As has been discussed in more detail elsewhere,[9] it is clear from the figure that those processes which allow quark exchange amplitudes tend to have cross sections about an order of magnitude greater than that observed in similar processes where quark exchange is not allowed.

Even before perturbative QCD was developed as a tool for hard exclusive scattering, the quark interchange mechanism[10] was proposed as a phenomenological mechanism for hadronic hard scattering. Now we have evidence that this mechanism may be realized in the scattering in this region of kinematics, $q^2 = 6$ $(\text{GeV/c})^2$.

3. New Generations of Color Transparency Measurements

Color transparency is a phenomena predicted by QCD as a consequence of the importance of the spatially compact intermediate state associated with the hard scattering process. Since a color neutral and spatially small state has a small interaction cross section with other strongly interacting matter, various nuclear effects can be observed.

The prediction is that at large q^2, hadron proton interactions in nuclei of the form

$$h + A \rightarrow hp(A - 1), \tag{2}$$

when measured in the kinematic region where the hp interaction is hard (near 90° in the CM), should exhibit reduced interactions between the hp initial and final states, and the A-1 nucleon system. Nominal initial and final state interactions can be calculated using very simple Glauber models. The most naive models of nominal absorption are probably correct to the 20% to 30% level.[11] More careful calculations should push the dependability of the nominal prediction to better than 10%.

The transparency ratio "T"is defined in terms of the observed pp differential cross section per target proton in (p,2p) interactions.

$$\frac{d^5\sigma}{dt\, d\vec{p}\, d\epsilon}_{(p,2p)} = TS(\vec{p}, \epsilon)\frac{d\sigma}{dt}_{pp} \tag{3}$$

In this expression $(\vec{p}, \epsilon)$ is the missing momentum and kinetic energy, carried away by the nucleus and $\frac{d\sigma}{dt}_{pp}$ is the pp cross section evaluated at energy appropriate in such an impulse approximation picture. $S(\vec{p}, \epsilon)$ is the spectral density normalized to unity. In Brookhaven E834, our group reported transparency ratios significantly above the Glauber prediction it certain kinematic regions.[5] The observation can be summarized as follows

- The transparency of (p,2p), corresponding to 90° pp in aluminum is seen to rise as the beam momentum is increased from 6 to 10 GeV/c. And falls from 10 to 12 GeV/c. This structure is anticorrelated with the cross section scaled by s^{10}, perhaps indicating that the effect of the nuclear filtering process is to remove components of th pp cross section which deviate the most from the pure power law dependence of dimensional scaling. Various models have been proposed to explain this structure.[12,13]

254

- The projected momentum distributions observed in this data are quite conventional in the central region, but have excess strength at large fermi momentum.[14] An uncertainty in these tails introduces a systematic uncertainty in the determination of the transparency ratio. A better understanding of the observed momentum distribution is interesting in it's own right and is essential in improving our understanding of the transparency ratio.

A new detector for color transparency related measurements has been constructed at Brookhaven. The new detector has advantages over the old one, including the ability to taking data with reduced background and 2-3 orders of magnitude greater sensitivity. In 1993, the first experiments with the fully operational detector are being done.

The primary reason for construction of the new apparatus is to increase the energy reach of these experiments, up to incident an incident momentum of 20 GeV/c. Since the cross section for these processes falls as the 10th power of incident energy, the increased sensitivity will be necessary at the higher energy.

The Eva detector has cylindrical symmetry and looks much like a collider detector with acceptance in the central region. Of course in this detector, the central region has been boosted forward, resulting in cylindrical elements telescoping forward. The detector uses a 1T solenoid magnet of 3 meter length and 1 meter radius. The detector elements are straw tube cylindrical chambers ranging in radius from .2 to 1.8 meters. The length of the chambers is 2 meters for all but the inner chamber. The detections of the 2 protons in a (p,2p) interaction is symmetric, with momentum measurement on both sides. The angular coverage of the detector becomes larger if one considers only the inner 2 chambers. While all chambers are needed for reconstruction of the highest energy tracks, the inner chambers will do a good job on tracks with momentum under 500 MeV/c.

The nature of this detector will allow a full reconstruction of the higher energy charged products in secondary interactions or in the breakup of the A-1 nucleon system. This allows the possibility of whole new classes of experiments, including the full reconstruction of interactions in the deuterium nucleus. For example, in

$$pd \to ppn$$

the process can be isolated from the missing mass or energy spectrum of the neutral candidates. In another example

$$pd \to ppn^*$$

with n^* decaying to a pair of charged particles, the event is fully analyzed in this detector.

An important issue in the understanding of transparency phenomena has been the question of the time scale for the system to evolve from a small state to the nominal hadron with full interaction. This issue has been addressed by several speakers at this conference, The ability to make measurements sensitive to secondary interactions in light systems like deuterium are important because the secondary interaction occurs almost immediately after or before the primary scatter.

During the next year, a variety of new results is expected using this detector.

4. Conclusion

It is seen that the perturbative picture of a hard hadron exclusive scattering has two very obvious features. The first is that the valence configurations of the proton(hadron) are the only ones that contribute to the scattering. The second is that this configuration is spatially compact.

The observation of 20 exclusive reactions at 90° indicates that those processes that can involve the exchange of quarks between the interacting hadrons tend to have larger cross sections that those that cannot. Because the valence quark flow is so important in these processes, it is tempting to infer that that the valence configurations are indeed important in these reactions.

Measurements of color transparency phenomena with the EVA detector at Brookhaven National Lab will be in progress during this next year.

1. G. P. Lepage and S. J. Brodsky. *Phys. Rev.* **D22** (1980) 2156.
2. A. H. Mueller. *Physics Reports* **73** 237 (1981).
3. A. Mueller, *Proceedings of the XVII Rencontre de Moriond*, ed. J. Tran Thanh Van, Editions Frontieres, Gif-sur-Yvette, France (1982) 13.
4. S. J. Brodsky. *Proceedings of the XIII International Symposium on Multiparticle Dynamics-1982*, eds. W. Kittel, W. Metzger and A. Stergiou, World Scientific, Singapore, (1983) 963.
5. Alan S. Carroll et al. *PRL* **61** (1988) 1968.
6. S.J.Brodsky and G.R. Farrar. *PRL* *31*(1973) 1153.
7. N. Isgur and C. H. Llewellyn Smith. *Nucl. Phys.* **B317**(1989) 526.
8. G. Farrar. *Phys. Rev. Lett.* **53**(1984) 28.
9. C. White et al. Preprint BNL-49059-Submitted to Physical Review (May 1993).
10. J.F. Gunion et al. *Phys. Rev.* **D8**(1973) 287.
11. G. R. Farrar, H. Liu, L.L. Frankfurt, and M.I. Strikman. *PRL* **61**(1988) 686 .
12. J. Ralston and B. Pire. *PRL* **61** (1988) 1823.
13. S.J. Brodsky and G.F. de Teramond. *PRL* **60**(1988) 1924.
14. S. Heppelmann et al. *Physics Letters* 232B(1989) 167.

COLOR TRANSPARENCY AND THE $(e, e'p)$ REACTION

A. LUNG

W. K. Kellogg Radiation Laboratory, California Institute of Technology
Pasadena, CA 91125, USA

Representing the NE18 Collaboration [1]

ABSTRACT

It has been suggested that nuclear matter would become "transparent" to hadrons involved in exclusive reactions at high momentum transfer. This prediction, known as "color transparency," is a consequence of arguments based on fundamental aspects of hadronic interactions in QCD. The quasielastic knockout of a proton in electron scattering, $(e, e'p)$, is a particularly clean process for the study of this predicted effect. Preliminary results from a recent experiment performed at SLAC will be discussed.

1. Introduction

The study of the transparency of the nuclear medium for exclusive hard-scattering reaction products is of central importance to understanding such processes. Mueller and Brodsky[2] have suggested that, at sufficiently high momentum transfer (Q^2), the final (and initial) state interactions of the hadrons with the nuclear medium should be reduced due to "color screening." The supporting arguments for this prediction, termed "color transparency," have been recently reviewed[3] and are briefly summarized below. The observation of this phenomenon would have substantial impact on our understanding of the mechanism of these hard-scattering processes and its relation to the fundamental theory of strong interactions, QCD.

Various models have been used to estimate the magnitude of the effect at experimentally accessible energies and the predictions differ considerably. Therefore it is important to experimentally establish the momentum scale required for the validity of the hard-scattering description.

We have performed an experiment at SLAC (NE18) to explore this issue in quasielastic proton knockout via electron scattering at high momentum transfer ($1 \leq Q^2 \leq 7 \, (\text{GeV/c})^2$). Preliminary results for H, D, C, and Fe are presented below.

2. Color Transparency

The existence of color transparency would require that three conditions be met in the scattering process.[3] Firstly, the high momentum transfer scattering should take place via selection of amplitudes in the initial and final state hadrons characterized by a small transverse size (much smaller than the hadron radius). Secondly, this small object participating in the reaction should be "color neutral" outside of this small radius in order not to radiate gluons (which would lead to inelasticity). This object, being small and color neutral, would then have reduced interactions with hadrons in the surrounding nuclear medium ("color screening"). Finally, it is necessary that this compact size be maintained for some distance in traversing the nuclear medium in order that the reduction in the interaction probablility be large enough to be observable.

It is interesting to note that such an effect is in fact observed in QED. The ionization density produced by high energy (~ 200 GeV) e^+e^- pairs in photographic emulsion has been observed to increase as a function of distance from the point of creation.[4] In QCD, the observation of Bjorken scaling in the low x (shadowing) region in deep inelastic muon scattering seems to indicate that $q\bar{q}$ pairs with small transverse size are created and that color screening takes place.[5]

The only presently available experimental data relevant to the question of color transparency in exclusive reactions are quasielastic $(p, 2p)$ data on nuclear targets taken at BNL.[6] There the measured "transparency" of the nuclear medium does increase with incident momentum, but also seems to decrease at the largest momenta. This effect appears to be correlated with the oscillating behavior observed in the free p-p elastic scattering cross section.[7] The correct interpretation of these data is not clear at the present time, and further measurements are planned at BNL.[8]

3. Quasielastic Electron Scattering

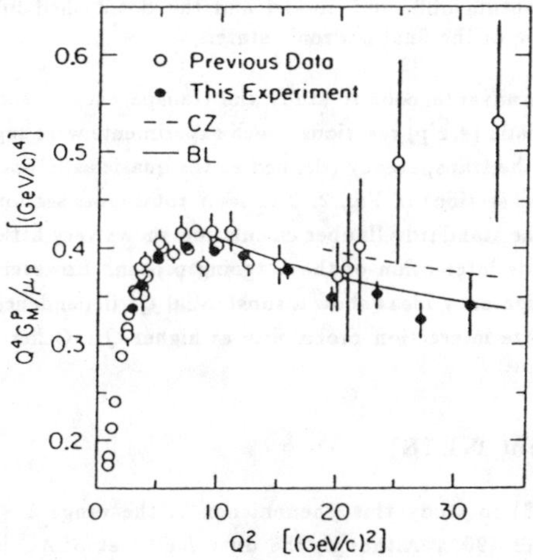

Figure 1 The magnetic form factor of the proton at high Q^2 from Ref. 9.

In contrast to the p-p elastic cross section, the elastic magnetic form factor of the proton[9] shown in Fig. 1 seems to exhibit the smooth behavior one would expect of a hard scattering process for $Q^2 > 5$ (GeV/c)^2. In fact, the Q^2 dependence follows the prediction based on quark counting rules, $G_M(Q^2) \sim Q^{-4}$. The interpretation of these data in terms of a perturbative amplitude is certainly controversial,[10] but at least there are no anomalous oscillations or other such obvious problems.

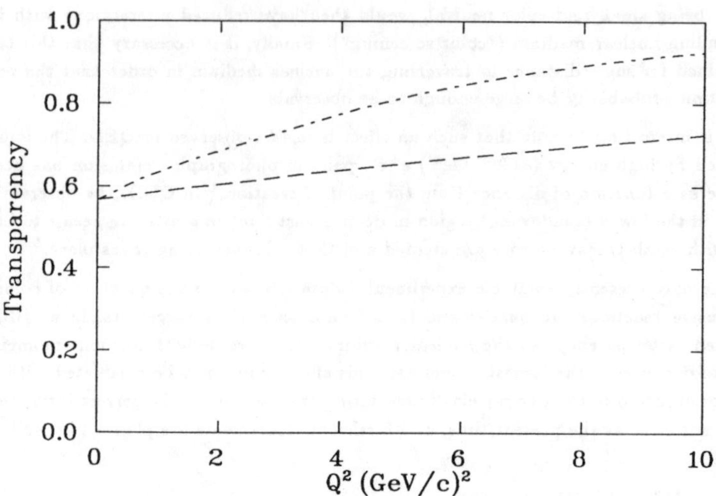

Figure 2 Predicted transparency[11] for quasielastic electron scattering from ^{12}C. The dashed line is a "quantum diffusion" model and the dot-dashed line is a naive parton model for the evolution of the final hadronic state.

Thus it would appear advantageous to study the transparency of the nuclear medium for the outgoing proton in quasielastic $(e, e'p)$ reactions. Such experiments were suggested in Ref. 11, and we show their predictions for the transparency (defined as the quasielastic nuclear cross section divided by A times the proton cross section) in Fig. 2. The N-N total cross section is more or less constant in this energy regime, so the standard Glauber calculation shows very little momentum dependence to the substantial final state interaction of the outgoing proton. However, their two models which incorporate the color transparency ideas show a substantial Q^2 dependence to the cross section due to the diminishing final state interaction probability at higher Q^2. Other theoretical models[12,13,14] yield similar predictions.

4. SLAC Experiment NE18

An experiment (NE18) to study this phenomenon in the range $1 \leq Q^2 < 7$ $(GeV/c)^2$ was performed at SLAC in Fall 1991. Although the duty factor at SLAC is very low ($\sim 10^{-4}$) for coincidence experiments, the $(e, e'p)$ kinematics are well determined for $q >> p_F$, where q is the three-momentum transfer and p_F is the Fermi momentum. That is, the protons are kinematically focussed into a small cone of angular radius $\delta\theta < p_F/q$ about the $\vec{q}$ direction. This feature of the reaction enhances the true to accidental ratio to an acceptable level even for the low duty factor at SLAC.

The experiment utilized the Nuclear Physics Injector at SLAC to provide electron beams of 1.9 to 5.1 GeV incident energy before scattering from a variety of nuclear targets (D, C, Fe, Au, and H for calibration). The 1.6 GeV spectrometer was employed at large scattering angles (35-60°) to detect the scattered electrons. A gas Čerenkov counter and Pb-glass calorimeter were used to identify the electrons and reject background pion events. The coincident proton was detected in the 8 GeV spectrometer using a time-of-flight measurement for identification. The experimental goal was to measure the integrated quasielastic $(e, e'p)$ cross section to $\sim 5 - 10\%$ precision as a function of A and Q^2 to test the predictions shown in Fig. 2.

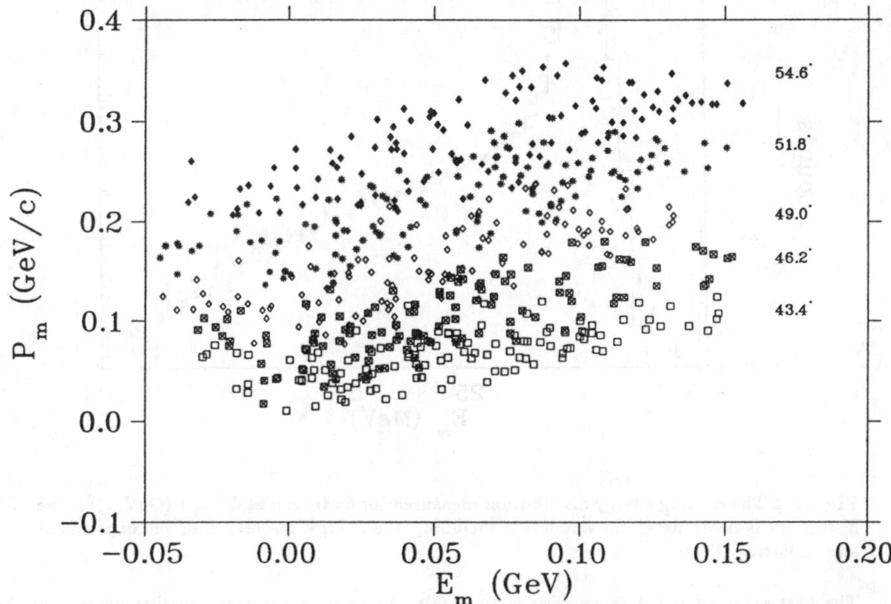

Figure 3 The missing energy versus missing momentum distribution at $Q^2 = 1$ (GeV/c)^2. Each symbol indicates the kinematic coverage for a different proton scattering angle, θ_p.

Experimental knowledge of the incident electron momentum and the momenta of the final state electron and proton allows the use of energy and momentum conservation to solve for the "missing" energy and momentum. The "missing" energy and momentum, E_m and P_m, are attributed to the initial state (bound) proton (assuming a quasielastic reaction).

To achieve the greatest kinematic coverage possible, the coincident protons were detected at several angles, θ_p, for each angular setting of the electron spectrometer, θ_e. Figure 3 shows the E_m versus P_m distribution for each of five proton scattering angles measured at $\theta_e = 35.5°$ for $Q^2 = 1$ (GeV/c)^2. Traditionally, one measures the distribution of these quantities (the nuclear "spectral function" $S(E_m, P_m)$) for various nuclei in quasielastic $(e, e'p)$ to study the single particle structure of nuclei. In our case, we wish to integrate over the spectral function in order to extract the total quasielastic $(e, e'p)$ cross section as a function of Q^2. The ratio of this total quasielastic cross section to that of the free proton is then a measure of the nuclear transparency. Table 1 shows the NE18 kinematics relevant to the "transparency" measurements.

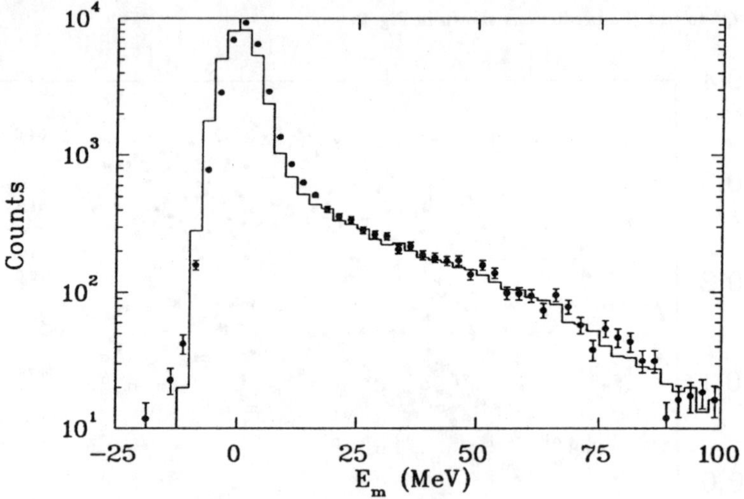

Figure 4 The missing energy distribution measured for hydrogen at $Q^2 = 1$ $(GeV/c)^2$. The histogram is a Monte Carlo simulation including the energy resolution of the experiment and radiative effects.

The first stage of the data analysis is to obtain the nuclear spectral function at each of the incident energies for each target. The hydrogen target allows one to test the experimental procedure. Here the initial proton is at rest and on-shell, so ideally there is no missing energy or momentum. In fact, radiative processes do generate tails; Fig. 4 shows the experimental missing energy distribution for hydrogen at $Q^2 = 1$ where the radiative tail is clearly evident. This tail is well-reproduced by a Monte Carlo simulation of the radiative processes.

In Fig. 5, the extracted momentum distribution for deuterium at $Q^2 = 1.2$ $(GeV/c)^2$ is shown along with a Monte Carlo calculation (radiative effects included) using the Bonn potential[15]. The agreement of the shape of the experimental distribution with the simulation is excellent. This demonstrates both that the experiment is capable of properly measuring the nuclear spectral function and that, at least for a simple light nucleus, the quasielastic description of the reaction is valid at these very high momentum transfers.

Table 1. Kinematics for SLAC experiment NE18.

Q^2	E_e	E'_e	P_p	θ_e	θ_p
$(\text{GeV/c})^2$	GeV	GeV	GeV/c		
6.8	5.12	1.47	4.49	56.6°	15.9, 17.3°
5.0	4.21	1.47	3.54	53.4°	20.9, 22.6°
3.0	3.19	1.47	2.45	47.7°	27.7, 30.5, 33.3°
1.0	2.01	1.39	1.20	35.5°	43.4, 46.2, 49.0, 51.8, 54.6°

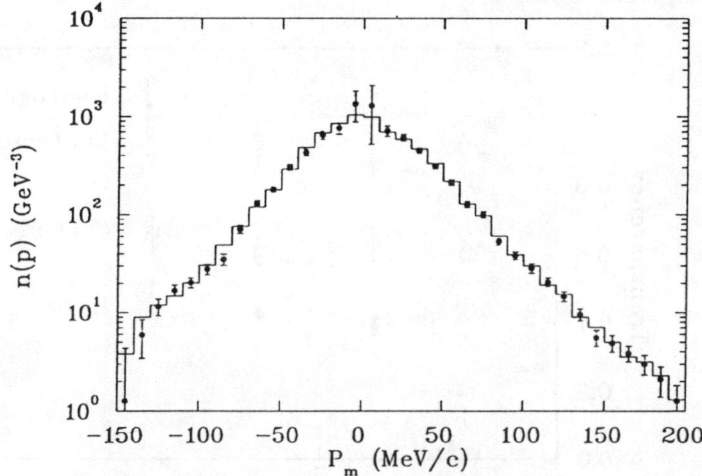

Figure 5 The momentum distribution measured for deuterium at $Q^2 = 1.2\ (\text{GeV/c})^2$. The histogram is a calculation using the Bonn potential[14] normalized to the experimental data.

In order to determine a measure of the nuclear transparency from these data, we have defined

262

transparency as the following ratio:

$$T = \frac{\int_{E_m,P_m} d^3P_m dE_m \ N_{exp}(P_m, E_m)}{\int_{E_m,P_m} d^3P_m dE_m \ N_{sim}(P_m, E_m)}$$

where N_{exp} is the experimentally measured distribution in E_m and P_m and N_{sim} is the corresponding distribution from a Monte Carlo simulation using an input spectral function. The input spectral function is a shell-model, single particle distribution normalized to $(1 - \kappa)$ where κ is the "wound" associated with high-momentum correlated nucleons. For Carbon, we use $\kappa = 0.09$ and for Iron we use $\kappa = 0.15$. The region of integration covers the area where the spectral function is large (typically $60 < P_m < 250$ MeV and $E_m < 100$ MeV). Preliminary results for the experimental transparency ratios are shown in Fig. 6. It is should be re-emphasized that these ratios do depend on the assumed spectral function used in the analysis. Note that we have also performed this analysis for Hydrogen and Deuterium. The Hydrogen results demonstrate that the normalization of the experiment is well-understood at each Q^2. The Deuterium transparency is 0.89 ± 0.07 independent of Q^2. The Carbon and Iron transparencies do not show the dramatic rise with Q^2 predicted in Ref. 11. Detailed comparisons with Glauber calculations of the final state effects as well as other models of color transparency are in progress.

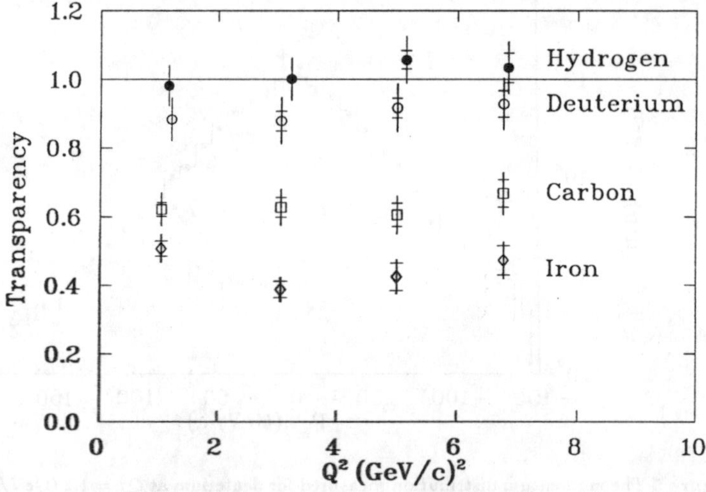

Figure 6 Preliminary results for the transparency ratios from the NE18 experiment for various nuclear targets as defined in the text. The uncertainties include systematic as well as statistical errors.

5. Conclusions

The NE18 experiment represents the first measurement of the $(e, e'p)$ reaction at high Q^2. The quasielastic reaction mechanism appears valid and the extracted momentum and missing-energy distributions are in good agreement with expectations based on lower energy data. The transparency ratio observed for Deuterium is 0.89 ± 0.07 independent of Q^2. The transparency ratios for Carbon and Iron are also approximately independent of Q^2, indicating that substantial color transparency effects are not present at these momentum transfers. The values of the transparency ratios are in rough agreement with several Glauber predictions, but further analysis is required to make accurate quantitative comparisons. Clearly the NE18 experiment is just beginning to explore the potentially interesting region of $Q^2 > 5$ $(GeV/c)^2$. One may need to reach $Q^2 > 10$ $(GeV/c)^2$ or even higher to clearly observe the color transparency phenomenon.

6. Acknowledgements

This work was supported by the National Science Foundation and the U.S. Department of Energy.

7. References

1. American University, Argonne, Caltech, CalState LA, CEBAF, Colorado, Illinois, Livermore, MIT, SLAC, Stanford, and Wisconsin.

2. A. Mueller, in *Proceedings of the Seventeenth Recontre de Moriond on Elementary Particle Physics,* (Les Arcs, France, 1982) ed. J. Tran Thanh Van (Editions Frontieres, Gif-sur-Yvette, France, 1982); S. J. Brodsky, in *Proceedings of the Thirteenth International Symposium on Multiparticle Dynamics* (Volendam, The Netherlands 1982) ed. by W. Kittel *et al.*, (World Scientific, Singapore, 1983).

3. L. Frankfurt, G. A. Miller, and M. Strikman, to be published.

4. D. Perkins, *Phil. Mag.* **46** (1955) 1146.

5. J. D. Bjorken, in *Proceedings of 1971 International Conference on Electron and Photon Interactions at High Energies* (Cornell Univ. Press), p. 282-298.

6. A. S. Carroll *et al.*, *Phys. Rev. Lett.* **61** (1988) 1698.

7. J. P. Ralston and B. Pire, *Phys. Rev. Lett.* **61** (1988) 1823.

8. S. Heppelmann, private communication.

9. R. G. Arnold *et al.*, *Phys. Rev. Lett.* **57** (1986) 174.

10. N. Isgur and C. H. Llewellyn Smith, *Phys. Lett.* *B* **217** (1989) 535.

11. G. Farrar *et al.*, *Phys. Rev. Lett.* **61** (1988) 686.

12. J. P. Ralston and B. Pire, *Phys. Rev. Lett.* **65** (1990) 2343.

13. B. K. Jennings and G. A. Miller, *Phys. Rev. D* **44**, (1991) 692.

14. O. Benhar *et al.*, *Phys. Rev. Lett.* **69** (1992) 881.

15. R. Machleidt, K. Holinde, and Ch. Elster, *Phys. Rep.* **149** (1987) 1.

Future Tests of Color Transparency using Leptons as a Probe

J. Jourdan

Institute of Physics, University of Basel, Klingelbergstrasse 82,
CH-4056 Basel, Switzerland,

Abstract

In this contribution an overview is given of existing plans to look for various effects of color transparency using leptons as a probe. Special emphasis is given to plans at a 15-30 GeV high duty factor electron facility.

1. Introduction

At large momentum transfer (Q^2) the relevant degrees of freedom of quasielastic $(e, e'p)$ scattering are the constituents of the nucleon. Measurements of exclusive processes like $(e, e'p)$ at these high Q^2 are thus intended to test theories of the strong interaction. They are in particular well suited to test the suppression of final state interactions (FSI) of the recoil proton in the nuclear medium as a result of the presence of color transparency (CT). The physical origin of CT is given by two basic properties of quantum chromodynamics: asymptotic freedom and the coherence of gluon radiation, leading to the vanishing of soft interaction cross sections for a small color neutral object. The arguments have been outlined by B.Jennings at this workshop [1] and are summarized as follows:

- At high Q^2 the dominant amplitude for the exclusive process is the one with the lowest number of quarks which consequently is of small transverse size. The final state can be a proton only if the absorbed momentum q from the virtual photon to the quark is distributed over the minimal number of constituents, the three quarks, through exchange of large transverse momentum gluons. As a consequence of the uncertainty principle, the transverse spatial extent of the struck proton is of order $1/q$, which at high Q^2, is much smaller than a normal size proton.

- Secondly, this small object, selected in the reaction, is color neutral outside of its small radius. Contributions arising from, individual closely spaced quarks tend to cancel. Thus this object, being small and color neutral has reduced interactions with hadrons in the surrounding nuclear medium ("color screening").

- This initial three quark state of the object is however not an eigenstate, and being small initially, its size will increase to a normal size proton in a characteristic time scale. To observe color transparency, it is necessary that the compact size be maintained over a finite distance in traversing the nuclear medium in order that the reduction in the interaction probability be large enough to be observable. To keep the object at small size for some time, a large momentum is required to make use of the consequently large time-dilation factor.

Before discussing planned experiments and the feasibility of future experiments at high Q^2 designed to test the concept of color transparency let me comment on other interesting observables that might show effects from CT:

- Above we have implicitly focused on the top of the quasielastic peak, *i.e.*, the kinematics where $x_{Bj} = 1$. Jennings and Kopeliovich [2] have noticed that the effect of CT depends on the Fermi momentum of the struck nucleon. By changing the value of $x_{Bj} \approx 1 + k_z/M_N$, one can control the composition of the ejectile wave packet, enhancing or diminishing its proton content, since the bound nucleon's Fermi momentum distribution is of finite width. Thus, Jennings and Kopeliovich predict an asymmetry in the transparency effect about $x_{Bj} = 1$.

- The measurement of $(e, e'N^*)$ in quasielastic proton kinematics is particularly interesting [3, 4]. The differing nodal structure of the nucleon and resonance wave functions allows both the magnitude and sign of the $(N \to N^*)$ matrix elements to differ from the "elastic" ones, measured in $(e, e'p)$ knock-out. Consequently, anomalous attenuation (shadowing) and enhancement (antishadowing) effects, similar to those discussed in the context of real and virtual photoproduction of vector mesons [5, 6, 7, 8, 9], may exist. This example illustrates the sensitivity of color transparency effects to the wave function of the initial ejectile and to those of the observed hadrons [5, 8].

- The resolution of nucleon knock-out from different nuclear shells is potentially interesting [10]. Both the FSI and the initial nucleon's Fermi motion are different for p and s shell knock-out. This admits the possibility that the shape of the nucleon spectral function – measured in $(e, e'p)$ knock-out – may change as Q^2 increases: CT need not manifest itself merely as a change in the total transmission.

- Studies of the polarization of the recoil proton in the $(e, e'p)$ experiment may also be a sensitive probe of CT. Recall that the full $(\vec{e}, e'\vec{p})$ cross section consists of 18 independent response functions [11]. For unpolarized electrons, and for measurements performed in the electron scattering plane, only the normal component (defined relative to the electron scattering plane) of the recoil proton's polarization survives. This polarization is induced by FSI, and thus, is a highly interesting observable for CT experiments [12]. The signature of CT, then, in these reactions is to observe a vanishing polarization at high Q^2.

Although the main discussion in this section has revolved around the $(e, e'p)$ reaction, it is clear that other reactions can be used to study CT as well. For example, pion electroproduction has been discussed in this context [13]. Other candidate reactions are the photo- and electroproduction of vector mesons [14, 15, 5, 8]. It is however important to emphasize the

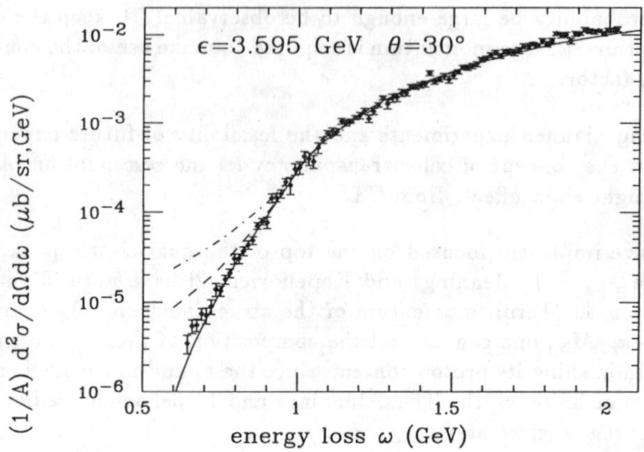

Figure 1: The nuclear matter data extrapolated from (e, e') data taken at SLAC compared to the nuclear matter calculation including FSI via a Glauber calculation (dash-dot). Taking into account the short range NN correlation results in the dashed line and in addition taking into account the effect of CT brings the calculation to a perfect agreement with data (solid).

intrinsic advantage of using electrons in probing this region of physics. In $(e, e'p)$ reactions, the kinematics is determined by the electron arm whereas in a quasielastic $(p, 2p)$ experiment this determination is *not* unambiguous as there are two strongly interacting particles in the final state. Consequently, the $(e, e'p)$ reaction is a cleaner probe of the desired physics.

Inclusive (e, e') data at moderate Q^2 and large values of x_{Bj} also may have a bearing on the question of CT. In the kinematical region, where the cross section is dominated by quasielastic scattering from individual nucleons, important contributions from FSI of the struck nucleon occur. It should be noted in particular that (e, e') is sensitive to CT at much lower Q^2 than exclusive $(e, e'p)$ reactions. In inclusive scattering one is sensitive only to the recoil proton interaction over the first $1/Q$ of its trajectory through the nucleus. To observe the full effect of CT, the lifetime of the "pointlike configuration" of the nucleon has to be of order c/Q--1 only.

This possibility has recently been explored in Benhar et al. [16] where an extrapolation of previous data from nuclei [17] is compared to a nuclear matter calculation. The calculation is based on a nuclear matter spectral function obtained from the correlated nuclear many body problem, and calculated by accounting for the recoil-nucleon FSI using correlated Glauber theory. The comparison with the extrapolated data suggest that CT effects can be important and that a study of the inclusive response at medium Q^2 thus can provide valuable information on CT (see figure 1).

The first exclusive $(e, e'p)$-experiment to look for a CT effect been done recently at SLAC [18] doing a measurement of the A and Q^2 dependence of the quasielastic cross section. In $(e, e'p)$ one needs c/R (R being the radius of the nucleus) to see the full effect of CT. This

Table 1: Assumed Beam Parameters

Maximal Beam Energy	15 GeV
Energy Resolution FWHM	3 10-4
Duty Factor	100%
Maximum Beam Current	50μA

much longer lifetime obtainable via time dilatation at large Q – is much harder to achieve. The first analyzed results are consistent with the conventional Glauber model. Thus no reduction of FSI has been observed in the Q^2-range of the experiment (1-7 GeV/c^2) [19]. An experiment in a similar Q^2-range is planned for CEBAF [20]. Also at CEBAF are plans to measure the recoil polarization of the knockout proton in an $(e, e'\vec{p})$ experiment [12]. Both CEBAF proposals are approved and will take data in a few years.

2. Plans at high Q^2

The result of the SLAC $(e, e'p)$-experiment indicated that the interplay between the perturbative and non-perturbative aspects of QCD cannot be easily explored by existing high energy machines. The SLAC electron machine is of a suitable energy, but its 10-4 duty factor is too low for high statistics coincidence measurements. CEBAF is capable of delivering the required beam characteristics, but its energy is incommensurate with the energies required to observe a significant effect in transparency experiments. The proposed European Electron Facility (ELFE), however, is well-suited to study this region.

As a result of a workshop held in Mainz in 1992 to discuss the physics case of ELFE a collaboration of physicists from 17 institutions has been formed to work out a proposal to study CT in the accessible Q^2-range [21]. To indicate the significance of data that can be taken at a high duty factor 15-30 GeV electron accelerator I would like to summarize the main content of this proposal.

Kinematics: We proposed to measure the quasi-free $(e, e'p)$ cross section and if possible the normal component of the polarization of the recoiling proton P_n on ^{2}H, C, Fe and Au targets. The kinematics and the targets have been selected to cover a large range of Q^2 and A such as to convincingly study the evolution of the CT effect. To work out the kinematics we assumed an electron beam with properties [22] as summarized in the following table.

The kinematics proposed is presented in table 2. In this table E' is the scattered energy and P the momentum of the knockout proton in parallel kinematics and T_p its kinetic energy. Θ_e is the scattering angle of the electron and Θ_p the scattering angle of the struck proton. The largest possible Q^2-point is limited by the beam energy and an assumed minimum scattering angle of the recoil proton of 10 degrees.

Spectrometers: The spectrometers needed for the experiment can have identical features for both the detection of the electron and the proton. The main characteristics are

Table 2: Kinematics

Q^2 $(GeV/c)^2$	P GeV/c	E' GeV	T_p GeV	Θ_e deg	Θ_q deg
6.0	4.03	11.80	3.20	10.57	32.43
9.0	5.66	10.20	4.80	13.93	25.68
12.0	7.27	8.60	6.39	17.54	20.86
15.0	8.88	7.00	7.99	21.78	16.99
18.0	10.49	5.41	9.59	27.24	13.63
21.0	12.09	3.81	11.19	35.29	10.46

Table 3: Spectrometer Parameters

Maximum Central Momentum	12 GeV/c
Angular Range	12° - 40°
$\Delta\theta$	60 mr
$\Delta\phi$	100 mr
Solid Angle	6 msr
Momentum Acceptance	$\pm 5\%$
Momentum Resolution	10^{-3}

summarized in table 3. The parameters result from the requirements of the kinematics outlined above and a solid angle that allows for an acceptable rate even at the highest Q^2-point. Additionally the Fermi cone of the quasi-free scattering, with opening angle given by $tg^{-1}(2k_F/q)$, has to be covered at all kinematic points. This is important in that it does not only maximize the coincidence rates but also makes the measurements less sensitive to details of the spectral function. The acceptance and resolution of the spectrometer assumed would also allow for a study of the x_{Bj} dependence as mentioned in the first section.

$(e, e'p)$ data are usually analyzed in terms of the missing momentum and energy defined as

$$\mathbf{Pm} = \mathbf{q} - \mathbf{p} \tag{1}$$

$$E_m = \nu - T - T_R \tag{2}$$

$(\mathbf{q}, \nu)$ is the four vector of the virtual photon and $(\mathbf{p}, T + M_p)$ the four vector of the knocked-out proton. The last term in the E_m-definition accounts for the recoil energy of the residual A-1 system. With the assumed momentum resolution of the two spectrometers and the beam energy resolution we reach a resolution in missing energy (all contribution added in quadrature) of 12-13 MeV full width which is adequate to discriminate against 2-nucleon knockout and allows for a limited study of shell effects.

Rates: The coincidence rate has been calculated using an impulse approximation model for the cross section integrated over the acceptance of the two spectrometers. Due to the fact that the entire Fermi cone is covered by the acceptance of the proton spectrometer the integrated cross section is not sensitive to details of the spectral function. The cross section is mainly determined by the off-shell e-p cross section were we use the deForest [23] prescription and a kinematical factor taking into account the recoil of the residual nuclear system. No transparency is assumed in the rate estimates. To account for FSI we reduce the quasi-free cross sections by 60% for C, 70% for Fe and 80% for Au.

Given that quasi-free $(e, e'p)$ scattering is a highly correlated process and a 100% duty factor is assumed the ratio of accidental to real coincidences should not be a problem. To check this point we estimate the inclusive rates for the Fe target as an example. The resulting ratio of signal to noise presents no difficulty, being better than 100 for all kinematics.

The coincidence rates have been calculated with a beam current of 50 μA and target thicknesses of 6% radiation length. Aiming for a statistical accuracy of <5% at the lower Q^2 values and <10% at all Q^2 results in a very resonable total running time of order 1100 hours for the entire experiment. One should note that in particular at high Q^2 the rates would increase considerably if transparency is present. Thus a decisive test of CT is possible at a 15-30 GeV electron accelerator such as ELFE.

Recoil Polarization Measurements: The normal component of the recoil polarization, P_n, can also be measured to study the onset of CT in nuclei [12]. In the absence of any nuclear medium effects, P_n, in an $(\vec{e}, e'\vec{p})$ reaction is independent of the polarization of the beam and should be zero in plane wave impulse approximation (PWIA) due to time reversal symmetry [24, 25]. This can be used to great advantage as an effective filter in the study of FSI phenomena in nuclei as it allows to isolate contributions of FSI to the $(e, e'p)$ cross section.

To measure the recoil polarization P_n, one would need to equip the hadron spectrometer with a focal plane polarimeter (FPP). The principle of measurement of P_n with a FPP is described in reference [12]. Without going into much detail, for an unpolarized beam, the statistical uncertainty in the measured polarization is given by:

$$\Delta P_n = \pi/2 (N_0 \epsilon)^{-1/2} \tag{3}$$

where the efficiency of the FPP is defined as $\epsilon = A_c^2 f$ and $f = N_f/N_0$ is the useful fraction of events accepted by the FPP. N_0 is the number of particles incident on the FPP after the first scattering at the experimental target.

The material for the analyzer is usually taken to be carbon for lower energies $(< 1 GeV)$ due to its large analyzing power in the forward angle cone $(\theta = 5^o$ to $20^o)$. Due to the lack of calibration data a measurement of the proton polarization for the elastic $(\vec{e}, e'\vec{p})$ reaction using values of the elastic form factors G_{Mp} and G_{Ep} from existing or forthcoming data pool is considered [12, 26].

For T_p as in the ELFE -proposal (T_p is in the range 3.2 GeV to 11.2 GeV), it is worthwhile to consider a lighter scatterer e.g. liquid hydrogen, instead of carbon, to increase the value of the efficiency parameter ε. In estimating the count rates needed to make an effective P_n

measurement we have used values of ε taken from a smooth extrapolation of the existing values at lower T_p values of order $3 \ 10^{-3}$. The experiment becomes increasingly difficult at Q^2 values above about $12(GeV/c)^2$ where the time required to obtain a precision of 20% or better in ΔP_n becomes inordinately long.

3. Conclusions

In this contribution I mentioned various proposals on $(e, e'p)$ on nuclei planned at low Q^2 and described the possibilities of $(e, e'p)$-experiments at large Q^2. These experiments are designed to provide data that allow us to test the concept of *color transparency*. This concept results from QCD calculations, and the systems studied are the *simplest Fock states* of hadrons. The quantitative understanding of these simplest states are fundamental for a study of the applicability of todays "accepted theory of the strong interaction (QCD)" to hadrons and the confinement region.

References

[1] B.K.Jennings. contribution to this workshop, 1993.

[2] B.K.Jennings and B.Z.Kopeliovich. *Phys. Rev. Lett.*, 70:3384, 1993.

[3] N.N.Nikolaev, A.Szczurek, J.Speth, J.Wambach, B.G.Zakharov, and B.R.Zoller. to be published, 1992.

[4] S. Gardner. Indiana University preprint IT/NTC 92–27.

[5] B.G.Kopeliovich and B.G.Zakharov. *Phys. Rev.*, D44:3466, 1991.

[6] N.N.Nikolaev. *Comments on Nuclear and Particle Physics*, 21:41, 1992.

[7] N.N.Nikolaev. IJMPE: Reports on Nuclear Physics, in press., 1992.

[8] S. Gardner. CEBAF preprint 92-PR-002, 1992.

[9] O. Benhar, B.Z. Kopeliovich, Ch. Mariotti, N.N. Nikolaev, and B.G. Zakharov. *Phys.Rev.Lett.*, 69:1156, 1992.

[10] L.L.Frankfurt, M.I.Strikman, and M.B.Zhalov. *Nucl. Phys.*, A515:599, 1990.

[11] T.W. Donnelly and A.S. Raskin. *Annals of Physics*, 169:247, 1986.

[12] A. Saha (spokesperson). CEBAF Proposal E-91-006, 1991.

[13] C.Carlson and J.Milana. *Phys. Rev.*, D44:1377, 1991.

[14] S.J.Brodsky and G.F. de Teramond. *Phys. Rev. Lett.*, 60:1924, 1988.

[15] N.N.Nikolaev and B.G.Zakharov. *Z.f.Phys.*, C49:607, 1992.

[16] O. Benhar, A. Fabrocini, S. Fantoni, V.R. Pandharipande, and I. Sick. *Phys.Rev.Lett.*, 69:881, 1992.

[17] D. Day, J.S. McCarthy, Z.E. Meziani, R. Minehart, R.M. Sealock, S. Thornton, J. Jourdan, I. Sick, B.W. Filippone, R.D. McKeown, R.G. Milner, D. Potterveld, and Z. Szalata. *Phys.Rev.C*, 40:1011, 1989.

[18] R. Milner and R. McKeown (spokespersons). SLAC-NPAS Proposal NE18, 1990.

[19] A. Lung. contribution to this workshop, 1993.

[20] R. Milner (spokesperson). CEBAF Proposal PR-91-007, 1991.

[21] H. Borel, J. Jourdan, and A. Saha (spokespersons). ELFE-Proposal, 1993.

[22] J. Arvieux. *Europhysics News*, Vol 23, 136, 1992.

[23] T. de Forest. *Nucl. Phys.A*, 392:232, 1983.

[24] A. Picklesimer *et al.* *Phys Rev.*, C32:1312, 1985.

[25] A. Picklesimer and J.W. van Orden. *Phys. Rev.*, C40:290, 1989.

[26] C.F. Perdrisat and V. Punjabi. CEBAF Proposal PR-89-014, 1989.

Fermi Momentum Bias of Color Transparency

BORIS Z. KOPELIOVICH
Joint Institute for Nuclear Research,
Laboratory of Nuclear Problems,
Head Post Office, P.O. Box 79, 10100 Moscow, Russia
e-mail: boris@bethe.npl.washington.edu

Abstract

Due to the specific kinematics of quasielastic scattering $(e, e'p)$, the final state interaction (FSI) itself generate color transparency (CT), and on the other hand, plays a role of a detector of the ejectile size. Even a small size ejectile displays a nonvanishing FSI: it transfers to nuclear matter a finite longitudinal momentum. This prevents, rather than helps, from the study of nuclear spectral function in $(e, e'p)$ at high Q^2.

A new sensitive observable of CT is found, which is an asymmetry of nuclear transparency with respect to the center of the quasielastic peak. It may provide the best way to study CT at low Q^2.

Color transparency (CT) is a phenomenon of suppression of nuclear attenuation[1,2] in some reactions. It is a consequence of color screening and the expected in PQCD smallness of hadronic fluctuations participating in the hard process. The examples are quasielastic scattering[3,4], and diffractive virtual photoproduction of vector mesons[5,6]. Results of experimental search for CT signal in the former case are rather disappointing. Measurement[7] of wide angle $(p, 2p)$ quasielastic scattering at BNL resulted in a puzzling energy dependence of teh nuclear transparency, contradicting theoretical expectations. Recent results of E18 experiment[8] at SLAC display no signal of CT in $(e, e'p)$ at $Q^2 < 7\ GeV^2$. However a confirmation of CT, which seems to be currently the best one, have just came from new data[9] of experiment E665 at Fermilab on Q^2-dependence of virtual photoproduction of ρ-mesons on nuclei, which are in an excellent agreement with the predictions of Ref.6. It worth mentioning also the signals of CT observed previously in the meson charge-exchange scattering on nuclei[10,11], and the inclusive production of hadrons in the deep-inelastic scattering on nuclei[12].

I would like to concentrate in this talk on the quasielastic $(e, e'p)$ scattering on nuclei, to demonstrate how rich is a quantum mechanical pattern of CT phenomenon[13], and particularly to show a failure of a classical treatment of CT as a suppression of final state interactions (FSI). On the contrary to naive expectations, just CT provides a substantial longitudinal momentum transfer from the ejectile to a nuclear medium during FSI, though the transmission coefficient may be unity.

A small-size ejectile have no definite mass. Such a wave packet is a superposition of many hadronic states which are eigenstates of mass matrix[1,14]. However in elastic $e - p$ scattering the ejectile has a definite mass of proton. This can be fixed either by the electron momenta (Bjorken variable, $x_B = 1$), or by a direct detection of the recoil proton. Therefore, in this case the ejectile is simply a proton of mean proton size, rather than a small size wave packet. So one should be careful at this point.

This simple observation does not contradict the PQCD expectation of the smallness of the ejectile size. According to quantum mechanics a measurement procedure itself affects the result, so one has to fix it. To minimize the influence of the measurement one may put the detector of size (FSI in the case under discussion) far apart from the the point, where the hard scattering occurs. However in this case the ejectile has enough time to develop its size up to the mean proton radius. Hence to test the initial ejectile size one should put the size detector as close to the hard interaction point as possible. However in this case the target proton cannot be treated more as being at rest. The uncertainty principle says that the closer is the detector, the larger is the uncertainty in the initial proton momentum. For this reason a value of x_B fixes nomore the missing mass of the ejectile, and heavier than a proton states can be produced in the final state (to avoid problems with off-mass-shellness we consider the imaginary part of the amplitude of $(e, e'p)$ included FSI, corresponding to all intermediate particles being on mass shell). The closer is the target to the size detector the more states are produced, the complete is the cancellation between amplitudes, provided CT.

Thus we are convinced that the procedure of measurement of ejectile size strongly affects the result. Actually the measurement, which is FSI, itself generates CT via Fermi motion. This is different from diffractive processes[5,6], where Fermi motion plays no role in formation of CT. The importance of Fermi motion for CT in quasielastic scattering was claimed independently in papers[15,16]. However the main effect under discussion, the Fermi momentum bias of CT, was missed in Ref.16. In fact this effect can be found also in formulas of paper[17], where it was missed as well.

Although being an important ingredient of CT, the Fermi motion restricts the amount of CT because the Fermi momentum spectrum in nuclei is concentrated within momenta of about $p_F \approx 0.2\ GeV/c$. If the Bjorken variable, $x_B = Q^2/(2m_p\nu)$, is fixed for example at $x_B = 1$, the electron knocks out a proton in the hard scattering on a bound proton, only if the latter was at rest. To produce an excited state of mass m^* in the hard scattering, the target nucleon must move in the direction opposite to the photon, $p_z \approx -(m^{*2} - m_p^2)/2\nu$. So the mass spectrum of produced states is restricted by:

$$m^{*2} < m_p^2 + 2\nu p_F \tag{1}$$

Consequently even if the ejectile in $(e, e'p)$ reaction has a very small size, $\rho^2 \approx 1/Q^2$, the nucleus, as a quantum size detector, cannot resolve it. Only that hadronic states, which can be created with available Fermi momenta, contribute to the production of the wave packet. Therefore a nucleus is sensitive only to a size larger than:

$$\rho^2 \geq \frac{m_p}{p_F}\frac{1}{Q^2}. \tag{2}$$

This restriction evidently reflects the fact that ejectile size cannot be measured at shorter distances after the hard interaction than the mean internucleon separation in nuclei. The latter is just related to the mean Fermi momentum.

Notice that at $x_B = 1$ the Fermi motion is used ineffectively: only the half of the quasielastic peak with $p_z < 0$ is used, no states are produced on mass shell at Fermi-momenta $p_z > 0$. Varying x_B (or the missing momentum, which is a difference between momenta of the photon and the detected proton), one can shift the mass spectrum of hadronic states along the p_z axis, increasing or suppressing the interval of mass covered by the Fermi momentum distribution. Thus one can effectively handle the transverse size of the produced wave packet, changing the Bjorken variable[15]. For instance, if we decrease x_B down to $x_B \approx 1 - p_F/m_p$, the mass squared interval, Δm^2, doubles in comparison with $x_B = 1$. This follows from the approximate relation between x_B, Fermi momentum $\vec{p}$ and the produced mass:

$$M^2 \approx m_p^2 + Q^2 \frac{1 - x_B}{x_B} - 2qp_z, \tag{3}$$

where $\vec{q}$ is the photon 3-momentum. As a consequence of the increase of the available mass interval, the nuclear transparency (Tr) grows. If one keeps decreasing x_B, he can arrive even at nuclear antishadowing, $Tr > 1$. Indeed, at some $x_B \leq 1 - p_F/m_p$, the lightest hadronic state, proton, will be pushed out of the Fermi distribution, i.e. its direct production will be extremely suppressed. However it can be still effectively produced via some heavier intermediate states, which use smaller Fermi momenta. The result depends much on the form of the edge of the Fermi momentum distribution and the mass spectrum of produced states. Nuclear transparency can even exceed unity. An example is given below.

One gets an opposite result increasing x_B. The larger is x_B, the narrower is the mass interval, which contributes to the ejectile wave packet. At last, at $x_B \geq 1 + p_F/m$ all the states except a proton are pushed out of the Fermi distribution peak. In this case the ejectile is simply a proton, the Glauber approximation is exact (up to usual inelastic corrections, which are as small as in the total hA cross section).

Thus we predict[15], an x_B-asymmetry of nuclear transparency which is the direct reflection of the deep quantum mechanical origin of the CT phenomenon.

We do not go into details of the hard interaction dynamics, but simply assume, that CT takes place, i.e. the amplitude of the diffractive interaction of the wave packet $|i\rangle$, produced in the hard interaction, is small.

$$\langle p|\hat{f}_D|i\rangle = O(1/Q^2), \tag{4}$$

where

$$|i\rangle = \hat{f}_H|p\rangle \tag{5}$$

Here $\hat{f}_H$ and $\hat{f}_D$ are the amplitudes of the hard interaction and the diffractive FSI respectively.

After decomposition of the produced wave packet $|i\rangle$ over hadronic states (proton and its diffractive excitations), relation (??) proves to be a condition of cancellation between many large diffractive amplitudes. Such a fine tuning is a manifestation of CT in the hadronic basis[1,14], called nowadays CT sum rule[18]. As we argued above different states need different Fermi momenta to be produced on mass shell, and thus acquire different weight factors caused, by restricted Fermi momentum distribution. Thus the CT sum rule proves to be strongly violated, dependent on x_B, as was explained above.

The first numerical estimate of the Fermi motion bias effect was done in Ref.15 in a simple two channel model. The complete set of hadronic states was replaced by two ones, a proton, $|p\rangle$, and a state $|p^*\rangle$, effectively reproducing the contribution of all diffractive excitations of a proton. Such a model contains four parameters, mass of $|p^*\rangle$, which was chosen at approximate center of gravity of mass distribution; two eigenvalues of the diffractive amplitude, fixed by CT sum rule and the value of pp total cross section; and one mixing angle, fixed by the ratio of forward diffractive to elastic cross sections. Evolution of the produced wave packet during propagation through nuclear matter was calculated exactly using the evolution equation of Ref.14. The Fermi distribution was taken in a simplest Gaussian form. Nucleon correlations in the nuclear density matrics were accounted in the first order. The results for x_B-dependence of nuclear transparency, which is a normalized ratio of nuclear quasielastic to elastic on a free proton cross sections, are shown in fig.1 vs values of $Q^2 = 7$, 15, and 30 GeV^2. One can be convinced that at $x_B > 1$ the nuclear transparency quickly approaches the Glauber approximation limit, whereas at $x_B < 1$ transparency rises and even exceeds unity.

After the Fermi bias effect was claimed and estimated in Ref.15, the calculations were repeated by other authors[19,20]. Nucleon correlations in nuclear density matrix were taken into account more carefully in Ref.20. This provides a distortion of the effective Fermi momentum distribution by the absorption of the recoil proton in nuclear matter. As a result a weak x_B dependence of nuclear transparency proves to appear even in the Glauber approximation. The form of the edge of Fermi momentum distribution, and the high-momentum tail, are most important for the magnitude of nuclear transparency at $x_B \leq 1 - p_F/m_p$. Depending on it, the interplay between the direct production of proton and production via intermediate excitations, can result in a nuclear antishadowing, like in Ref.15, or a shadowing, like in Ref.20. Unfortunately the high-momentum tail of the Fermi distribution is badly known. The statement of Ref.20, that no nuclear antishadowing, they got, is due to the coherency constraint, is incorrect, since the latter was completely included in the evolution operator of Ref.15.

The negative result of search for CT in experiment $E18$ at SLAC[8], as well as many theoretical expectations[18,15], show that CT effect, integrated over x_B,is miserable at $Q^2 \leq 10$ GeV^2. Such a weak CT signal on the top of the Glauber background is doubtful to be extracted, even by a high statistics measurement, having unavoidable uncertainties related to usual experimental cuts, which is difficult to incorporate with theoretical calculations. The Fermi momentum bias of nuclear transparency provides a new exciting opportunity to observe a CT signal even in this range of Q^2. Since the main effect is expected at the tail of Fermi momentum distribution, high statistics measurements are desired. It seems that among the forthcoming electron beam facilities only CEBAF is relevant to this problem.

To conclude, a few comments in order.

- CT in quasielastic scattering on nuclei is possible only due to Fermi motion. However the finiteness of Fermi momenta strongly violates the CT sum rule. This depends on Bjorken x_B (or missing momentum), which allows to handle the amount of CT. The predicted x_B-asymmetry of nuclear transparency is a new observable of CT, reflecting its a deep quantum mechanical nature.

- The finiteness of available Fermi momenta and energy conservation cut off most of hadronic states, which contribute to the CT sum rule. One might arrive at a wrong conclusion, forgetting about it. An example is the proposal[21] for CEBAF to study CT effects in $(e, e'2p)$ on light nuclei. Even at low $Q^2 \approx 4\ GeV^2$ the authors predict about 50% deviation from Glauber model. They use three channel model with masses $m = m_p$, $m^* = 1.6\ GeV$ and $m^{**} = 3\ GeV$. It is easy to check however, that two latter states are too heavy to be produced with reasonable Fermi momentum at $Q^2 = 4\ GeV^2$. Thus this model, corrected for the conservation of energy, predicts no deviation from the Glauber model in the CEBAF range of Q^2.

- Although different hadronic states produced in the hard interaction use different Fermi momenta, the final proton momentum is fixed. The difference between the longitudinal momenta of hadronic states after the hard interaction, is compensated by longitudinal momentum transfer in diffractive FSI. Thus a small size ejectile, consisted of many hadronic waves, transfer to the nuclear matter a positive (relative to the photon direction) longitudinal momentum of the order of mean Fermi momentum. This keeps being valid at any Q^2. Therefore CT does not mean an overall suppression of FSI, but only strong cancellation in the transmission amplitude.

- According to the wide spread opinion, the high Q^2 quasielastic scattering, provides a clean tool for measurement of the nuclear spectral function, because CT suppresses FSI. Such a believe is based on the classical treatment of CT. It is amazing that a correct quantum mechanical analyses leads to the opposite conclusion: CT does not help, but just spoils any opportunity to measure the nuclear spectral function. Firstly, according to the previous comment, CT does mean that FSI exists, provided an additional longitudinal momentum transfer, escaping detection. It is unavoidable property of CT. Secondly, if one wants to measure the large momentum tail of Fermi distribution he faces the Fermi bias of nuclear transparency, which cannot be calculated reliably. One is safe of these problems only at $x_B < 1 - p_F/m_p$, where the Glauber approximation is correct. One may wonder however in this case, why does he need high Q^2 at all.

- We see that the $(e, e'p)$ reaction has a poor kinematics for CT studies: a nucleus is a too rough analyser of the ejectile transverse size, unable to resolve a small structure at $\rho \approx 1/Q^2$. One has to increase Q^2 which decreases the cross-section while one really just wants to increase the laboratory frame momentum of the final particle. The same concerns wide-angle quasi-elastic $(p, 2p)$ scattering.

- The observed phenomenon of Fermi bias might shed light[15] on the puzzling results of measurement of nuclear transparency in quasielastic $(p, 2p)$ scattering in the BNL experiment[7]. Actually the data were distributed over missing momentum, so the asymmetry predicted here for such a distribution has a direct concern to interpretation of the data. Of course situation is more complicate than in $(e, e'p)$: there are three proton legs in $(p, 2p)$, of different momenta. Intermediate state excitations in incoming and outgoing legs produce opposite effects on missing momentum dependence. All these contributions were taken into account in Ref.22 within the same two-component model, as used in Ref.15. It is found that the missing momentum asymmetry nicely explains the observed drop of transparency at 12 GeV/c beam momentum, which have just confused theorists, expecting a rising energy dependence. At the same time the effect of missing momentum asymmetry is weaker at lower beam momenta, due to stronger influence of time evolution. So the calculations[22] agree with the the the data[7] at 6 and 10 GeV/c momenta as well.

A successful attempt to explain the data[7] by the Fermi-bias effect was also undertaken in Ref.23. It was found that the effect[24] of energy-dependent oscillations of the cross section, if any, is very small. Notice that the model of evolution with distributed mass, used in Ref.23, ignores main sources (rapidity and transverse momentum distributions) of suppression of the overlap between DIS and diffraction amplitudes. Such a model cannot be realistic.

References

1. A.B. Zamolodchikov, B.Z. Kopeliovich and L.I. Lapidus, *JETP Lett.* **33**, (1981) 595

2. G. Bertsch, S.J. Brodsky, A.S. Goldhaber and J.G. Gunion, *Phys. Rev. Lett.* **47**, (1981) 297

3. A.H. Mueller, "Proceedings of XVII Recontre de Moriond", Moriond, 1982, ed J. Tran Thanh Van, p13.

4. S.J. Brodsky in Proceedings of XIII Symposium on Multiparticle dynamics, ed W. Kittel, W. Metzger and A. Stergiou (World Scientific, Singapore 1982)

5. B.Z. Kopeliovich and B.G. Zakharov, *Phys. Rev.* **D44**, (1991) 3466

6. B.Z. Kopeliovich, N.N. Nikolaev, J. Nemchik and B.G. Zakharov, preprint TRIUMF, TRI-PP-93-5, 1993. *Phys. Lett. B*, in press.

7. A.S. Carroll et al, *Phys. Rev. Lett.* **61**, (1988) 1698

8. The NE18 Collaboration, B.W Fillipone, PANIC XIII, Book of Abstract, p.177, 1993

9. The E665 Collaboration, Guang Y. Fang, PANIC XIII, Book of Abstracts, p.223, 1993

10. B.G. Zakharov and B.Z. Kopeliovich, *Sov. J. Nucl. Phys.* **46** (1987) 911

11. B.Z. Kopeliovich and B.G. Zakharov, *Phys. Lett. B* **264** (1991) 434

12. B.Z. Kopeliovich and J. Nemchik, INFN-ISS 91/3, Sezione Sanita Rome, 1991

13. B.Z. Kopeliovich, DFTT 24/93, Torino University, 1993; to be published in the Proceedings of the VIII Rencontre de Moriond conference.

14. B.Z. Kopeliovich and L.I. Lapidus, *JETP Lett.* **32**, (1980)595

15. B. Jennings ans B. Kopeliovich, *Phys. Rev. Lett.* **70**, (1993)3384

16. A. Biamconi, S. Boffi, D.E. Kharzeev, preprint FNT/T-92/38, (1992)

17. B.K. Jennings and G.A. Miller, *Phys. Rev.* **D44**, (1991)692

18. N.N. Nikolaev et al., Jülich preprint KFA-IKP(Th)-1992-16(1992)

19. A. Biamconi, S. Boffi, D.E. Kharzeev, FTN/T-92/47,
20. N.N. Nikolaev et al., Jülich preprint KFA-IKP(Th)-1993-16(1992)
21. K. Egiyan et al., PANIC XIII, Book of Abstracts, p.178, 1993
22. B. Jennings and B. Kopeliovich, PANIC XIII, Book of Abstracts, p.165
23. B. Jennings and G. Miller, preprint UW, Seattle, (1993)
24. B. Pire and J. Ralston, *Phys. Rev. Lett.*, **61** (1988) 1823

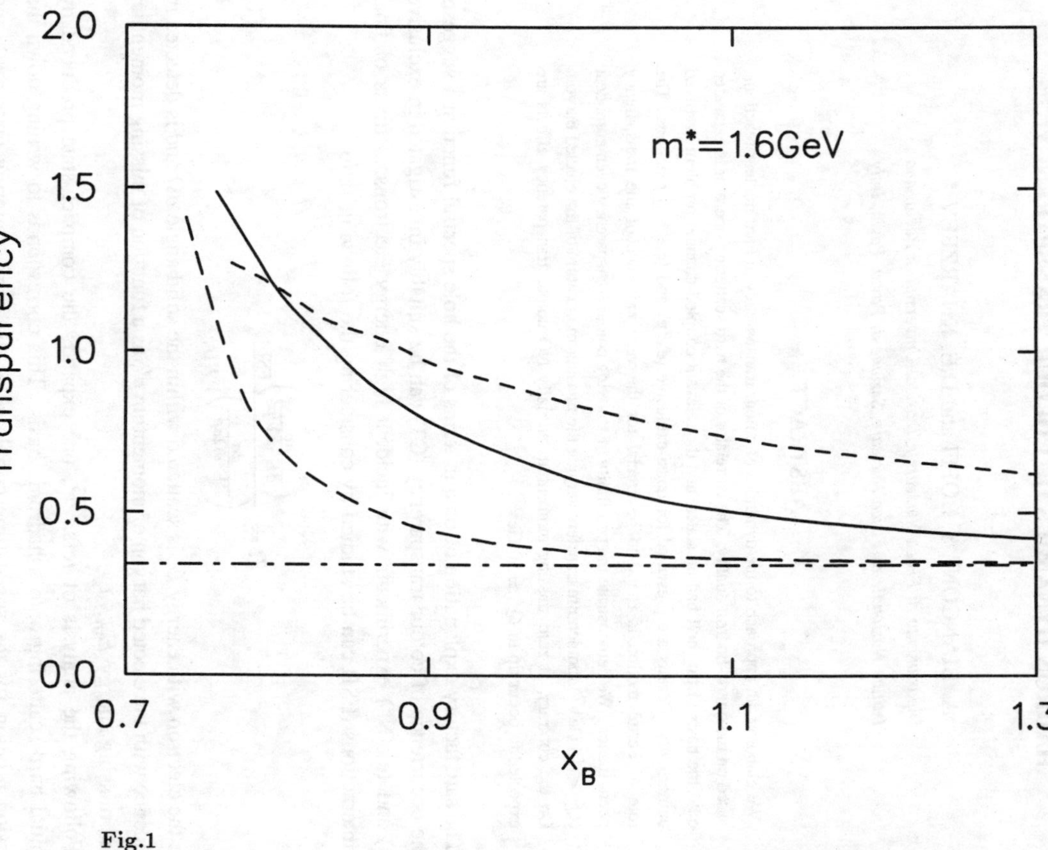

Fig.1
The nuclear transparency in $(e, e'p)$ as function of Bjorken variable x_F, versus $Q^2 = 7, 15$ and $30\ GeV^2$

HADRON DYNAMICS OF COLOUR TRANSPARENCY

A. BIANCONI, S. BOFFI and D.E. KHARZEEV *

Dipartimento di Fisica Nucleare e Teorica, Università di Pavia, and

Istituto Nazionale di Fisica Nucleare, Sezione di Pavia, Pavia, Italy

ABSTRACT

We elaborate an approach to the dynamics of colour transparency in hard nuclear reactions, based on the quark–hadron duality. The correlator of the hard scattering operator is expanded onto the basis of on-shell hadron states, and the relevant coupled-channel problem is solved numerically. Results are presented for quasi-exclusive (e,e'p) and (e,e'N*) reactions. The role of Fermi motion is found to be crucial for the very existence of the transparency phenomenon. We demonstrate the possibility of studying colour transparency even at modest $(Q^2 < 15\,\mathrm{GeV}^2)$ momentum transfers varying the missing momentum of the ejected baryon. On the contrary, at zero missing momentum virtually NO colour transparency effects are expected to occur up to $Q^2 \simeq 50\,\mathrm{GeV}^2$.

The satisfactory explanation of data in terms of the hole spectral function [1] suggests that the occurrence of colour transparency (CT) can be suitably investigated in exclusive (e,e'p) and (e,e'N*) experiments with incident high–energy electrons. Effects of final state interactions (FSI) can be studied by considering the following ratio

$$R = \frac{\left(\frac{d\sigma}{dE'\,d\Omega\,dp'}\right)_{\mathrm{FSI}}}{\left(\frac{d\sigma}{dE'\,d\Omega\,dp'}\right)_{\mathrm{PWIA}}}, \tag{1}$$

where the electron with energy E' is scattered within the solid angle $d\Omega$ and is detected in coincidence with the ejected baryon of momentum p', as a function of missing momentum p_m and missing energy E_m.

Following the analysis of refs. [2,3], one expands the compact state produced by the initial hard scattering onto a hadronic basis. This corresponds to writing down the dispersion relation for the space–time correlator of the hard scattering operator [4] and reflects the quark–hadron duality, stating that both languages – quark and hadronic – can always be used. The different hadron components of the wave packet are just the usual on-shell hadrons of normal hadron size. In nuclear matter they undergo elastic and diffractive FSI, which cause distortion of the wave packet. Imposing the CT behaviour

---

* On leave from Moscow State University, Moscow 119899 Russia

required by QCD at high momentum transfers fixes the relative phases of elastic and diffractive amplitudes [5,3,4]; as a result, elastic and diffractive FSI interfer destructively leaving the initial wave packet unaffected by the presence of the nucleus. At moderate momentum transfers only a few hadron states contribute; FSI with the nuclear medium will turn the intermediate state into the physical final state. Thus a suitable formulation of the problem is in terms of a coupled-channel approach [3]. In this work we have considered (apart from the proton) two different baryon states abundantly produced in hard ep scattering [6] – N(1535) and N(1680). In the hard scattering the hit proton can be excited into N* which is then de–excited by FSI into an outgoing proton with momentum p'.

The Fermi motion plays a fundamental role in exclusive knock-out reactions, allowing for the very existence of CT [3,7]. In a hypothetical nucleus where nucleons have fixed zero three-momentum (i.e. no Fermi motion), the four-momentum of nucleons in the laboratory is simply $P_0 = (m_N, 0)$. If the kinematics of the process allows to fix the transferred momentum Q, then the invariant mass of the state produced by hard scattering is strictly determined: $s = s_0 = (P_0 + Q)^2$. One finds therefore that the wave packet is composed by only one hadron state of mass $m_0 = \sqrt{s_0}$ (if it exists). This state obviously will be attenuated in the nuclear medium in a usual way since its wave function is of the normal hadron size – CT cannot appear. The existence of Fermi motion smears the distribution in the nucleon momentum and therefore causes an intrinsic uncertainty in the invariant mass of the produced state. However, at low Q the Fermi motion is still unable to allow for a composition of a sufficiently coherent wave packet. At large momentum transfers the situation changes; uncertainty in the invariant mass of the produced state becomes larger than a characteristic mass spacing between physical hadronic states, and coherent compact wave packet can be produced. The excited components of the wave packet then undergo in the nucleus multiple diffractive transitions. The cancellation of the corresponding amplitudes with the amplitude of elastic rescattering causes CT.

Numerical calculations have been performed for the quasi–elastic ^{27}Al(e,e$'$p) and ^{27}Al(e,e$'$N(1535)) reactions. The values of the corresponding transparency coefficients R are reported in fig. 1. For $p_m = 0$, R attains its maximum at $q \to \infty$, as expected. Also, NO CT is virtually present up to $Q^2 \sim 50\,\text{GeV}^2$. The non trivial result is however that CT is present even at modest momentum transfer at non-zero and positive values of p_m. Thus at intermediate momentum transfer the only possibility to observe CT is to study the domain where the missing momentum of the ejected baryon is $p_m > 0$.

Cutting the surface at fixed values of $p_m > 0.2 \div 0.4\, p_F$ the resulting curve versus q starts at low q from a value manifesting ordinary nuclear attenuation. Then increasing

282

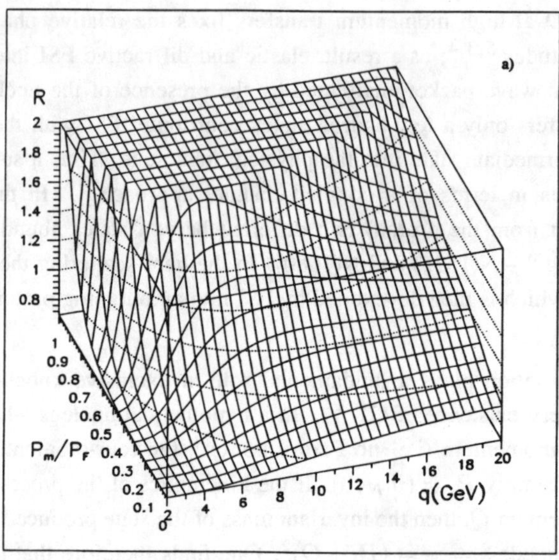

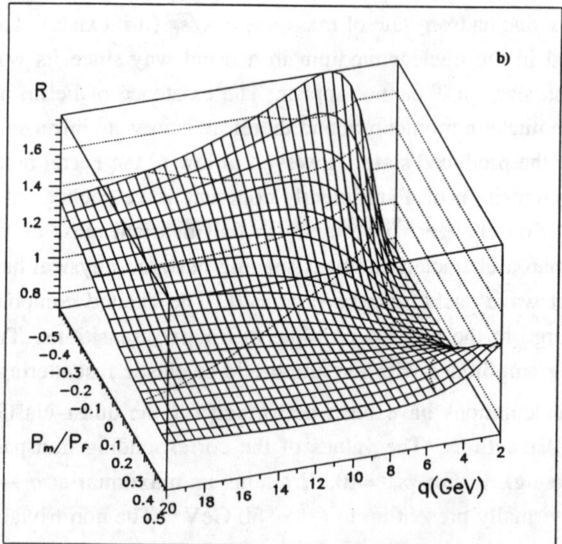

Fig. 1. The transparency coefficient R for the reactions: a) ^{27}Al(e,e'p) and b) ^{27}Al(e,e'N(1535)), as a function of three-momentum transfer q (GeV) and missing momentum p_m (in units of Fermi momentum $p_F = 234$ MeV).

q, R increases towards a maximum and then starts decreasing again towards asymptotic values (for $q \to \infty$) which are again expression of normal nuclear attenuation. The maxima of R are strictly related to the maxima of the probability of exciting the baryonic resonance in the hard scattering vertex.

So the non trivial conclusion is that only for $p_m \sim 0$ transparency is attained at asymptotic momentum transfer. At non-asymptotic q transparency maxima manifest themselves at other values of p_m. This trend is reminiscent of the features of the data in the BNL experiment [8], which were taken at different values of p_m.

The above model takes into account only two specific resonances which dominate the second and third resonance regions at high Q^2. Since at moderate Q^2 there is a considerable admixture of the other resonances, we consider the two above states as effectively summing up the whole contribution. This is not a limitation because the relevant parameters are the resonance energy and width.

Basing on our results we conclude that an interesting region to be explored by (e,e'p) and (e,e'N*) experiments is for $Q \sim 2 \div 10$ GeV and p_m up to a considerable fraction of p_F. Such an investigation could already start at SLAC and at CEBAF [9].

The counting rate for such momenta will take advantage from an increased beam energy as well as from a high duty cycle. Thus, to explore the p_m dependence of R with sufficient accuracy one needs good resolution spectrometers [10] and high luminosity beams, such as e.g. the one proposed for the European Electron Facility. A more extensive range of missing momenta could then be explored with a possible access to the CT regime. Based on the simulation of ref. [11] reasonable counting rates can be predicted.

References

1) S. Boffi, C. Giusti and F.D. Pacati, Phys. Rep. **226** (1993) 1

2) B.K. Jennings and G.A. Miller, Phys. Lett. **B 236** (1990) 209; Phys. Rev. **D44** (1991) 692

3) A. Bianconi, S. Boffi and D.E. Kharzeev, Phys. Lett. **B 305** (1993) 1; in: *Proc. of Diquarks Workshop* (Torino, 1992), eds. M. Anselmino and E. Predazzi (World Scientific, Singapore), in press; Nucl. Phys. **A** (1993) in press

4) D.E. Kharzeev, Inst. Phys. Conf. Series **124** (1992) 229; Nucl. Phys. **A** (1993), in press

5) L.L. Frankfurt, G.A. Miller and M.I. Strikman, Comm. Nucl. Part. Phys. **21** (1992) 1

6) P. Stoler, Phys. Rep. **226** (1993) 103

7) B.K. Jennings and B.Z. Kopeliovich, Phys. Rev. Lett. **70** (1993) 3384

8) A.S. Carrol *et al.*, Phys. Rev. Lett. **61** (1988) 1698

9) J. van de Brand *et al.*, SLAC–NPAS proposal NE18 (1989); R.D. McKeown, Nucl. Phys. **A532** (1991) 285c; R.G. Milner *et al.*, Letter of intent, SLAC (1991); R.G. Milner *et al.*, CEBAF proposal (1991)

10) H.P. Blok *et al.*, in: *The European Electron Facility*, eds. J. Arvieux and E. De Sanctis (Editrice Compositori, Bologna, 1993) to be published

11) H. Borel, S. Fleck, J. Marroncle, F. Staley and C. Vallet, Nucl. Phys. **A532** (1990) 291c

Physics with a $15 \div 30$ GeV Electron Accelerator (ELFE)

Bernard Frois

DAPNIA, Centre d'Etudes Nucléaires de Saclay,

91191 Gif sur Yvette, France.

Bernard Pire

CPT, Ecole Polytechnique,

F-91128 Palaiseau, France

Abstract

This paper presents a brief overview of the physics with a 15-30 GeV continuous beam electron facility proposed in Europe.

1 INTRODUCTION

In the last two decades, we have seen the emergence of a theory that identifies the basic constituents of matter and describes the strong interaction. The elementary building blocks of atomic nuclei are colored quarks and gluons. The theory describing their interactions is Quantum Chromodynamics (QCD) which has two special features, asymptotic freedom and color confinement. Asymptotic freedom means that color interactions are weak at short distances. Color confinement results in the existence of hadrons and in the impossibility to observe quarks and gluons as single particles. Color confinement and asymptotic freedom lead to the existence of two regimes. At short distances, quarks and gluons are in the regime of asymptotic freedom and behave in essence as free particles. At large distances, color interactions are strong and confine quarks and gluons in hadrons (mesons and baryons). A nucleus appears then to be built of nucleons interacting through the exchange of mesons.

Until now, the study of nuclear structure has mostly focused on the distributions of nucleons and mesons in nuclei. Experimental results have shown that even in the dense nuclear interior, nucleons keep their identity. A coherent description of nuclei at the fermi scale has been achieved according to this concept of nuclei made of nucleons. At shorter distances, nucleons start to overlap and one must take their internal structure into account. Mesonic theory provides an efficient and economical description of nuclear reactions involving momentum transfers up to about 1 (GeV/c)2, but for higher momentum transfers the situation becomes much more complex. In order to describe very short range processes, one must introduce more and more mesons. The limits of the description of nuclei in terms of nucleons and mesons will be studied in Europe with the accelerators of Amsterdam, Bonn and Mainz. In the United States research at CEBAF, the 4 GeV continuous beam electron accelerator built at Newport News (Virginia), will start operation in 1995. CEBAF will explore in details the structure of the nucleon and its resonances.

Although one knows the microscopic theory for the strong interactions, *one does not understand how quarks build up hadrons*. After twenty years of theoretical developments we still lack reliable, analytic tools for this problem. This is one of the most important problem of contemporary physics. Therefore a common goal of nuclear, particle and astrophysics is today to understand the formation of hadrons from quarks and gluons. This is the central problem to solve if one wants to understand the formation of matter.

The advantages of electron scattering

Electron scattering is the most appropriate probe to attack this problem. Electrons are pointlike charges and their interaction with other elementary particles is well understood. This interaction is sufficiently weak to allow electrons to penetrate in the heart of a nucleus without significant perturbation of its structure. Electron beams probe matter with a spatial resolution that depends on their energy. The higher the energy, the better is their resolution.

Exclusive reactions: A new tool

Nearly all existing data on quark distributions in hadrons have been obtained by inclusive scattering of high energy particles. In such reactions, one strikes quarks with considerable momentum and energy and reconstruct quark distributions from scattering data. The experimental observation amounts to an average over all the possible quark configurations in the nucleus. Therefore, it is impossible to get precise information on specific configurations and to follow their evolution which is controlled by the confining mechanisms. One needs a different type of data sensitive to the time evolution of a system of correlated quarks. This is the domain of exclusive reactions where scattered particles emitted in a specific channel are observed in coincidence.

In inclusive scattering, the energy ν and the four momentum $Q2$ of the virtual photon are the only two independent kinematic variables. In exclusive scattering, t the momentum transfer between the photon and the detected hadron, is a new variable to tune the size of the interaction volume. The production of heavy flavors and the determination of spin observables open new possibilities to disentangle various aspects of the dynamics.

286

The study of exclusive reactions is not possible with existing accelerators because of their technical limitations. Muons and neutrino beams have too low an intensity. Existing high energy electron accelerators have too small a duty factor.

The ELFE project

During the last five years, several conferences and workshops have discussed the best experimental approach to understand the evolution from quarks to hadronic matter. Proposals using ELFE (An Electron Laboratory For Europe): a $15 \div 30$ GeV high luminosity, continuous beam electron accelerator have been discussed and collaborations have been formed at the Mainz workshop in 1992. This project will be presented to NUPECC at the end of 1993. These proposals form an extensive research program on exclusive reactions to probe the evolution of correlated quarks systems. Using the nucleus itself as a microscopic detector is one of the important ideas of this program. One measures the same reaction using nuclei of different sizes and thus observes the differences in the evolution from quarks and gluons to hadrons in the nuclear medium. This is possible only in the $15 \div 30$ GeV energy range. One must have sufficiently high energy to describe the reaction in terms of electron-quark scattering. However, the energy transfer should not be too high since one is interested in the formation of hadrons inside the nuclear medium and not outside of the nucleus.

This research program lies at the border of nuclear and particle physics. Most of the predictions of QCD are only valid at very high energies where perturbation theory can be applied. In order to understand how hadrons are built, however, one is in the domain of confinement where the coupling is strong. Up to now there are only crude theoretical models of hadronic structure inspired by QCD. One hopes that in the next ten years major developments of nonperturbative theoretical methods such as lattice gauge theory will bring a wealth of results on the transition from quark to hadron. It is fundamental to guide theory by the accurate, quantitative and interpretable measurements obtained by electron scattering experiments.

The research program of ELFE addresses the questions raised by the quark structure of matter: the role of quark exchange, color transparency, flavor and spin dependence of structure functions and differences between quark distributions in the nucleon and nuclei, color neutralization in the hadronization of a quark...All these questions are some of the many exciting facets of the fundamental question:

"How do color forces build up hadrons from quarks and gluons? "

PUZZLING EFFECTS OBSERVED IN HIGH ENERGY EXPERIMENTS

- Deep inelastic scattering experiments on nuclei have revealed a significant variation of structure functions with the density of the nucleus. This effect was discovered by the EMC collaboration using a high energy muon beam and subsequently investigated in detail by the NMC collaboration also at CERN. Many different explanations have been proposed in terms of shadowing, mesons in nuclei, effects of binding or modification of the nucleon size in the nuclear medium.

- Hadron production at high transverse momentum in hadron-nucleus collisions have revealed a puzzling $A\alpha$ dependence with α varying up to 1.3. Explanations of this effect involve the successive scattering of a naked quark from quarks and gluons bound in nearby nucleons, before the formation of a hadron.

- Suppression of charmonium production has been observed in high energy heavy ion collisions. To isolate the possible signals from a quark gluon plasma, it is necessary to understand the formation and propagation of a $c\bar{c}$ pair in a dense medium.

- Proton-proton elastic scattering data at large angle measured at Brookhaven seem to be compatible with an effect of color transparency. The interpretation of these data is still controversial.

- Contrary to perturbative QCD predictions, helicity non conservation has been observed in several hard exclusive reactions.

2 ELFE EXPERIMENTAL PROGRAM

The central idea of the ELFE project is to use exclusive reactions and the nucleus as a microscopic detector to determine the time evolution of the elementary quark configurations in the building up of hadrons. Two typical examples of this research program are color transparency in quasi-elastic (e,e'p) reactions and in charmonium production, and hadronization in the nuclear medium. For these processes, the nucleus is used as a medium of *varying length*.

The typical time scales to build up a hadron is $\tau_o \sim 1\text{fm/c}$ in its rest frame. This is the time needed by a quark to travel through distances characteristics of confined systems. Due to the Lorentz dilation factor $\gamma = E/M$, the time scale τ, in the laboratory frame, is several fm/c's.

At this scale, the only available detector is the nucleus.

ELFE will focus on the following research topics:

- *Exclusive processes.* Exclusive electroproduction processes, including polarization experiments, are needed to study the spatial structure of hadrons. Because they require coherent scattering of the quarks, exclusive observables are sensitive to the quark gluon wave function of the hadrons. Typical examples are virtual Compton scattering and form factors of mesons or baryons.

- *Nucleus as a detector.* A central idea is to use the nucleus as a microscopic detector to determine the time evolution of the elementary quark configurations in the building up of hadrons. A typical example of this research program is color transparency in quasi-elastic reactions and in charmonium production. Another one is hadronization in the nuclear medium. For these processes, the nucleus is used as a medium of *varying length*.

- *Heavy Flavors.* The study of the production and the propagation of strangeness and charm provides us with an original way to understand the structure of hadronic matter. The corresponding reactions do not involve the valence quarks of the target and probe its sea quark (intrinsic strange or charm content) and gluon distributions.

- *Short Range Structure of Nuclei.* At short distances, nuclear structure cannot be reduced to nucleons or isobar configurations. To unravel such exotic configurations dedicated experiments (large x structure functions and ϕ production) are proposed.

3   ACCELERATOR AND DETECTOR REQUIREMENTS

The choice of the energy range of 15 to 30 GeV for the ELFE accelerator is fixed by three constraints:

- Hard electron-quark scattering: one must have sufficiently high energy and momentum transfer to describe the reaction in terms of electron-quark scattering. The high energy corresponds to a very fast process where the struck quark is quasi-free. High momentum transfers are necessary to probe short distances.

- Nuclear sizes: The energy of the incident electron beam is determined to match the characteristic interaction time τ to the diameter of the nucleus. Starting from the rest frame time $\tau_o \sim 1$ fm/c and taking into account a typical Lorentz dilation factor $\gamma = E/M$ this means a time τ of several fm/c's in the laboratory. If the energy transfer is too large, the building-up of hadrons occurs outside the nucleus which can then no longer be used as a microscopic detector.

- Charm production requires a minimum electron beam energy of 15 GeV to have reasonable counting rates.

Exclusive and semi-inclusive experiments are at the heart of the ELFE project. To avoid a prohibitively large number of accidental coincident events a high duty cycle is imperative. The ELFE experimental program also requires a high luminosity because of the relatively low probability of exclusive processes. Finally a good energy resolution is necessary to identify specific reaction channels. A typical experiment at 15 GeV (quasielastic scattering for instance) needs a beam energy resolution of about 5 MeV. At 30 GeV the proposed experiments require only to separate pion emission. These characteristics of the ELFE accelerator are summarized in table 1.

Beam Energy	$15 \div 30 GeV$
Energy Resolution FWHM	$3 \times 10{-}4$ @ 15 GeV
	$10{-}3$ @ 30 GeV
Duty Factor	$\simeq 100$ %
Beam Current	$10 \div 50 \mu$A
Polarized Beams	$P > 80$ %

Table 1: ELFE Accelerator Parameters

Due to the very low duty cycle available at SLAC and HERA (HERMES program) one can only perform with these accelerators inclusive experiments and a limited set of exclusive experiments.

ELFE will be the first high energy electron beam beyond 10 GeV
with both high intensity and high duty factor.

The various components of the ELFE experimental physics program put different requirements on the detection systems that can be satisfied only by a set of complementary experimental equipment. The most relevant detector features are the acceptable luminosity, the particle multiplicity, the angular acceptance and the momentum resolution. High momentum resolution ($5 \times 10{-}4$) and high luminosity (1038 nucleons/cm2/s) can be achieved by magnetic focusing spectrometers. For semi-exclusive or exclusive experiments with more than two particles in the final state, the largest possible angular acceptance ($\sim 4\pi$) is highly desirable. The quality and reliability of large acceptance detectors have improved substantially in the last two decades. The design of the ELFE large acceptance detectors uses state of the art developments to achieve good resolution and the highest possible luminosity.

ELFE belongs to a coherent long term strategy

Complementary experiments to study the quark structure of matter have been proposed with ultra high energy heavy ions. Thus, the experimental strategy follows two distinct paths.

- Study of the evolution from quasi free quarks or correlated quark systems to massive particles (hadrons) where they are confined. We propose in this report to build a 15-30 GeV continuous beam electron accelerator to study specific reaction channels to well identified final states. ELFE: an Electron Laboratory For Europe, dedicated to the study of the quark and gluon structure of matter.

- Study of quark deconfinement in heavy ion collisions at ultra high energies to discover the quark gluon plasma. The United States are now building the relativistic heavy ion collider (RHIC) at Brookhaven. In Europe, an exploratory program has started at CERN. An ambitious program proposes to use the future large hadron collider (LHC) at CERN.

4   COLOR TRANSPARENCY: A TYPICAL EXAMPLE

Color transparency has been extensively discussed at this workshop. This phenomenon illustrates the power of exclusive reactions to isolate simple elementary quark configurations. The experimental technique to probe these configurations is the following:

- For a hard exclusive reaction, say electron scattering from a proton, the scattering amplitude at large momentum transfer $Q2$ is suppressed by powers of $Q2$ if the proton contains more than the minimal number of constituents. This is derived from the QCD based quark counting rules, which result from the factorization of wave-function-like distribution amplitudes. Thus protons containing only valence quarks participate in the scattering. Moreover, each quark, connected to another one by a hard gluon exchange carrying momentum of order Q, should be found within a distance of order $1/Q$. Thus , at large $Q2$ one selects a very special quark configuration: all connected quarks are close together, forming a small size color neutral configuration sometimes referred to as a *mini hadron*. This mini hadron is not a stationary state and evolves to build up a normal hadron.

- Such a color singlet system cannot emit or absorb soft gluons which carry energy or momentum smaller than Q. This is because gluon radiation — like photon radiation in QED — is a coherent process and there is thus destructive interference between gluon emission amplitudes by quarks with "opposite" color. Even without knowing exactly how exchanges of soft gluons and other constituents create strong interactions, we know that these interactions must be turned off for small color singlet objects.

An exclusive hard reaction will thus probe the structure of a *mini hadron*, i.e. the short distance part of a minimal Fock state component in the hadron wave function. This is of primordial interest for the understanding of the difficult physics of confinement. First, selecting the simplest Fock state amounts to the study of the confining forces in a colorless object in the "quenched approximation" where quark-antiquark pair creation from the vacuum is forbidden. Secondly, letting the mini-state evolve during its travel through different nuclei of various sizes allows an indirect but unique way to test how the squeezed mini-state goes back to its full size and complexity, *i.e.* how quarks inside the proton rearrange themselves spatially to "reconstruct" a normal size hadron. In this respect the observation of baryonic resonance production as well as detailed spin studies are mandatory.

To the extent that the electromagnetic form factors are understood as a function of $Q2$, $eA \rightarrow e'(A - 1)p$ experiments will measure the color screening properties of QCD. The quantity to be measured is the transparency ratio T_r which is defined as:

$$T_r = \frac{\sigma_{Nucleus}}{Z\,\sigma_{Nucleon}} \tag{1}$$

At asymptotically large values of $Q2$, dimensional estimates suggest that T_r scales as a function of $A^{\frac{1}{3}}/Q2$. The approach to the scaling behavior as well as the value of T_r as a function of the scaling variable determine the evolution from the pointlike configuration to the complete hadron. This highly interesting effect can be measured in an $e, e'p$ reaction that provides the best chance for a *quantitative* interpretation.

The interplay between the perturbative and non-perturbative aspects of QCD cannot be easily explored by existing high energy machines. The SLAC electron machine is of a suitable energy, but its $10-4$ duty factor is too low for high statistics coincidence measurements. CEBAF is capable of delivering the required beam characteristics, but its energy is too low to observe a significant effect in transparency experiments.

5 CONCLUSIONS

The ELFE research program lies at the border of nuclear and particle physics. Most of the predictions of QCD are only valid at very high energies where perturbation theory can be applied. In order to understand how hadrons are built, however, one is in the domain of confinement where the coupling is strong. It is fundamental to guide theory by the accurate, quantitative and interpretable measurements obtained by electron scattering experiments, in particular in exclusive reactions.

This research domain is essentially a virgin territory. There is only a limited amount of experimental data with poor statistics. It is not possible to make significant progress in the understanding of the evolution from quarks to hadrons with the available information.

This lack of data explains to a large extent the slow pace of theoretical progress. The situation will considerably improve due to technical break-throughs in electron accelerating techniques. ELFE will be the first high energy machine offering the high luminosity and high duty cycle demanded by the exclusive reaction program.

A few topics of the experimental program proposed at ELFE can be covered by existing or planned facilities at the price of considerable efforts. This is the case of the proton electric form factor at SLAC. Also the proton transverse spin structure function can be studied at RHIC through dilepton pair production in polarized proton-proton collisions. These topics are but a small part of the extensive ELFE research program. The exploratory program on color transparency at Brookhaven with protons and at SLAC with electrons did strengthen the need for dedicated experiments with high energy resolution and high duty cycle electron beam. The HERMES program at HERA proposes a first detailed study of semi-inclusive reactions. ELFE experiments will increase the statistics by orders of magnitude thus allowing a much more detailed understanding of color neutralization.

The goal of the ELFE research program, starting from the QCD framework, is to explore the coherent and quark confining QCD mechanisms underlying the strong force. It is not to test QCD in its perturbative regime, but rather to use the existing knowledge of perturbative QCD to determine the reaction mechanism and access the hadron structure.

ELFE will use the tools that have been forged by twenty years of research in QCD, to elucidate the central problem of color interaction: color confinement and the quark and gluon structure of matter.

Ackowledgements

This overview is based on a collective contribution to the book on the ELFE project edited by J. Arvieux and E. de Sanctis to be published by Nuovo Cimento. This book contains the original research proposals and a complete list of references. We would like to give our warmest thanks to our colleagues who have contributed to develop the ideas presented at Mainz in a coherent research program: J. Arvieux, P. Bertin, G. Chanfray, J. Y. Grossiord, P. Guichon, P. Kroll, J.-M. Laget, J.-F. Mathiot, J. Mougey, A. Mueller, P. Mulders and P. Stoler.